Engineering Metrology and Measurement

Engineering Metrology and Measurement

K Duraivelu

BE, MBA, ME, PhD, FIE
Professor & Dean, Faculty of Engineering & Technology,
Vadapalani Campus, SRM University, Chennai

S Karthikeyan

ME, PhD
Professor & Head, Department of Mechanical Engineering,
Vadapalani Campus, SRM University, Chennai

Universities Press

ENGINEERING METROLOGY AND MEASUREMENT

Universities Press (India) Private Limited

Registered Office
3-6-747/1/A & 3-6-754/1. Himayatnagar, Hyderabad 500 029 (Telangana), India
e-mail: info@universitiespress.com; www.universitiespress.com

Distributed by
Orient Blackswan Private Limited

Registered Office
3-6-752, Himayatnagar, Hyderabad 500 029 (Telangana), India

Other Offices
Bengaluru / Bhopal / Bhubaneshwar / Chennai / Guwahati / Hyderabad /
Jaipur / Kolkata / Lucknow / Mumbai / New Delhi / Noida / Patna / Vijayawada

ISBN 978-93-86235-52-7

501410

06122017

Typest in Palatino 11/14 by
Bookcraft Publishing Services (I) Pvt Ltd, Chennai

Printed at
Yash Printographics, Noida

Published by
Universities Press (India) Private Limited
3-6-747/1/A & 3-6-754/1, Himayatnagar,
Hyderabad 500 029, Telangana, India

Foreword

It is with great pleasure that I write this foreword for the book *Engineering Metrology and Measurement*, authored by K Duraivelu and S Karthikeyan. The authors have done a remarkable job in covering all the essential aspects of the subject in an interesting manner.

In this current era of mass production, with high precision and assured quality, understanding the concepts of metrology and measurement is essential. Further, the measured data should be suitably interpreted using statistical tools for effective quality control. This book covers all these aspects in 11 chapters, and each chapter is followed by a summary, multiple choice questions and review questions.

This will be helpful both to students as well as to practicing engineers who wish to upgrade their knowledge.

This book will surely be of great support to anyone who wants to understand the subject without much difficulty, as the language is clear and jargon is mostly avoided in the text. I believe this book will provide its readers very valuable information about the field of Metrology and Measurement. I heartily recommend this book to everyone.

Dr. R Velraj
Professor & Director, Institute for Energy Studies,
Anna University, Chennai

Foreword

I am very pleased to write a foreword for the book *Engineering Metrology and Measurement*, written by K Duraivelu and S Karthikeyan.

This textbook provides comprehensive coverage of syllabi in the areas of Metrology, Measurement and Statistical Quality Control. The language is simple and can be easily read and understood by undergraduate engineering students.

The book includes numerous illustrations that help explain theoretical concepts in an elegant manner. Apart from several solved problems, I find that there are many multiple choice questions, which are quite useful for students preparing for competitive exams.

Due to its wide coverage of the syllabi of several Indian universities, I am sure that this book will have a huge audience. It is no wonder that a renowned publisher such as Universities Press has agreed to publish this very useful textbook.

Dr. V Raghavan
Associate Professor, Mechanical Engineering,
Indian Institute of Technology Madras, Chennai

Preface

Metrology and Measurement form an integral part of our everyday life, ranging from the daily measurement of the weight of vegetables and grocery items to the measurement of body temperature, and so on. In any engineering field, it plays a vital role in the processes of production, inspection and quality control of a physical product. Moreover, accurate measurement in real time is crucial for monitoring and controlling the operations of process industries, power plants, meteorology departments and so on. Knowledge of metrology and practice of its application is of increasing importance in industry.

Though the field of Metrology and Measurement is as vast as an ocean, this book is written in such a way that it covers the syllabi of most Indian Universities in the given context. The fundamentals of Statistical Quality Control are also covered to the required depth. Primarily, the fundamentals have been dealt with in detail to make the book compact and readable. Every attempt has been made to make this book more interesting and useful to scholars of Mechanical Engineering, Production Engineering, Industrial Engineering, Instrumentation and allied branches. This book will have achieved its purpose if it offers students the basic knowledge upon which further development may be built. It is our firm conviction that, in order to satisfactorily learn even the simpler aspects of Metrology and Measurement, students must undertake extensive practical experiments.

We express our sincere gratitude to Mr. M Subramaniyan, Director (Vadapalani Campus), and to Dr. C Muthamizhselvan, Director (E&T) of SRM University, for their constant encouragement in publishing this textbook. We also thank the publisher, Universities Press – in particular, Mr. Thomas Rajesh, Vice-President, and Ms. Madhavi Sethupathi, Senior Editor, for her intensive editing, which has helped shape this textbook.

We thank Dr. S Richard, Associate Professor of Mechanical Engineering of Dr. Sivanthi Aditanar College of Engineering, for his kind contribution to the chapter on temperature measurement. We also thank Mr. S Thamilarasu, Assistant Professor, SRM University, and Mr. P Manikandan, Assistant Professor, Mahendra College of Engineering, for their help with some of the figures.

We personally dedicate this work to our respective parents, spouse and child(ren). Any suggestions towards further improvement of the text will be gratefully acknowledged and incorporated in the next edition.

K Duraivelu and **S Karthikeyan**

Contents

Engineering Metrology

1

Basic Concepts of Measurement

1.1 INTRODUCTION

Metrology is the science of measurement. The word metrology comes from Greek – *metron* means measure and *logos* means study. It covers both the theoretical and the practical aspects of the measurement of any physical quantity. The International Bureau of Weights and Measures (BIPM – Bureau International des Poids et Mesures) defines metrology as, "the science of measurement, embracing both experimental and theoretical determinations at any level of uncertainty in any field of science and technology". Engineering metrology is restricted to the measurement of length and angle.

Metrology can be further classified as follows:

- **Scientific or fundamental metrology**: It concerns the establishment of quantity systems, unit systems, units of measurement, development of new measurement methods, realisation of measurement standards and the transfer of traceability from these measurement standards to end-users in society.
- **Applied, technical or industrial metrology**: It concerns the application of measurement science to production and other processes and their use in society, ensuring the suitability of measuring instruments, the calibration of measuring instruments and the quality control of measurements.
- **Legal metrology**: It concerns activities that result from statutory requirements and pertain to measurement, units of measurement, measuring instruments and methods of measurement performed by competent bodies.

A measuring system comprises (i) a measuring standard, (ii) a measuring instrument, (iii) a work-piece and (iv) a measuring environment.

1.2 SIGNIFICANCE OF MEASUREMENT

Human beings come across many things in their life that need to be measured. For example, before purchasing clothes, the required size of the cloth must be measured first. Any quantity of study in physics is called a **physical quantity**. Every accepted standard quantity is a standard of measurement, and is also called a **unit**. The measurement of a particular quantity is the same at all places and at all times. Some standard units for making physical measurements have been introduced for this purpose.

The following are the **purposes** for which any physical quantity is measured:

- To ensure that the part to be measured conforms to the established standard.
- To satisfy interchangeability of manufacture.
- To provide customer satisfaction by ensuring that no faulty product reaches the customers.
- To coordinate the functions of quality control, production, procurement and other departments of the organisation.
- To judge the possibility of making some of the defective parts acceptable after minor repairs.

1.3 PRECISION AND ACCURACY OF MEASUREMENT

Precision and accuracy (Fig. 1.1) are two important terms used in the measurement of any physical quantity. The following paragraphs define the meaning of and differences between these two terms.

Precision: It is the degree which determines how well identically performed measurements agree with each other. It is the repeatability of the measuring process. It refers to the ability of a measurement to be consistently reproduced. Precision does not carry any meaning for only one measurement. It exists only when a set of observations is gathered for the same quantity under identical conditions. In such a set, the observations will scatter about a mean. The smaller the scattering, the more precise the measurement.

Accuracy: It is the degree of agreement between the measured value and its true value. It is the degree of exactness with which the final product corresponds to the measurement standard. The difference between the measured value and the true value is known as **error of measurement.** Accuracy is the quality of conformity.

To distinguish between precision and accuracy, the following simple example can be used. A non-operational needle-clock will show the same time (precision reading) all the time, but will show the correct time (accurate reading) only twice in a day.

Of the two, only precision is required in a measuring instrument, though accuracy is usually sought for in a measuring process. Achieving high precision is easier and cheaper than achieving high accuracy. If the measuring instrument is of

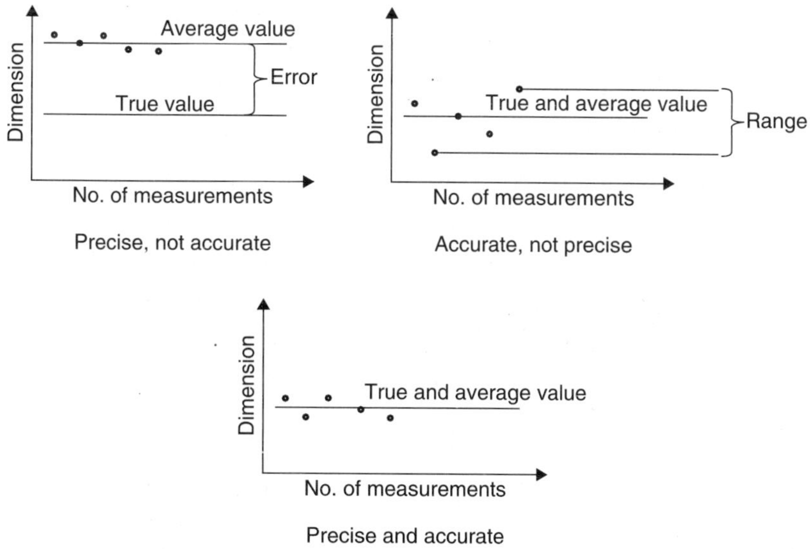

Figure 1.1 Precision and accuracy

high precision and is calibrated for error, then the true value can be easily obtained from the measured average value after deducting the instrument error. So, a high-precision rather than a high-accuracy instrument is required, considering the cost and reliability of the measuring instrument. The differences between precision and accuracy are given in Table 1.1.

The question of which of the two is more vital depends on the situation. For example, for a carpenter entrusted with the job of fitting a shelf in a cupboard, precision is more important. This can be achieved only when he uses the same scale

Table 1.1 Precision vs. accuracy

Precision	Accuracy
Repeated similar measured values	Measured value is closer to true value
The degree which determines how well identically performed measurements agree with each other	The degree of agreement between the measured value and its true value
It exists only when a set of observations is gathered for the same quantity under identical conditions	One observation is sufficient to know accuracy
It is concerned with the reliability of measurement	It is concerned with the quality of conformity

to measure the cupboard and the board for the shelf. It hardly matters whether his scale is accurate or not. If, however, such a board is ordered for purchase from a pre-cut board, accuracy becomes more vital. He must measure the size of the cupboard very accurately before placing the order.

Interchangeability: Interchangeability is the ability to select components for assembly at random and fit them together within proper tolerances. Not only should a nut from a lot fit any appropriate bolt from its lot – both being manufactured in the same plant by the same men – but, it should also fit a bolt from some other manufacturer. The simplest way to maintain compatibility of parts for interchangeable manufacture is by adopting accuracy in measurement everywhere.

1.3.1 Factors Affecting the Accuracy of the Measurement System

The following are the factors that affect the accuracy of measurement systems.

- Factors affecting the standard of measurement
  - Coefficient of thermal expansion
  - Elastic properties
  - Stability with time
  - Geometric compatibility
- Factors affecting the work-piece to be measured
  - Coefficient of thermal expansion
  - Elastic properties
  - Arrangement of supporting work-piece
  - Hidden geometry
  - Surface defects such as scratches and waviness
- Factors affecting the inherent characteristics of the instrument
  - Repeatability and readability
  - Calibration errors
  - Effect of friction, backlash, etc.
  - Inadequate amplification for accuracy objective
  - Deformation in handling or use
- Factors affecting the person
  - Improper training/skill
  - Inability to select proper standards/instruments
  - Lower commitment towards achieving accuracy during the process of measurement
- Factors affecting the environment
  - Temperature, humidity, atmospheric pressure, etc.
  - Cleanliness

- Adequate illumination
- Heat radiation from lights/heating elements

1.4 RELIABILITY OF MEASUREMENT

Reliability of measurement refers to the consistency of accurate results over consecutive measurements over time. If a measuring instrument is not precise, it will give different values for the same dimension, when it is measured again and again. Such an instrument is considered untrustworthy. The first and fundamental requirement for the effectiveness of any good measuring instrument is that it should have adequate repeatability or precision. A measuring instrument that gives precise (same) values all the time is far more reliable than one that gives accurate (true) values rarely but is not precise all the time. A precise value can be converted into an accurate value by taking into account the constant error of the precision instrument.

If the precision measuring instrument is highly calibrated for its error of measurement and the constant error of measurement is known in advance, then the accurate (true) value can be obtained as follows

$$\text{True value} = \text{Measured value} \pm \text{Error}$$

A calibrated and precise measuring instrument is more reliable and is hence used in metrological laboratories.

1.5 METHODS OF MEASUREMENT

The following are the different methods of measurement:

- **Direct measurement**: The value of the quantity to be measured is obtained directly without the necessity of carrying out supplementary calculations based on a functional dependence of the quantity to be measured in relation to the quantities actually measured. For example, the weight of a substance is measured directly using a physical balance.
- **Indirect measurement**: The value of the quantity is obtained from measurements carried out by the direct method of measurement of other quantities that are connected with the quantity to be measured by a known relationship. For example, the weight of a substance is measured by measuring the length, breadth and height of the substance directly and then by using the relation, Weight = Length × Breadth × Height × Density.
- **Measurement without contact**: The sensor is not placed in contact with the object whose characteristics are being measured.

- **Combination measurement:** The results of direct or indirect measurement or different combinations of those values are made use of and the corresponding system of equations is solved.
- **Fundamental measurement**: This is based on the measurement of the base quantities that enter into the definition of the quantity.
- **Measurement by comparison**: The value of the quantity to be measured is compared with a known value of the same quantity (direct comparison) or a known value of another quantity which is a function of the quantity to be measured (indirect comparison).
- **Measurement by substitution**: The value of a quantity to be measured is replaced by a known value of the same quantity, selected so that the effects produced in the indicating device by these two values are the same (a type of direct comparison).
- **Measurement by transposition**: The value of the quantity to be measured is, in the beginning, balanced by a known value A of the same quantity. Then the value of the quantity to be measured is put in place of this known value and is again balanced by another known value B. If the position of the element indicating equilibrium is the same in both the cases, the value of the quantity measured is equal to A and B.
- **Differential measurement**: The quantity to be measured is compared with a quantity of the same kind which has a value known to be slightly different from that of the quantity to be measured, and the difference between the values of these two quantities is measured.
- **Measurement by complement:** The value of the quantity to be measured is complemented by a known value of the same quantity, selected in such a way that the sum of these two values is equal to a certain value of comparison fixed in advance.
- **Measurement by interpolation:** It consists of determining the value of the quantity measured on the basis of the law of correspondence and known values of the same quantity, the value to be determined lying between two known values.
- **Measurement by extrapolation**: It consists of determining the value of the quantity measured on the basis of the law of correspondence and known values of the same quantity, the value to be determined lying outside the known values.

1.6 IMPORTANT TERMS IN MEASUREMENT

A few terms commonly used by scientists and engineers in the field of measurement are defined as follows:

Constant of a measuring instrument: The factor by which the indication of the instrument shall be multiplied to obtain the result of measurement.

Nominal value of a physical measure: The value of the quantity reproduced by the physical measure and which is indicated on that measure.

Conventional true value of a physical measure: The value of the quantity reproduced by the physical measure, determined by measurement carried out with the help of measuring instruments, which shows a total error that is negligible.

Standard: The physical embodiment of a unit. For every kind of quantity to be measured, there should be a unit to express the result of the measurement and a standard to enable the measurement.

Calibration: The process of determining the values of the quantity being measured corresponding to a pre-established arbitrary scale. It is the measurement of the measuring instrument. The quantity to be measured is the input to the measuring instrument. The input affects some parameter, which is the output, and is read out. The amount of output is governed by that of the input. Before we can read any instrument, a scale must be framed for the output by successive applications of some previously standardised (input) signals. This process is known as calibration.

Traceability of measurement: The ongoing validations that the measurement of the final product conforms to the original standard of measurement.

Resolution of the instrument: The smallest change in a measured value that the instrument will be able to detect perfectly. It is also known as the sensitivity of an instrument. It can be defined as the ratio of the magnitude of the output (response) to the magnitude of the input (quantity being measured). The deflection/scale factor is the reciprocal of the sensitivity of the instrument.

Example Problem 1: A Wheatstone bridge requires a change of 6 Ω in the unknown arm of the bridge to produce a change in deflection of 3 mm of the galvanometer. Find the sensitivity and scale factor of the bridge.

Solution:

$$\text{Sensitivity} = \frac{\text{Magnitude of output}}{\text{Magnitude of input}}$$

$$= 3 \text{ mm} \,/\, 7\Omega = 0.5 \text{ mm}/\Omega$$

$$\text{Scale factor} = \text{Reciprocal of sensitivity}$$

$$= 1/0.5 = 2 \;\Omega/\text{mm}$$

Readability of the instrument: The susceptibility of a measuring instrument to having its indications converted to a meaningful number. It implies the ease with which observations can be made accurately.

Repeatability of the instrument: The variation in measurement taken by a single person or instrument on the same product and under the same conditions. It is also called test-retest reliability. A measurement is said to be repeatable when this variation is smaller than some agreed limit.

Reproducibility of the instrument: The degree of agreement between measurements conducted on replicated specimens in different locations by different people using the same instrument.

Least count of the instrument: The minimum value that can be measured using the instrument.

Uncertainty of measurement: The expression of the result of a measurement that states the range of values within which the true value is estimated to lie.

Hysteresis: A phenomenon that depicts different output effects when a system is loaded and unloaded. It is the non-coincidence of the loading and unloading curves. It arises from the fact that the energy put into the stressed parts during loading is not recoverable upon unloading. This property obeys the Second Law of Thermodynamics, ruling out any perfectly reversible process in the world.

Backlash: The maximum distance or angle through which a part of a mechanical system may be moved in one direction without causing the motion of the next part.

Fidelity: The degree to which a measurement system indicates changes in the measured quantity without any dynamic error.

Dynamic error: The difference between the true value of the quantity that changes with time and the value indicated by the measurement system, if no static error is assumed. It is also called measurement error.

1.7 STANDARDS OF MEASUREMENT

The main standards of measurement systems followed across various countries are given below:

- **FPS system**: In this system, the units of length, mass, time and temperature are foot (or yard), pound (or slug), second and Rankine (or Fahrenheit), respectively. This is common in English-speaking countries and was developed by Britain.
- **Metric system**: This is a decimal system of weights and measurement is based on the metre as the unit of length. It was first used in France.

- **CGS system:** This prescribes centimetre, gram and second for length, weight and time, respectively.
- **MKS system:** This prescribes metre, kilogram and second for length, weight and time, respectively.
- **MKSA (Giorgi) system:** This added ampere, the unit of electric current, to the MKS system.
- **SI system:** In 1960, the General Conference on Weights and Measures (CGPM) formally gave the MKSA system the title, Le Systeme International d'Unites, with the abbreviation SI (also called International System of units). In SI, the main departure from the traditional metric system is the use of newton as the unit of force. India, by Act of Parliament No. 89, 1956, switched over to the SI system.

1.8 BASIC UNITS IN THE SI SYSTEM

For length: By the wavelength standard, 1 metre (m) is equal to 1,650,763.73 wavelengths in vacuum of the red-orange radiation of krypton Isotope-86 gas at 60 K. This standard overcomes all the practical problems of maintaining, handling and transferring the physical material standard, besides being highly stable and accurate. In 1982, the definition of the metre was changed to the distance light travels in 1/299,792,458ths of a second. For measurement, the light from a helium-neon laser illuminates iodine, which fluoresces at a highly stable frequency.

By the line standard, 1 metre is the distance between the axes of two lines engraved on a polished surface of a platinum–iridium bar M (90% platinum and 10% iridium) kept at the Bureau of Weights and Measures (BIPM) at Sevres near Paris (Fig. 1.2). This bar is kept at 0°C, under normal atmospheric pressure,

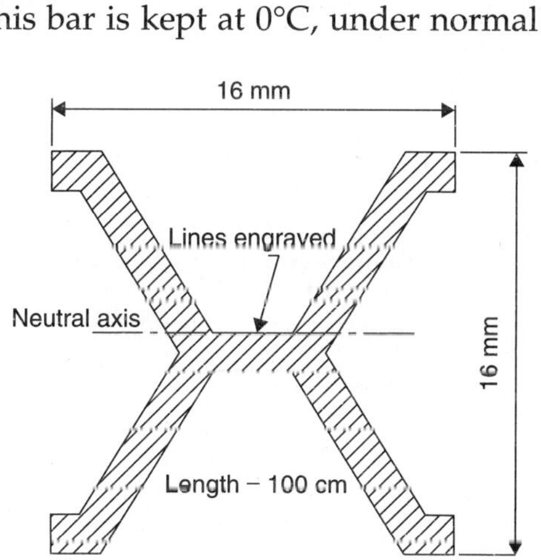

Figure 1.2 International prototype metre cross-section

and supported by two rollers of at least 1 cm diameter symmetrically situated in the same horizontal plane at a distance of 588.9 mm (Airy points) so as to give minimum deflection.

For mass: The SI unit of mass is the kilogram (kg). It is defined in terms of the platinum–iridium mass maintained under very accurate conditions at the International Bureau of Weights and Measures in Sevres, France.

For time: The SI unit of time is the second (s), which is equal to the duration of 9,192,631,770 periods of the radiation corresponding to the transition between the hyper fine levels of the ground state of the Caesium 133 atom.

For current: Ampere (A) is the constant current which, if maintained in two straight parallel conductors of infinite length of negligible circular cross-section and placed one metre apart in vacuum, would produce between these conductors, a force equal to 2×10^{-7} Newton per unit length.

For temperature: Kelvin (K) is the fraction 1/273 of the thermodynamic temperature of the triple point of water.

For luminous intensity: Candela (cd) is the luminous intensity in the perpendicular direction of a surface of 1/600,000 m^2 of a black body at the temperature of freezing platinum under a pressure of 101,325 N/m^2.

For the amount of a substance: Mole (mol) is the amount of substance of a system which contains as many elementary entities as there are atoms in 0.012 kg of Carbon-12.

Supplementary SI units

- For plane angle: radian (rad)
- For solid angle: steradian (sr)

Table 1.2 gives the seven base units in the SI system.

Table 1.2 Base units in the SI system

Quantity	Unit	
	Name	Symbol
Length	Metre	m
Mass	Kilogram	kg
Time	Second	s

(*Continued*)

Table 1.2 Continued

Quantity	Unit	
	Name	Symbol
Electric current	Ampere	A
Temperature	Kelvin	K
Amount of substance	Mole	mol
Luminous intensity	Candela	cd

Derived SI units: Other dimensions can be derived from the seven base units by multiplying or dividing them. A few such derived units are assigned special names; others are not. For example, the unit of force, newton, is derived from kilogram, metre and second as kg s^2/m. In contrast, area is simply mentioned as m^2. Some derived units carrying special names are listed in Table 1.3.

Table 1.3 Derived units

Quantity	Unit	Symbol	Equivalent other units
Plane angle	radian	rad	$m \times m^{-1}$
Solid angle	steradian	sr	$m^2 \times m^{-2}$
Frequency	hertz	Hz	s^{-1}
Force	newton	N	$kg \times m/s^2$
Pressure, stress	pascal	Pa	N/m^2
Energy	joule	J	$N \times m$
Power	watt	W	J/s
Electric charge	coulomb	C	$A \times s$
Electric potential difference	volt	V	W/A
Electric capacitance	farad	F	C/V
Electric resistance	ohm	Ω	V/A
Magnetic flux	weber	Wb	$V \times s$
Magnetic flux density	tesla	T	Wb/m^2
Inductance	henry	H	Wb/A

Some derived quantities without special units, but expressed in other units, are listed in Table 1.4.

Table 1.4 Derived quantities

Derived quantity	Symbol
Area	m^2
Acceleration	m/s^2
Angular acceleration	rad/s^2
Angular velocity	rad/s
Density	kg/m^3
Dynamic viscosity	Pa.s
Heat flux	W/m^2
Moment of force	$N \times m$
Specific heat capacity	$J/kg \times K$
Velocity	m/s
Volume	m^3

1.9 CLASSIFICATION OF STANDARDS

Line and end standards: In the line standard, length is the distance between the centres of engraved lines, whereas in the end standard, it is the distance between the end faces of the standard. For example, the line standard is a measuring scale, whereas the end standard is a block gauge.

Primary, secondary, tertiary and working standards

- **Primary standard:** There is only one material standard and it is preserved under the most careful conditions and is used only for comparison with the secondary standard.
- **Secondary standard:** It is similar to the primary standard, as nearly as possible, and is distributed to a number of places for safe custody. It is used for occasional comparison with tertiary standards.
- **Tertiary standard:** It is used for reference purposes in laboratories and workshops and is used for comparison with the working standard.
- **Working standard:** It is used daily in laboratories and workshops. Low grades of materials may be used.

1.10 ERRORS IN MEASUREMENT

Error in measurement is the difference between the measured value and the true value of the measured dimension. It can be expressed as follows:

$$\text{Error in measurement} = \text{Measured value} - \text{True value}$$

The error in measurement may be expressed as an absolute error or as a relative error.

- **Absolute error:** It is the algebraic difference between the measured value and the true value of the quantity measured. It is also called **static error**. It is further classified as
  - **True absolute error:** It is the algebraic difference between the measured average value and the conventional true value of the quantity measured.
  - **Apparent absolute error:** It is the algebraic difference between one of the measured values of a series of measurements and the arithmetic mean of all the measured values in that series.

  Static correction is the error to be corrected, and hence, the negative value of absolute error.
- **Relative error:** It is the quotient of the absolute error and the value of comparison (which may be the true value, the conventional true value or the arithmetic mean value of a series of measurements) used for the calculation of that absolute error. It is also called **relative static error.**

Example Problem 2: If the actual (true) value is 5,000 and the estimated (measured) value is 4,500, find the absolute and relative errors.

Solution: Absolute error = True value – Measured value

$$= 5{,}000 - 4{,}500 = 500 \text{ units}$$

$$\text{Relative error} = \frac{\text{Absolute error}}{\text{True value}} = \frac{500}{5{,}000}$$

$$= 0.1, \text{ i.e., } 10\%$$

1.10.1 Types of Errors in Measurement

This section discusses five major types of errors in measurement, namely, errors of measurement, instrumental errors, observation errors, errors based on nature and errors based on control.

Measurement errors

- **Systematic error:** It is the error that, during several measurements made under the same conditions of the same value of a certain quantity, remains constant in absolute value and sign or varies in a predictable way in accordance with a specified law when the conditions change. The causes of these errors may be known or unknown. The errors may be constant or variable. Systematic errors are regularly repetitive in nature and are avoidable.

- **Random error:** This error varies in an unpredictable manner in absolute value and in sign when a large number of measurements of the same value of a quantity are made under practically identical conditions. Random errors are non-consistent and are accidental in nature. Random errors are normally of limited time duration and are unavoidable. The differences between systematic and random errors are tabulated in Table 1.5.

Table 1.5 Systematic error vs. random error

Systematic error	Random error
Repetitive in nature	Non-consistent
Associated with validity	Associated with reliability
Results from improper conditions or procedures that are consistent in action	Inherent in the measuring instruments
Remains constant and varies in a predictable way in value and in sign	Varies in an unpredictable manner in value and in sign.
Can be controlled in magnitude and sense	Cannot be determined from knowledge of the measuring system or condition
Can be reduced or eliminated	Cannot be eliminated; only the results can be corrected
Examples include calibration error, environmental error, stylus pressure error, reading errors, etc.	Examples include errors due to variation in the position of the setting standard and work-piece, errors due to displacement of lever joints of instruments, errors resulting from backlash/friction, etc.

- **Parasitic error:** It is the error, often gross, which results from incorrect execution of measurement.

Instrument errors

- **Error of a physical measure:** It is the difference between the nominal value and the conventional true value reproduced by the physical measure.
- **Error of a measuring mechanism:** It is the difference between the value indicated by the measuring mechanism and the conventional true value of the measured quantity.
- **Zero error:** It is the indication of a measuring instrument for the zero value of the quantity measured.

- **Calibration error of a physical measure:** It is the difference between the conventional true value reproduced by the physical measure and the nominal value of that measure.
- **Complementary error of a measuring instrument:** It is the error of a measuring instrument arising from the fact that the values of the influence quantities are different from those corresponding to the reference conditions.
- **Error of indication of a measuring instrument**: It is the difference between the measured values of a quantity, when an influence quantity successively takes two specified values, without changing the quantity measured.
- **Error due to temperature**: It is the error arising from the fact that the temperature of instrument does not maintain its reference value.
- **Error due to friction**: It is the error due to the friction between the moving parts of the measuring instrument.
- **Error due to inertia**: It is the error due to inertia (mechanical, thermal or otherwise) of the parts of the measuring instrument.

Observation errors

- **Reading error**: It is the error of observation resulting from incorrect reading by the observer of the indication of a measuring instrument.
- **Parallax error**: It is the reading error that is produced, when, with the index at a certain distance from the surface of the scale, the reading is not made in the direction of observation provided for the instrument.
- **Interpolation error**: It is the reading error resulting from the inexact evaluation of the position of the index with regard to two adjacent graduation marks between which the index is located.

Errors based on nature

- **Systematic error** (discussed earlier)
- **Random error** (discussed earlier)
- **Illegitimate error**: As the name implies, it should not exist. These include mistakes and blunders, computational errors and chaotic errors. Chaotic errors are random errors but, unlike the latter, they create chaos in the final results.

Errors based on control

- **Controllable errors:** The sources of error are known and it is possible to control these sources. These can be calibration errors, environmental errors or errors due to non-similarity of conditions while calibrating and measuring.
  - **Calibration errors:** These are caused by variations in the calibrated scale from its normal value. The actual length of standards such as slip gauges will vary from the nominal value by a small amount. This will cause an error of constant magnitude.

- **Environmental (ambient/atmospheric condition) errors:** An international agreement has been reached on the ambient condition, which is a temperature of 20°C, pressure of 760 mmHg and humidity of 10 mmHg. Instruments are calibrated at these conditions. If there is any variation in the ambient condition, errors may creep into the final results. Of the three, the effect of temperature is the most considerable.
- **Stylus pressure errors**: Though the pressure involved during measurement is generally small, it is sufficient to cause an appreciable deformation of both the stylus and the work-piece. This will cause an error in measurement.

✦ **Avoidable errors**: These errors may occur due to parallax in the reading of measuring instruments. They occur when the scale and pointer are separated relative to one another. The two practices commonly used to minimise avoidable errors are

- Reduce the separation between the scale and pointer to minimum.
- Place a mirror behind the pointer to ensure normal reading of the scale in all cases.

Avoidable errors also occur also due to non-alignment of work-piece centres, improper location of measuring instruments, etc.

✦ **Non-controllable errors**: These are random errors that are not controllable.

1.10.2 Causes of Errors

✦ **Errors due to deflection (Errors of supports)**: When long bars are supported as a beam, they become deformed or deflected (Fig. 1.3). This elastic deformation occurs because long bars, supported at the ends, sag under their own weight. The amount of deflection depends upon the positions of the supports. This problem was considered by Sir GB Airy, who showed that the positions of the supports can be arranged to give minimum error. Slope and deflection at any point can be calculated from the theory of bending.

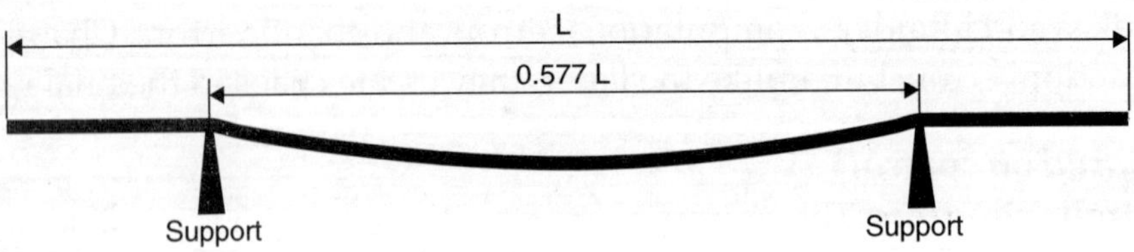

Figure 1.3 Slope at ends is equal to zero

Two conditions are considered (Fig. 1.4): For a bar of length L, supported equidistant from the centre on supports distance l apart, for no slopes at the ends, l = 0.577 L (suitable for line standards and end bars). For minimum deflection of the beam, l = 0.554 L (suitable for straight edges).

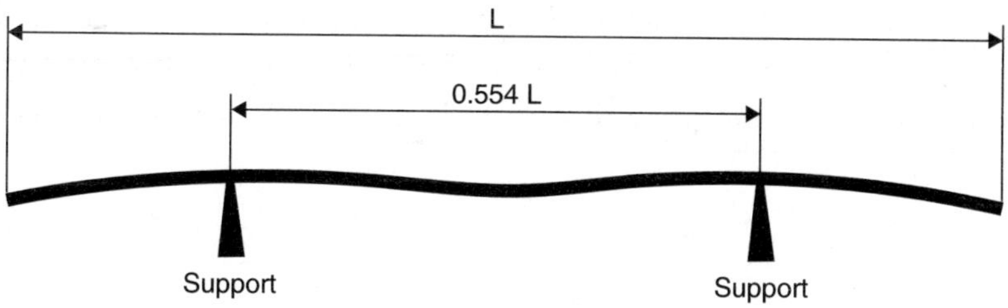

Figure 1.4 Deflection at ends is equal to deflection at centre

- **Errors due to misalignment**: Abbe's principle of alignment should be followed in measurement to avoid cosine errors, sine errors, etc. According to Abbe's principle, "the axis or line of measurement of the measured part should coincide with the line of the measuring scale or the axis of measurement of the measuring instrument".

  While measuring the length of any part, if the measuring instrument or scale is inclined to the line of dimension being measured, the error induced will be called a cosine error or a sine error. The cosine error is shown in Fig. 1.5. The combined cosine and sine error occurs if the micrometer axis is not truly perpendicular to the axis of the work–piece, as shown in Fig. 1.6.

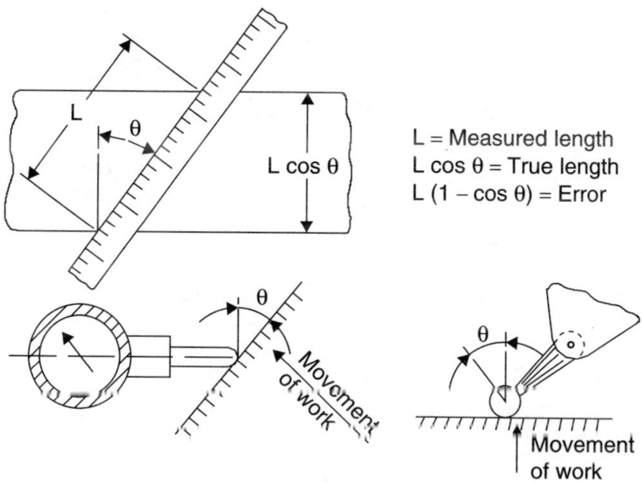

Figure 1.5 Cosine errors of measurement

Similarly, the same error occurs when measuring an end gauge in the horizontal comparator. If the gauge is not supported so that its axis is parallel to the axis of the measuring anvils, or if its ends – though parallel to each other – are not square with the axis, an error occurs. This is shown in Fig. 1.7.

The above combined cosine and sine errors can be avoided by using gauges with spherical ends. If the axes of the two spherical end gauges are not aligned, an error in length will occur. This is equal to $(a + b)$, as shown in Fig. 1.8.

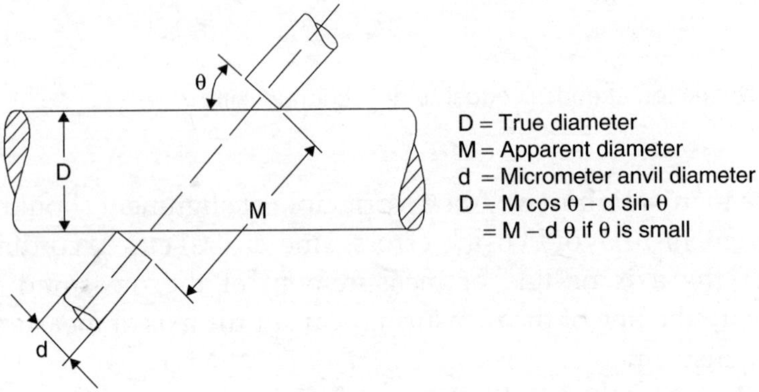

Figure 1.6 Combined cosine and sine error

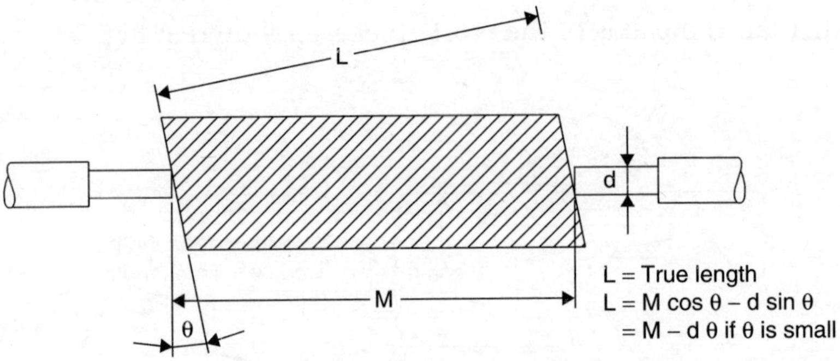

Figure 1.7 Combined cosine and sine error (in end gauge)

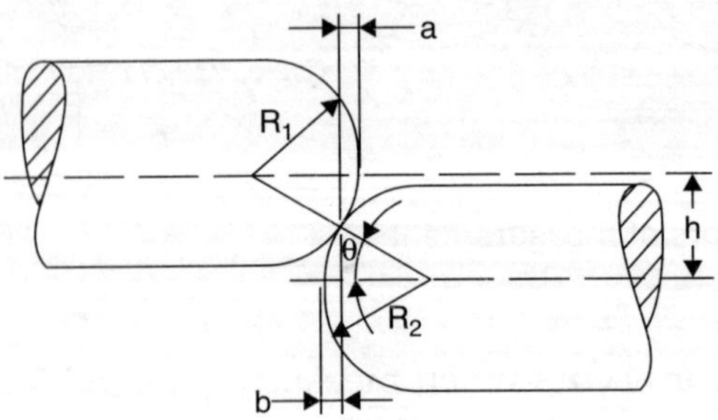

Figure 1.8 Alignment of spherical end gauges

From Fig. 1.8, it is known that

$$a = R_1(1 - \cos\theta)$$

$$= R_1(1 - \sqrt{1 - \sin^2\theta}) = R_1\left[1 - \sqrt{1 - \frac{h^2}{(R_1 + R_2)^2}}\right]$$

$$= R_1\left[1 - 1 + \frac{h^2}{2(R_1 + R_2)^2}\right] \text{ very nearly } = \frac{R_1 h^2}{2(R_1 + R_2)^2}$$

Similarly, $b = \dfrac{R_2 h^2}{2(R_1 + R_2)^2}$

Therefore, $(a + b) = \dfrac{(R_1 + R_2)h^2}{2(R_1 + R_2)^2} = \dfrac{h^2}{2(R_1 + R_2)}$

- **Error due to contact pressure**: Variations in the contact pressure between the anvils of the instrument and the work-piece being measured produce considerable differences in the reading (Fig. 1.9). The deformation of the work-piece and the anvils of the instrument depend upon the contact pressure and the shape of the contact surfaces.

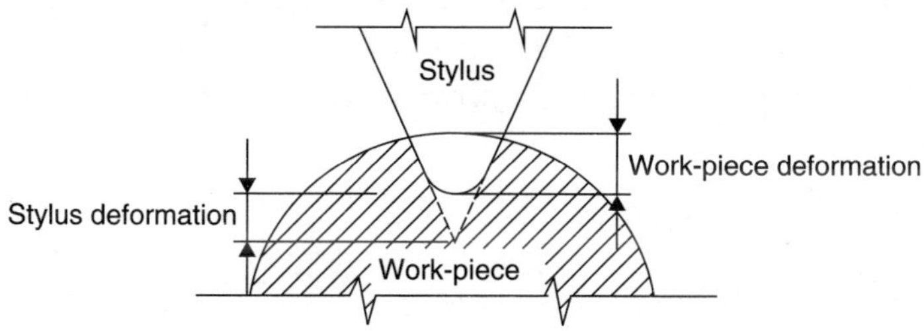

- **Error due to vibrations**: The errors due to vibrations can be avoided by
  - Locating the laboratory away from the sources of vibration
  - Placing slipping cork, felt and rubber pads under the gauge
  - Mounting the gauge pedestal or floor sections on tar mastic
  - Placing a gauge on a surface plate resting, in turn, on a heavy plate
- **Error due to dirt**: If the datum surface is not flat or if foreign matter such as dirt, chips, etc. are present between the datum and the work-piece surface, then errors will be introduced in the reading taken (Fig. 1.10).
- **Error due to poor contact**: To avoid this type of error, the gauge with the smaller area of contact should be used while measuring irregular or curved surfaces and the correct pressure should be applied while making the contact (Fig. 1.11).

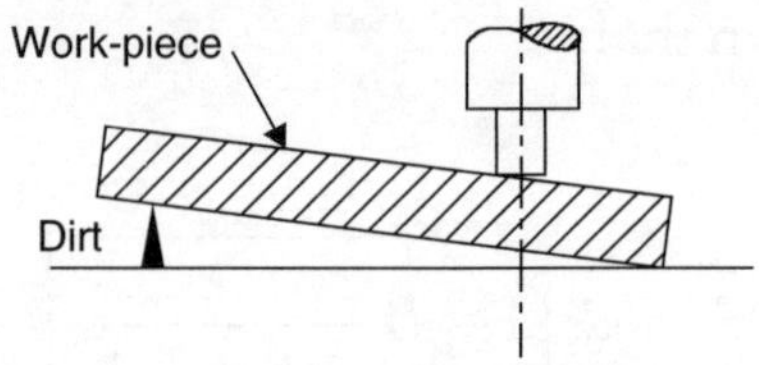

Figure 1.10 Error due to dirt

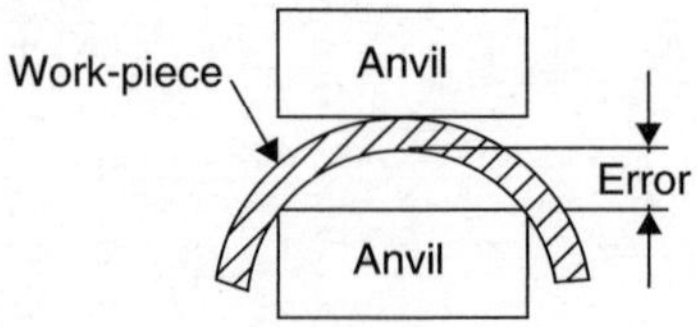

Figure 1.11 Error due to poor contact

- **Error due to wear in gauges:** Wear of the measuring surfaces of the instrument occurs due to repeated use. This error can be avoided by hardening the surfaces or by using chrome-plated parts for contact surfaces.
- **Error due to looseness**: Looseness can be tested by setting the gauge contact on the gauge anvil, zeroing the metre, applying finger pressure or a light tab to each location, where looseness might be expected, and then noting the reading again.
- **Error due to location**: Figure 1.12 shows how imperfect location of the surface causes errors.

$$\text{Reading} = a + b = c \tan \theta + t \sec \theta$$

$$\text{Error} = a + b - t = c \tan \theta + t (\sec \theta - 1)$$

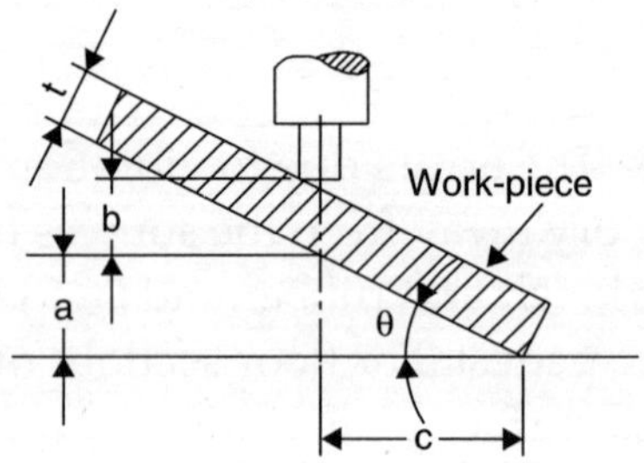

Figure 1.12 Error due to location

- **Error due to parallax effect**: Parallax error occurs when
  - The line of vision is not directly in line with the measuring scale
  - The scale and the pointer are separated from each other

In Fig. 1.13, d = separation of scale and pointer; D = distance between the pointer and the observer's eye; and θ = angle that the line of sight makes with the normal to the scale.

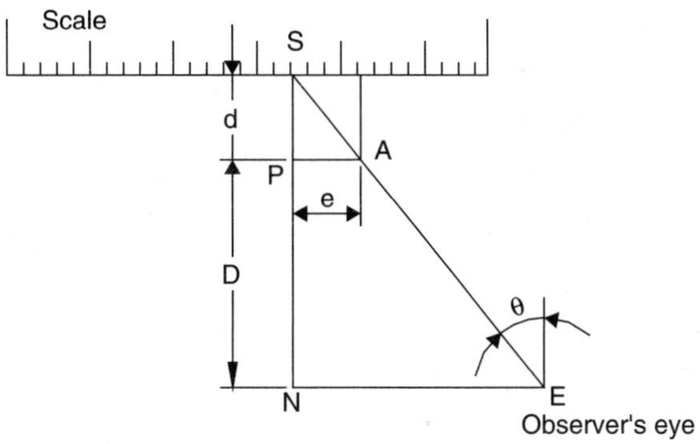

Figure 1.13 Error due to parallax effect

$$\frac{PA}{NE} = \frac{d}{d+D}; \; PA = \frac{d}{d+D} \times NE$$

As $\quad \cot(90 - \theta) = \dfrac{NE}{d+D}$

$$NE = \cot(90 - \theta)\,(d + D) = (d + D)\tan\theta]$$

Therefore, $\quad PA = \dfrac{d}{d+D}(d+D)\tan\theta$

Hence, error $\quad PA = d\tan\theta; \; e = d\,\theta$, if θ is small.

SUMMARY

- Metrology is the study of the measurement of length and angle. It has three basic subfields: scientific or fundamental metrology, applied or industrial metrology, and legal metrology.
- The significance of measurement is to ensure that the part to be measured conforms to the standard; to meet interchangeability; to provide customer satisfaction; to coordinate the functions of the quality control, production and

purchase departments; to judge the possibility of making some defective parts acceptable after minor corrective measures.

- Precision is the repeatability of the measuring process, whereas accuracy is the degree of agreement between the measured value and its true value.
- Interchangeability is the ability to select components for assembly at random and fit them together within proper tolerances.
- The various factors affecting the accuracy of a measuring system are those affecting the standard of measurement; the work-piece to be measured; the inherent characteristics of the instrument; the measuring person; and the environment.
- Reliability of measurement refers to the consistency of accurate results over consecutive measurements over time.
- The various methods of measurement are direct measurement; indirect measurement; measurement without contact; combination measurement closed series; fundamental measurement; measurement by comparison; measurement by substitution; measurement by transposition; differential measurement; measurement by complement; measurement by interpolation; and measurement by extrapolation.
- Calibration of an instrument is the process of determining the values of the quantity being measured corresponding to a pre-established arbitrary scale.
- Resolution of an instrument refers to the smallest change in a measured value that the instrument will be able to detect perfectly.
- The least count of an instrument is the minimum value that can be measured using the instrument.
- Standards are classified as line and end standards; and primary, secondary, tertiary and working standards.
- Types of error in measurement include systematic error, random error, parasitic error, instrumental error, observation error, zero error, reading error, parallax error, interpolation error, illegitimate error, calibration error, environmental error, stylus pressure error, etc.
- The causes of errors are deflection, misalignment, contact pressure, vibrations, dirt, poor contact, wear in gauges, looseness, location and parallax effect.
- Abbe's principle states that, "the axis of line of measurement of the measured part should coincide with the line of the measuring scale or the axis of measurement of the measuring instrument".

EXERCISES

Multiple Choice Questions

1. ________________ concerns the establishment of quantity systems, unit systems and units of measurement.
   a) Technical metrology
   b) Industrial metrology
   c) Scientific metrology
   d) Legal metrology

2. Each accepted standard quantity is called a ________________
   a) Unit
   b) Measurement
   c) Dimension
   d) Value

3. ________________ is the degree that determines how well identically performed measurements agree with each other.
   a) Accuracy
   b) Unit
   c) Value
   d) Precision

4. ________________ is the degree of agreement between the measured value and its true value.
   a) Precision
   b) Measurement
   c) Accuracy
   d) Limit

5. The difference between average value and true value is called ________________
   a) Accuracy
   b) Precision
   c) Error
   d) Limit

6. ________________ is concerned with the reliability of measurement.
   a) Accuracy
   b) Precision
   c) Error
   d) Limit

7. ________________ is concerned with the quality of conformity.
   a) Precision
   b) Error
   c) Accuracy
   d) Limit

8. ________________ is the ability to select components for assembly at random and fit them together within proper tolerances.
   a) Precision
   b) Interchangeability
   c) Accuracy
   d) Error

9. The following is NOT a factor that affects the standard of measurement:
   a) Coefficient of thermal expansion
   b) Elastic properties
   c) Hidden geometry
   d) Stability with time

10. ________________ of measurement refers to the consistency of accurate results over consecutive measurements over time.
a) Accuracy
b) Precision
c) Reliability
d) Limit

11. True value is
a) Measured value / Error
b) Measured value × Error
c) Measured value + Error
d) Measured value – Error

12. ________________ is a type of measurement where the value of the quantity is obtained directly.
a) Indirect measurement
b) Fundamental measurement
c) Direct measurement
d) Substitution measurement

13. Measuring the weight of a substance using a physical balance is an example of
a) Indirect measurement
b) Direct measurement
c) Fundamental measurement
d) Substitution measurement

14. ________________ is a method of measurement where the value of the quantity is measured on the basis of law of correspondence and known values.
a) Direct measurement
b) Indirect measurement
c) Fundamental measurement
d) Interpolation measurement

15. ________________ is the process of determining the values of the quantity being measured corresponding to a pre-established arbitrary scale.
a) Standardisation
b) Calibration
c) Traceability
d) Measuring

16. ________________ of the instrument refers to the smallest change in a measured value that the instrument will be able to detect perfectly.
a) Standard
b) Calibration
c) Traceability
d) Resolution

17. Sensitivity of an instrument is given as
a) Response × Quantity being measured
b) Response / Quantity being measured
c) Response – Quantity being measured
d) Response + Quantity being measured

18. ________________ is the susceptibility of the measuring instrument to having its indications converted to a meaningful number.
a) Sensitivity
b) Readability
c) Repeatability
d) Reproducibility

19. ________________ is the variation in measurement taken by a single person (or) instrument on the same product and under the same conditions.
   a) Sensitivity
   b) Reproducibility
   c) Readability
   d) Quantity

20. The minimum value that can be measured using an instrument is termed
   a) Readability
   b) Reproducibility
   c) Least count
   d) Readability

21. ________________ is a phenomenon that depicts different output effects when a system is loaded and unloaded.
   a) Uncertainty
   b) Hysteresis
   c) Least count
   d) Repeatability

22. ________________ is defined as the degree to which a measurement system indicates changes in the measured quantity without any dynamic error.
   a) Uncertainty
   b) Hysteresis
   c) Fidelity
   d) Least count

23. ________________ is the error which, during server measurement made under the same conditions, remains constant (or) varies in a predictable way in accordance with a specified law.
   a) Random error
   b) Systematic error
   c) Absolute error
   d) Relative error

24. ________________ error varies in an unpredictable manner in absolute value and sign when a large number of measurements of the same value of a quantity are made.
   a) Systematic
   b) Random
   c) Absolute
   d) Relative

25. ________________ is the indicator of a measuring instrument for the zero value of a quantity measured.
   a) Calibration error
   b) Corresponding error
   c) Zero error
   d) Temperature error

26. By reducing the separation between scale and pointer to minimum, error is
   a) Equalised
   b) Minimised
   c) Maximised
   d) Remains unchanged

27. Combined cosine and sine errors can be avoided by using gauges with ________________ ends.
   a) Rectangular
   b) Circular
   c) Taper pin
   d) Spherical

28. The degree of closeness of a measured value of a certain quantity with its true value is ____________
    a) Standard
    b) Accuracy
    c) Sensitivity
    d) Precision

29. The ability by which a measuring instrument can detect small differences in the quantity being measured by it is called ____________
    a) Sensitivity
    b) Damping
    c) Accuracy
    d) Flexibility

30. To compare an unknown with a standard through a calibrated system is called ____________
    a) Direct comparison
    b) Indirect comparison
    c) Both a and b
    d) None of the above

31. The following is an internationally recognised and accepted unit system:
    a) MKS
    b) FPS
    c) SI
    d) All of the above

32. The following is a line standard of measurement:
    a) Measuring tape
    b) Slip gauge
    b) Micrometer
    d) End bars

33. The principle of interchangeability is used for
    a) Mass production
    b) Production of identical parts
    c) Parts within the limits of sizes
    d) All of the above

Answers

1. c	2. b	3. d	4. c	5. c
6. b	7. c	8. b	9. c	10. c
11. c	12. c	13. b	14. d	15. b
16. d	17. b	18. b	19. b	20. c
21. b	22. c	23. b	24. b	25. c
26. b	27. d	28. b	29. a	30. b
31. c	32. a	33. d		

Review Questions

1. Describe the significance of measurement.
2. Write the differences between precision and accuracy. Give a few examples.

3. What are the general characteristics affecting the selection of measuring instruments?
4. What are the factors affecting the accuracy of a measuring system?
5. Define calibration of a measuring instrument. What is the need for calibration?
6. Define the following terms with respect to a measuring instrument:
   i) Sensitivity
   ii) Readability
   iii) Repeatability
   iv) Reproducibility
   v) Interchangeability
   vi) Reliability
   vii) Resolution
   viii) Least count
   ix) Uncertainty
7. Define the static and dynamic characteristics (static and dynamic response) of a measuring instrument.
8. Why are standards of measurement necessary?
9. What are the major standards of a measuring system?
10. Give the standard for length, mass, time and temperature.
11. Define metre, kilogram and second in the SI system.
12. Discuss the fundamental and derived units in detail.
13. What is the difference between primary, secondary and tertiary standards?
14. Distinguish between line and end standards. Give examples.
15. How is an error in measurement expressed?
16. A 10.00-mm job is measured with an instrument and the following readings were obtained: 10.02, 9.97, 10.03, 9.96 and 9.98 mm. What is the error of the instrument?
17. Explain the types of errors in measurement, with examples.
18. Discuss the following errors in measuring instruments and methods of preventing the same: a) Loading error b) Systematic error c) Illegitimate error
19. Differentiate random errors from systematic errors.
20. Explain the causes of errors, with suitable examples.
21. Define Abbe's principle of measurement.
22. What is cosine error?
23. Derive the expression to find the value of combined cosine and sine error in measurement.
24. Derive the expression to find the value of error due to parallax effect during measurement.
25. Discuss the classification of measuring methods.
26. Distinguish between the direct and indirect methods of measurement. Give examples to support your answer.

2

Linear Measurement

2.1 INTRODUCTION

Measurement systems are mainly used in industries for quality inspection and control. Quality control engineers often use some measuring system, such as linear and angular measuring instruments, for quality inspection. These measuring instruments are used to compare the actual dimensions with existing standard dimensions.

Linear measurement is concerned with the measurement of length, diameter, height and thickness. It is used to find the distance between two specific points on a straight line. The principle of linear measurement is to compare the actual measurements with standard dimensions that have been accurately marked on the instrument.

Measuring instruments are designed for two purposes:

- **Line measurement**: The distance between two lines is measured. For example, the steel rule.
- **End measurement**: The distance between two surfaces is measured. For example, the block gauge.

Some of the common measuring instruments used for linear measurements are

- Vernier calipers
- Micrometer (Screw gauge)
- Vernier height gauge
- Depth gauge

Vernier scale: When two scales that are slightly different in size are used, the accuracy of measurement is enhanced.

Least count of Vernier instruments: Vernier instruments always have two scales, namely, the main scale and the Vernier scale. When the zeros of the main scale and the Vernier scale are made to coincide, the Vernier scale will have one more division than that of the main scale. Hence, the value of a division on the Vernier scale is slightly smaller than that of the main scale. This difference is the least count

(LC) of the instrument. In general, the least count is the minimum value that can be measured using the instrument.

$$LC = \begin{matrix}\text{Value of one division} \\ \text{on the main scale}\end{matrix} - \begin{matrix}\text{Value of one division} \\ \text{on the Vernier scale}\end{matrix}$$

LC for Vernier calipers, Vernier height gauge and Vernier depth gauge can be written as

$$\text{Least count (LC)} = 1\ \text{MSR} - 1\ \text{VSD}$$

where MSR is the value of the main scale reading and VSD is the Vernier scale division (value in terms of the main scale).

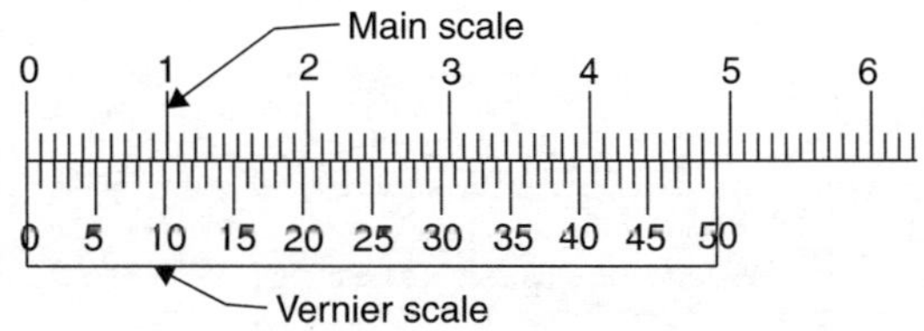

Figure 2.1 Vernier scale

For the Vernier scale shown in Fig. 2.1, the LC is calculated as

$$LC = 1\,\text{mm} - \frac{49}{50}\,\text{mm} = 0.02\,\text{mm}$$

LC can also be calculated as

$$\text{Least count} = \frac{\text{One main scale division value}}{\text{Number of divisions in the Vernier scale}}$$

Accordingly, for Fig. 2.1 $LC = \dfrac{01}{50}\,\text{mm} = 0.02\,\text{mm}$

Range of the instrument: The minimum and the maximum values that can be measured using the instrument.

2.2 LINEAR MEASUREMENT INSTRUMENTS

Gauges are used to check engineering components produced on a large scale, where the job is usually handled by semi-skilled workers. This type of measurement cannot be relied upon if accuracy is more important. The different methods and instruments used for precise and accurate measurements are discussed in this chapter. The conventional instruments for linear measurement like Vernier calipers, micrometer, Vernier height gauge and depth gauge are discussed in this section.

Vernier calipers: This is used to measure the width, depth and diameter of components (internal and external). The main scale carries a fixed jaw and the movable scale carries a movable jaw (Fig. 2.2). Both the jaws have measuring tips and the measuring faces are called 'nibs'.

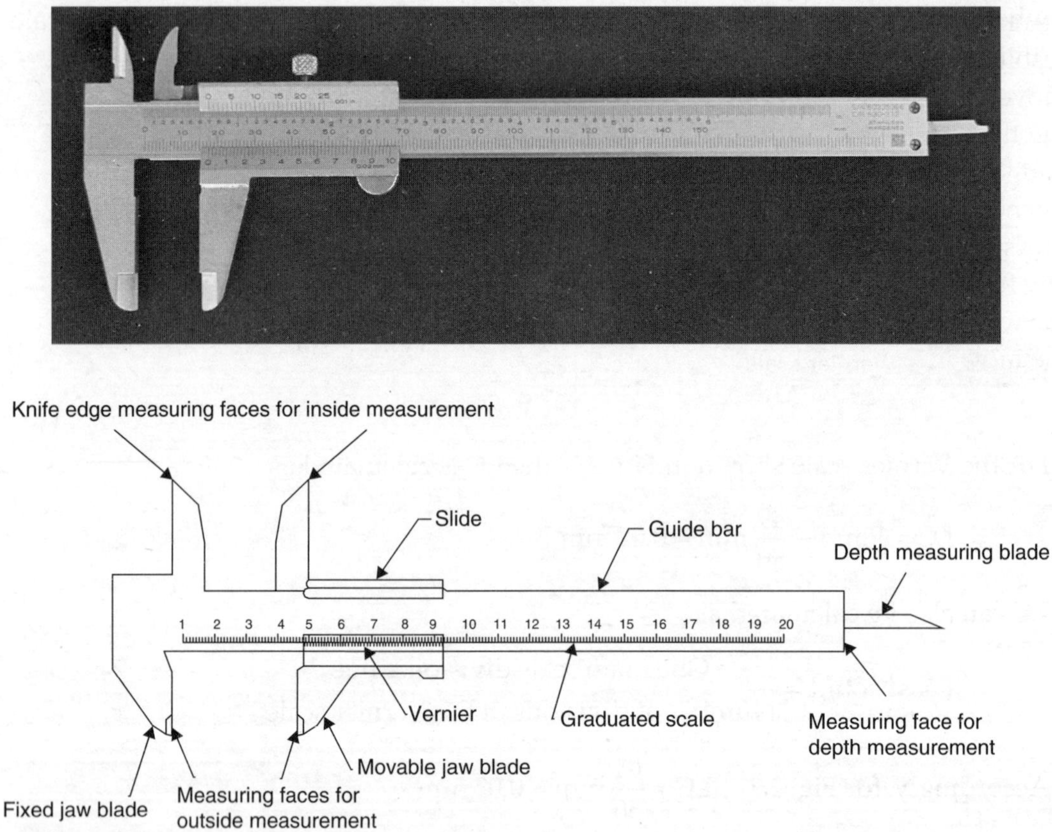

Figure 2.2 Vernier calipers

For external measurement, the nibs are flat, while for internal measurement, they are curved. When both the jaws are closed, the zero on the main scale should coincide with the zero on the Vernier scale. Adjustment screws are provided for precise setting of the movable jaw. A clamping screw is provided to lock the movable Vernier scale. While using the Vernier calipers, the zero line on the Vernier scale should coincide with the zero line on the main scale.

The work-piece whose dimension is to be measured should be held between the measuring jaws. The reading on the main scale to the left of the zero on the Vernier scale is noted. Also, one of the lines on the Vernier scale that exactly coincides with a line on the main scale is noted. Then the total reading is calculated as:

Total reading = Main scale reading + [Number of the Vernier scale division that coincided with the main scale division × Least count]

Micrometer (Screw gauge): The working principle of the micrometer is that the distance moved by the nut along the screw is proportional to the number of revolutions made by the nut (Fig. 2.3). Hence, by controlling the number of resolutions and fractions of the revolution made by the nut, the distance the nut moves along the screw can be accurately determined. The ratchet at the end of the thimble is used to ensure that uniform pressure is applied between the anvil and spindle faces, after touching the ends of the work-pieces to be measured. This ratchet arrangement is used to avoid errors due to variations in the measuring pressure. The clutch in the ratchet screw arrangement will start slipping to prevent any further movement of the spindle when the spindle is brought into contact with the work-piece at the correct measuring pressure. The locknut provided will arrest the movement of the spindle to ensure a correct reading.

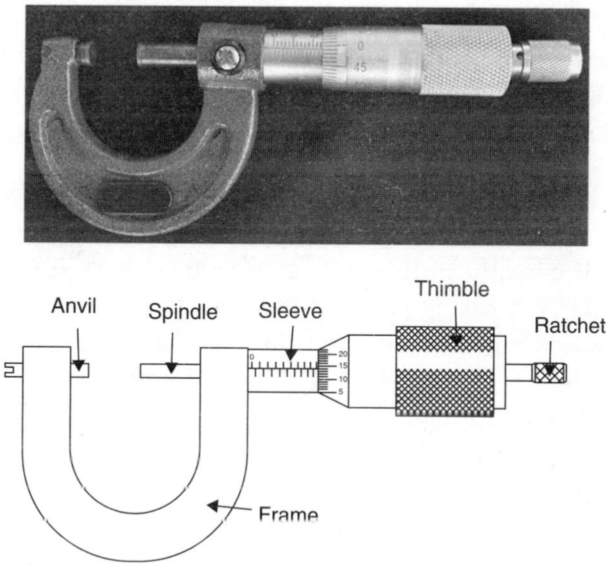

Figure 2.3 Micrometer

The micrometer has a precision screw nut assembly at one end of the frame. The pitch of the screw is usually 0.5 mm, and hence, the flat end of the spindle extending from the screw which forms one of the measuring tips moves by 0.5 mm for every rotation of the screw. The other fixed reference is an anvil at the other end of the frame of the micrometer. The spindle is a slide fitted over a barrel and the barrel is attached to the frame. The barrel is usually graduated in units of 0.5 mm. The thimble has 50 divisions around its periphery on the circular portion. Thus, each division on the thimble corresponds to 0.01 mm.

The accuracy and precision of the micrometer may be affected by the following important errors:

- **Zero error**: This is the reading indicated when there is no gap between the anvil and spindle faces. This error could be due to wear of the measuring faces of the spindle and anvil or due to slackness of the end float of the spindle screw. The error due to wear of the measuring faces can be rectified by rotating the thimble relative to the spindle. The error due to slackness can be rectified by tightening the taper nut on the outside of the partially slotted hub nut.
- **Screw pitch error**: The screw of the micrometer is generally subjected to progressive or periodic pitch errors. Progressive pitch error is due to constant pitches, but differs slightly from the correct pitch. Periodic error occurs periodically, say, for every rotation of the screw. If the wear in the micrometer screw is not even, erratic pitch errors may occur.

Laser micrometer: When the object is placed between a scanning laser beam and a sensor, the diameter/dimension is sensed by the sensor and digitally displayed on the monitor. This set-up is called a laser micrometer, and is discussed in detail in Chapter 5.

Vernier height gauge: This is also a Vernier calipers but is attached to a special base block and has other attachments (Fig. 2.4). The whole assembly is composed in such a way that the height of the parts can be measured. A removable clamp is attached between the measuring jaw and the Vernier scale. A scriber is fitted with the measuring jaw to scribe the lines on the parts, whenever it is required. The measuring jaws of the Vernier height gauge can be replaced by a dial gauge. The surface plate is used as a datum surface. A slider is attached to the Vernier gauge to make fine adjustments to the Vernier scale against the main scale. To fit the slider perfectly, a clamp is provided. The procedure to calculate the total reading is the same as for the Vernier calipers.

Vernier depth gauge: This is also a type of Vernier calipers. The graduated scale directly slides through the base, but the Vernier scale remains fixed (Fig. 2.5). This instrument is used for measuring the depth of holes, recesses and the distance from a surface. During measurement, the base and the anvil should be firmly rested on the surface of the work-piece to be measured. The process for obtaining

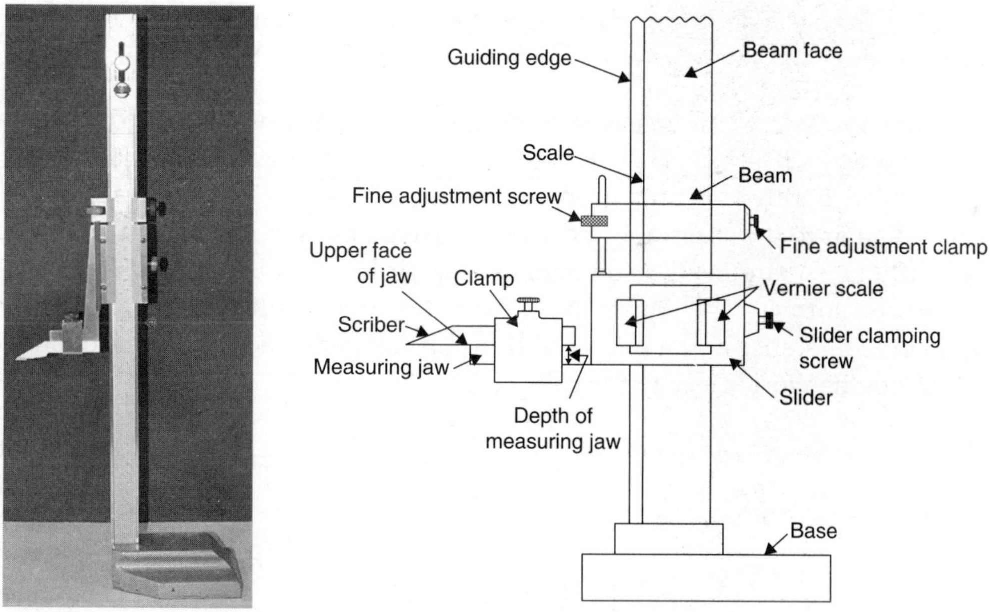

Figure 2.4 Vernier height gauge

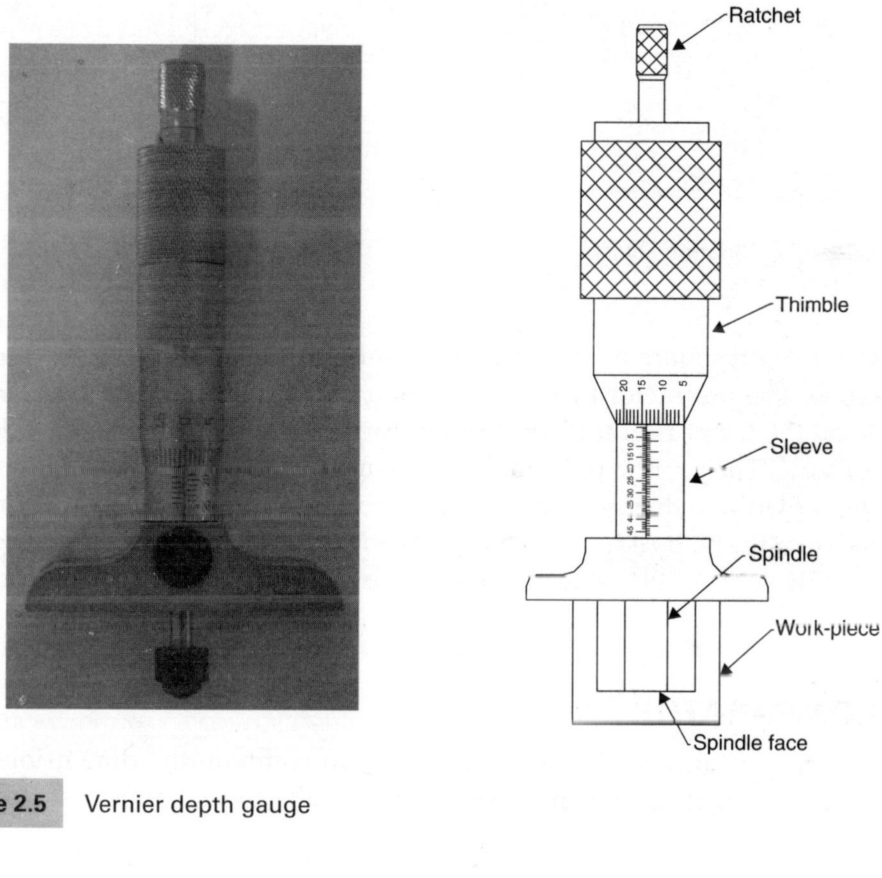

Figure 2.5 Vernier depth gauge

a reading from the depth micrometer is the same as for an external measurement in a micrometer. However, the scale is generally calibrated in the reverse direction.

Combination set: This consists of a steel rule, square head, protractor and centre head (Fig. 2.6). The heavy steel rule has grooves all along its length. A sliding square head is fitted on this groove. One surface of the square head is always perpendicular to the scale, and it can be adjusted at any point using a locking bolt and nut. The square head also contains a spirit level which is used to check the level of the surface. The square head provides two additions to the steel rule: a right-angle reference and a means to transfer the measured point or the reference point from the work-piece to the rule.

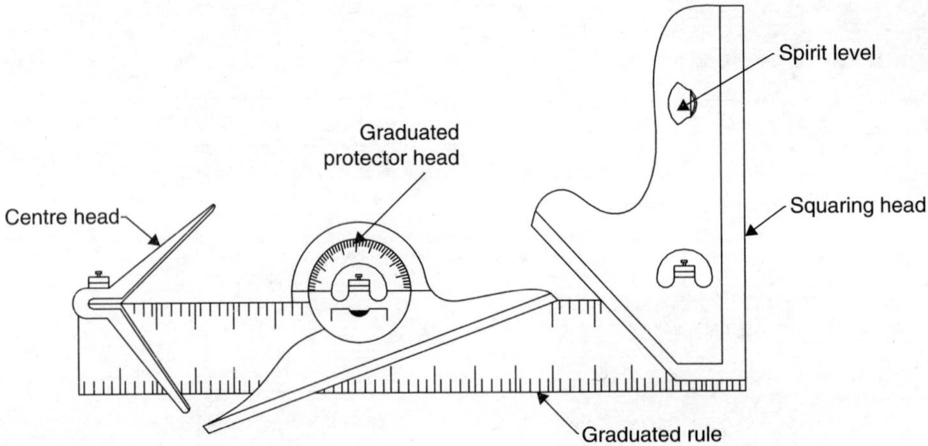

Figure 2.6 Combination set

The steel rule with square head is called a **combination square**. This is used as a depth gauge. The protractor head consists of a turret that rotates within a stock. The scale on the turret is graduated from 0° to 180° in either direction – a double protractor scale. The protractor head is used to measure the angle. When the blade (steel rule) is placed in the centre head, one edge of the blade bisects the V-angle of the head so that the blade edge lies along the centre line of any circle placed against the faces of the V. The centre head is widely used for measuring the diameter of a circle.

2.3 COMPARATOR

This is a precision measuring instrument used to compare the dimensions of a given component with a working standard (generally block gauges). It does not measure the actual dimension but indicates by how much the component differs from the basic dimensions (working standard).

Uses

- For calibrating working gauges
- As a working gauge
- As a final inspection gauge

Essential characteristics of a good comparator

- Robust design and construction
- Linear characteristics of scale
- High magnification
- Quick results
- Versatility
- Minimum wear of contact point
- Free from backlash
- Quick insertion of work-piece
- Provision for compensation from temperature effects
- Means to prevent damage to the work-piece during use

2.3.1 Classification of Comparators

Comparators can be broadly classified as given below:

- Mechanical comparator
  - Dial indicator
  - Johansson Mikrokator comparator
  - Sigma comparator
  - Reed type mechanical comparator
- Optical comparator
  - Zeiss ultra optimeter
  - Zeiss optotest comparator
- Mechanical–Optical comparator
- Pneumatic comparator
  - Back pressure comparator
  - Flow–velocity pneumatic comparator
  - Differential comparator
- Electrical comparator
- Electronic comparator
- Fluid displacement comparator

In addition, the comparators used for calibration of gauges – are:

- Brooke's level comparator
- Eden-Rolt millionth comparator

2.3.2 Basic Working Principle of a Comparator

The comparator is first adjusted to zero on its dial or recording device with a gauge block in position. The gauge block (slip gauge) is of the dimensions that the work-piece should be. The work-piece to be checked is then placed in the position of the gauge block and the comparator gives the difference in dimensions in relation to the gauge block. The dimensions of the work-piece may be less than, equal to or greater than the standard dimensions. The difference in dimensions will be shown on the dial or on the recording device of the comparator.

2.3.3 Mechanical Comparators

Various mechanical comparators are discussed in this section.

Dial indicator: This is the simplest type of mechanical comparator. It consists of a base with a rigid column and an arm carrying a dial gauge (dial indicator). The arm can be adjusted vertically up and down along the column. The arm can be swivelled and the dial gauge can be locked in any position along its arm. The whole set-up is placed on the surface plate, which is used as a datum surface (Fig. 2.7).

Initially, the slip gauges are built for the required height (standard dimensions) for which the work-piece needs to be checked and placed beneath the ball anvil of the plunger of the dial gauge, and the dial gauge is manually adjusted for the zero reading for this known height of the slip gauges. The slip gauges are then replaced by the work-piece, which is compared with the standard dimensions (height of the slip gauges is set).

If the dial indicator shows zero deflection and the plunger is in contact with the work-piece, the work-piece height is said to be the height of the slip gauges set (standard dimensions). If a deflection is shown on the dial indicator, the deflection reading gives the increase in the height of the work-piece from the standard dimensions. If the plunger does not touch the work-piece, then a set of slip gauges is built and attached on top of the work-piece so as to bring the plunger into contact with the work-piece, and the dial indicator shows a zero reading. The height of the slip gauges set gives the decrease in the height of the work-piece from the standard dimensions.

Johansson Mikrokator comparator: This consists of a thin metal strip that carries at the centre of its length a very light glass tube pointer. From the centre, the metal strip is permanently twisted to form right- and left-hand helices. One

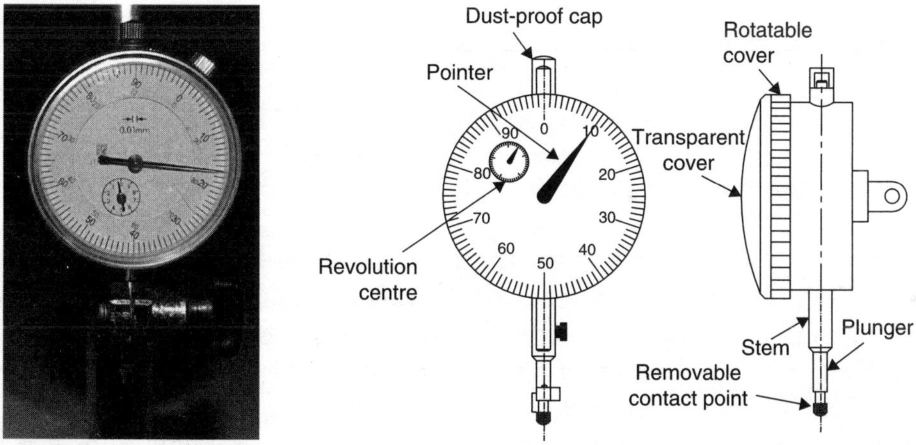

Figure 2.7 Dial gauge

end of the metal strip is fixed to the adjustable cantilever strip and the other end is anchored to the spring strip elbow. One arm of the spring strip elbow is carried on the measuring plunger.

The measuring plunger moves through a slit washer which constraints the moving plunger to move only in the vertical direction. As the measuring plunger moves, upwards or downwards, the elbow acts as a bell crank lever and causes the twisted strip to change its length, and thus, further twist or untwist. Hence, the pointer at the centre of the twisted strip rotates an amount ($d\theta$) against a circular scale proportional to the change in length ($d\ell$) of the strip.

The set-up of the Johansson Mikrokator comparator is shown in Fig. 2.8.

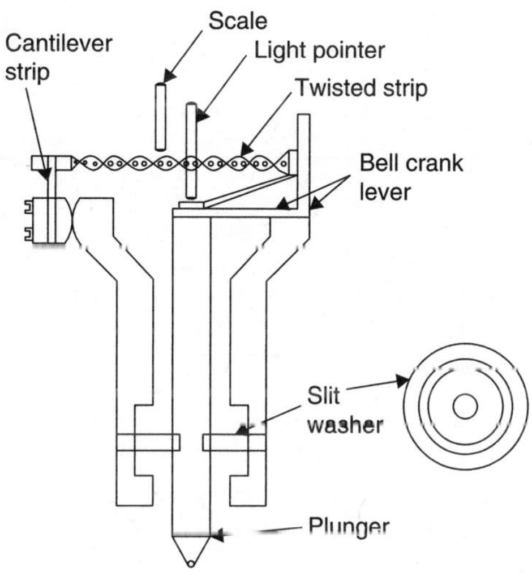

Figure 2.8 Johansson Mikrokator comparator

It can be shown that the ratio $\frac{d\theta}{d\ell} = \text{Amplification} = -\frac{9.1\ell}{W^2 n}$

where ℓ is the length of the twisted strip, measured along its neutral axis; W is the width of the twisted strip; n is the number of turns and θ is the twist of the midpoint of the strip with respect to the end.

Sigma comparator: It consists of a plunger, mounted on a pair of slit diaphragms to enable frictionless linear movement. A knife edge is mounted on the plunger. The knife edge bears upon the face of the moving member of a cross-strip hinge. The cross-strip hinge consists of a moving block and a fixed member, connected by thin flexible strips alternately at right angles to each other. It can be shown that if an external force is applied to the moving member, it will pivot as a hinge, about the line of intersection of the strips. Attached to the moving member is an arm which divides into a Y form. The constructional details of a typical Sigma comparator are shown in Fig. 2.9.

If the effective length of the Y arm is L and the distance from the hinge pivot to the knife edge is x, then the first stage of the magnification is L/x. To the extremities of the Y arm are attached a phosphor-bronze strip which is passed around a drum of radius r attached to the pointer spindle. If the pointer length is R, then the second stage of magnification is R/r, and the total magnification is $LR/(xr)$. The magnification can be adjusted by slackening one screw and tightening the other screw attaching the knife edge to the plunger, and thus, adjusting the distance x.

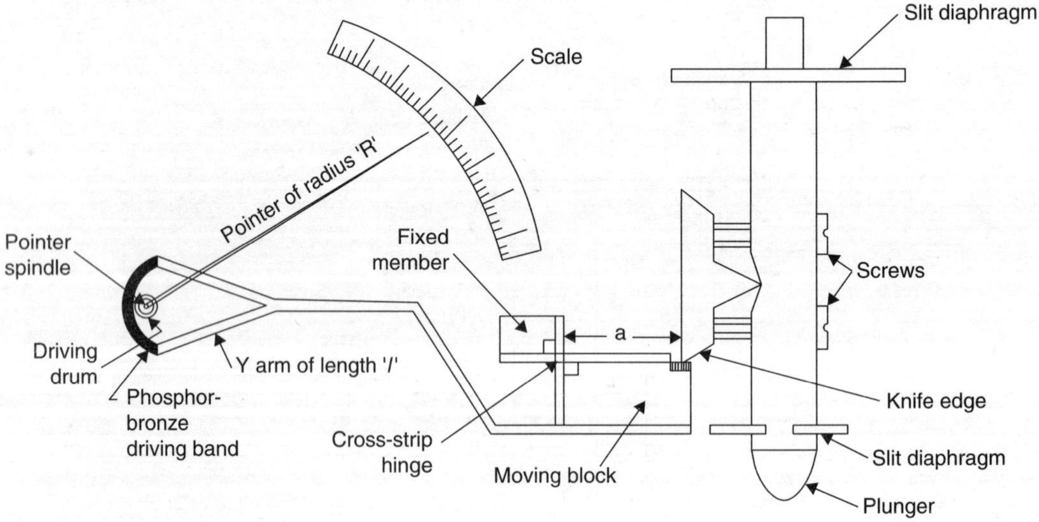

Figure 2.9 Sigma mechanical comparator

A range of instruments of differing magnification can be produced with drums of different radii *r* and suitable strips.

As the plunger moves up, it will cause the knife edge to move, which in turn will make the moving block of cross-strip hinge pivot. Hence, the entire Y arm swings about at the line of intersection of the strips, making the phosphor-bronze strip pass around the drum *r*. The drum *r* will rotate proportionally, causing the pointer to swing against the circular scale. The circular scale is already calibrated with the known length of movement of the plunger, using block gauges. Thus, the linear movement of the plunger is converted to the angular rotation of the pointer to obtain a direct reading from the circular scale.

The other interesting features of this instrument are:

- Better safety
- Deadbeat readings can be obtained
- Fine adjustment is possible
- Parallax error is avoided
- The measuring pressure is kept constant by the magnet plunger

Reed type mechanical comparator: It consists of a sensitive, high-quality dial indicator mounted on a base supported by a sturdy column. It comprises two blocks, namely, a fixed block and a moving block. The fixed block is rigidly fastened to the gauge head while the moving block carries the gauging spindle and is connected horizontally to the fixed block with reeds. There is a pointer at the joint block of the vertical reeds. When the work-piece is placed below the spindle, the movement of the spindle causes the movable block to move vertically. The vertical reed on the movable block thus slides past the vertical reeds on the fixed block. This makes the pointer swing against the circular scale. The amount of swing is proportional to the distance that the moving block has moved. The scale against the pointer is calibrated by means of slip gauges. The set-up is shown in Fig. 2.10.

2.3.4 Optical Comparator

In this system, a small displacement of the measuring plunger is amplified by a mechanical system consisting of pivoted levers. This amplified mechanical movement is further amplified by a single optical system involving the projection of an image. The scale is calibrated using slip gauges. The set-up is shown in Fig. 2.11.

In such a mechanical optical comparator, as shown in Fig. 2.11,

$$\text{Mechanical amplification} = \frac{L_2}{L_1} \text{ units}$$

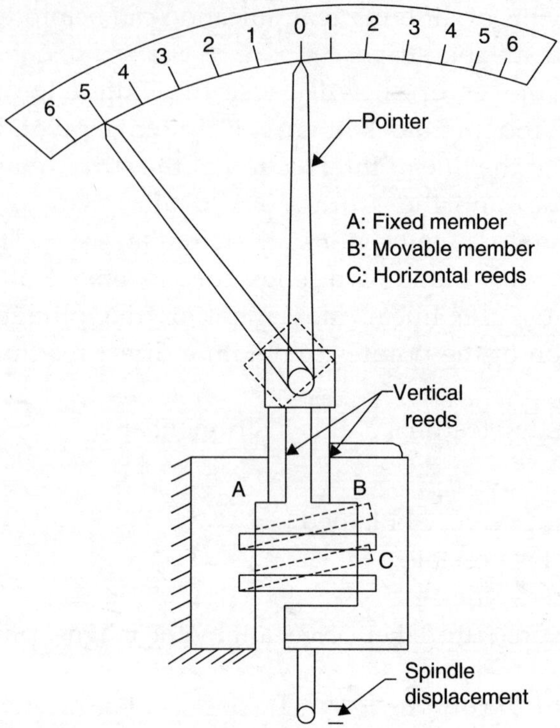

Figure 2.10 Reed type mechanical comparator

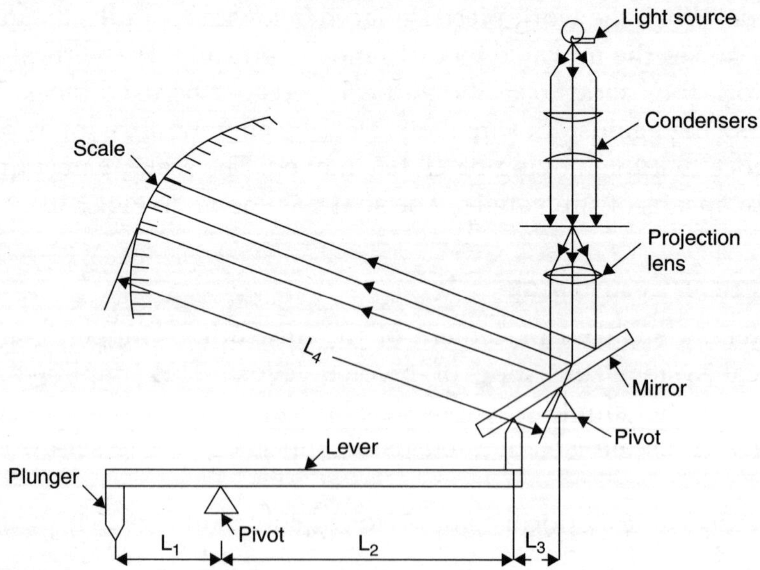

Figure 2.11 Optical comparator

Optical amplification = $(L_4 / L_3) \times 2$ units (as the reflected ray is deflected at an angle of 2θ from the original line and gives the corresponding reading on the scale, and hence, the optical amplification that is obtained is two times the original value on the scale)

$$\text{Total amplification} = \left(\frac{L_2}{L_1}\right) \times \left(\frac{L_4}{L_3}\right) \times 2 \text{ units}$$

From the value of total amplification, it is obvious that optical comparators are capable of a high degree of measuring accuracy owing to high magnification and reduction of the moving members to the minimum. Also, this system has fewer wear problems as the only wearing members are the plunger, its guide and mirror pivot bearing. It is important that the mirror used in this system be of the front reflection type.

2.3.5 Pneumatic Comparator

In this type of comparator, no physical contact is made with the setting gauge or with the part being measured. Internal dimensions may be readily measured, not only with respect to tolerance boundaries, but also to geometric form. Further, the system lends itself to the inspection of a single or a number of dimensions simultaneously, either during or immediately after the operating cycle of a machine tool.

Principle: When the measuring probe is directed towards measurement, the back pressure varies when the gap or distance between the work and the head varies. Back pressure variation indirectly measures the dimension.

Back pressure pneumatic comparator: This uses a water manometer to indicate the back pressure. It consists of a vertical metal cylinder filled with water to a certain level and a dip tube immersed in it to a depth corresponding to the air pressure required. A calibrated manometer tube connects the cylinder and the control orifice, as shown in Fig. 2.12.

The air from its normal source of supply is filtered and passes through a flow valve. Its pressure is then reduced and maintained at a constant value by a dip tube immersed in a water chamber, the pressure value being determined by the head of the water displaced, excess air escaping into the atmosphere. The air at reduced pressure then passes through the control orifice and escapes from the measuring (jet) orifice. The back pressure in the circuit is indicated by the head of the water displaced in the manometer tube. The tube is graduated linearly to show the change in pressure resulting from changes in dimension d of the internal hole of the work-piece being measured.

Amplifications of up to 50,000 are obtainable with this system. The set-up of a typical Solex pneumatic comparator is shown in Fig. 2.12.

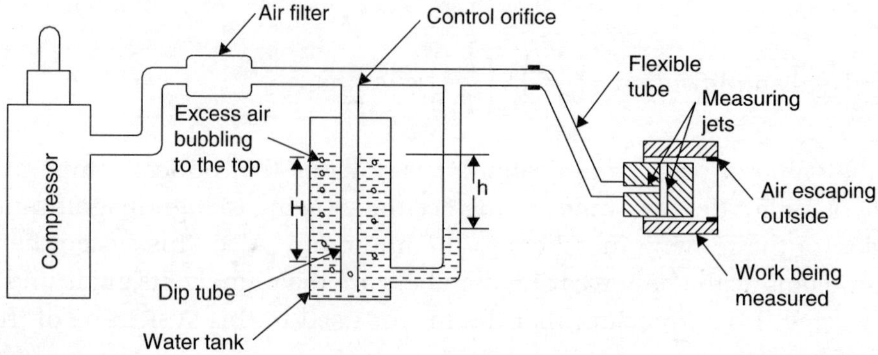

Figure 2.12 Back pressure pneumatic (Solex) comparator

Flow–Velocity type comparator: It consists of a vertical glass column (straight or tapered) with a scale engraved on the glass wall. A float is kept inside the glass column. The set-up of a flow velocity comparator is shown in Fig. 2.13.

Pressurised air is supplied to the vertical glass column through a filter and a pressure regulator. A shut-off valve is used to allow the filtered air to the pressure regulator when air supply is required. The top end of the vertical glass column is connected to the gauging head through a plastic tube. When the pressurised air passes through the glass column, the float floats based on the pressure difference in the gauging head.

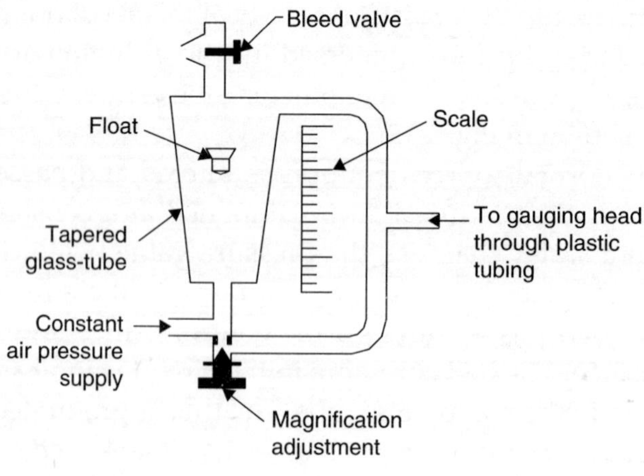

Figure 2.13 Flow–velocity pneumatic comparator

The set-up is calibrated using standard holes. The greater the gap between the outer wall of the gauging head and the inner wall of the hole (work-piece), the smaller the pressure difference in the vertical glass tube, and hence, the lower the position of the float against the scale.

Differential comparator: This consists of a differential pressure–indicating mechanism connected across two air channels, with a built-in gauge zeroing valve. Regulated compressed air is supplied to the two channels, each of which is fitted with a separate control orifice (also called master jet). From Control Orifice 1, air flows to the measuring head, where it meets further restriction in the form of the work-piece whose dimension is to be measured. This restriction builds up back pressure in Channel A. The regulated air flows simultaneously through Control Orifice 2 to the reference jet. By adjusting the valve of the reference jet, the pressure in the space between Control Orifice 2 and the reference jet is adjusted to match the back pressure from the measuring jet, which is sensed by the pressure-indicating device fixed across the two air channels, as shown in Fig. 2.14. After the reference jet is adjusted, the pressure indicator indicates equal pressure in both the air channels, and hence, shows a zero reading on the scale. This zero reading is achieved for the master work-piece whose dimension is exactly equal to the

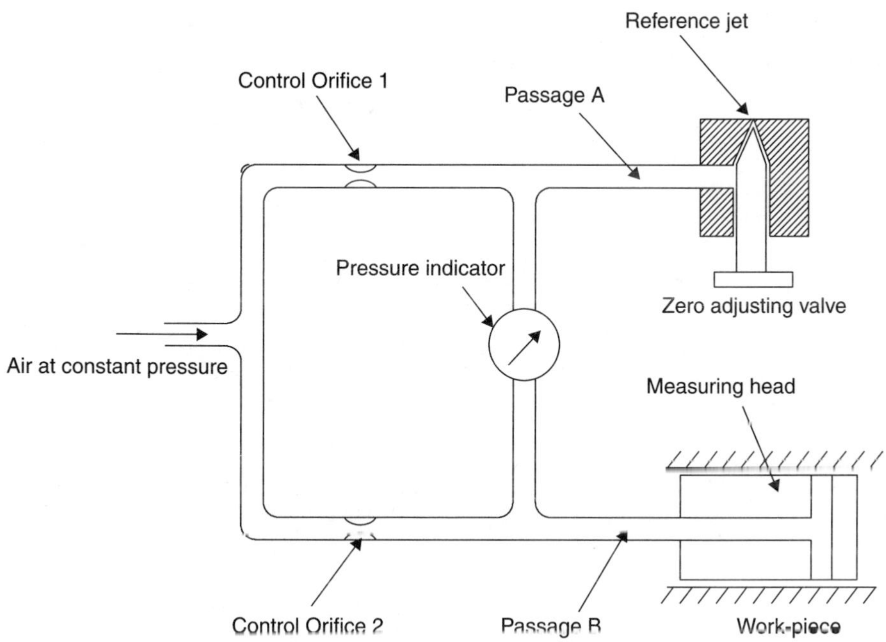

Figure 2.14 Differential comparator

nominal size. Now, variation in the dimensions at the measuring head will cause a change in the back pressure in Channel A. This pressure will be different from the mean pressure which has already been set in channel B. The difference in pressure between the two air channels will be indicated by the pressure-indicating device, which can be directly calibrated in terms of variation in the dimension from the mean dimension. As the instrument is based on the measurement of differential pressure, it is called a differential comparator.

2.3.6 Electrical Comparator

It consists of a plunger, steel armature and four inductances. The plunger rests on the surface to be checked. The steel armature rests on a diamond ball which, in turn, rests on the top of the plunger. C and D are inductance elements and form the two arms of a bridge. The steel armature is held on one side by a spring and by C and D on the other side. When electromagnets are placed near the armature, there will be higher magnification of inductance, and vice versa. The screws B are adjusted to change the magnification of inductance. Two similar inductance arms are outside and these four inductance arms form a Wheatstone bridge. The set-up is shown in Fig. 2.15.

During measurement, the armature may be lifted up or lowered down by the movement of the plunger. This change in the position of the armature will induce a change in the current or potential, according to the displacement of the plunger. This induced current is amplified before it is displayed in the output meter. The readings are first calibrated using the slip gauges. Also, the screws B are used for

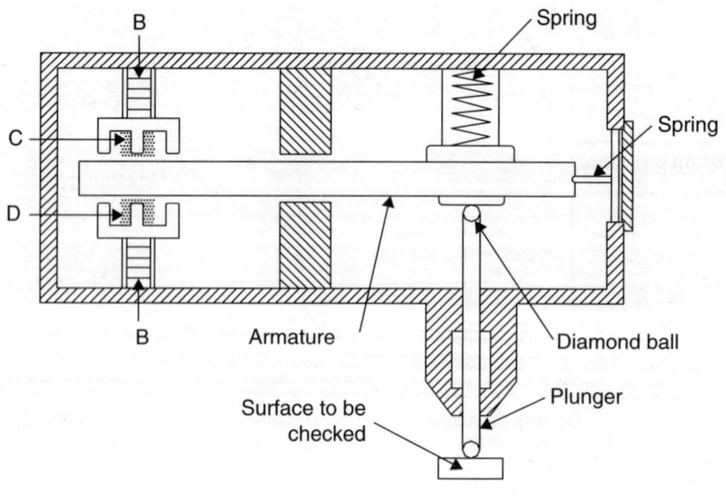

Figure 2.15 Electrical comparator

initial adjustment of the bridge to give a zero reading when there is no irregularity, i.e., for the plane surface.

2.3.7 Electronic Comparator

Frequency modulation or radio oscillation is used in the working principle of this comparator (Fig. 2.16). It consists of the following:

- **Transducer**: It converts the movement of the plunger (kinetic energy) into an electrical signal (electrical energy). It is connected to the oscillator.
- **Oscillator**: It receives the electrical signal from the transducer and raises the amplitude of the frequency wave by adding carrier frequency. This process is called **modulation.**
- **Amplifier**: It is connected to the oscillator and the demodulator. The signal from the oscillator is amplified to the required level.
- **Demodulator**: It cuts off the external carrier wave frequency. It converts the modulated wave into the original wave as an electrical signal.

The work-piece to be measured is placed under the plunger of the electronic comparator. The linear movement of the plunger is converted into an electrical signal by the transducer. The oscillator then modulates this electrical signal. The modulated signal is amplified by means of an amplifier and then sent to the

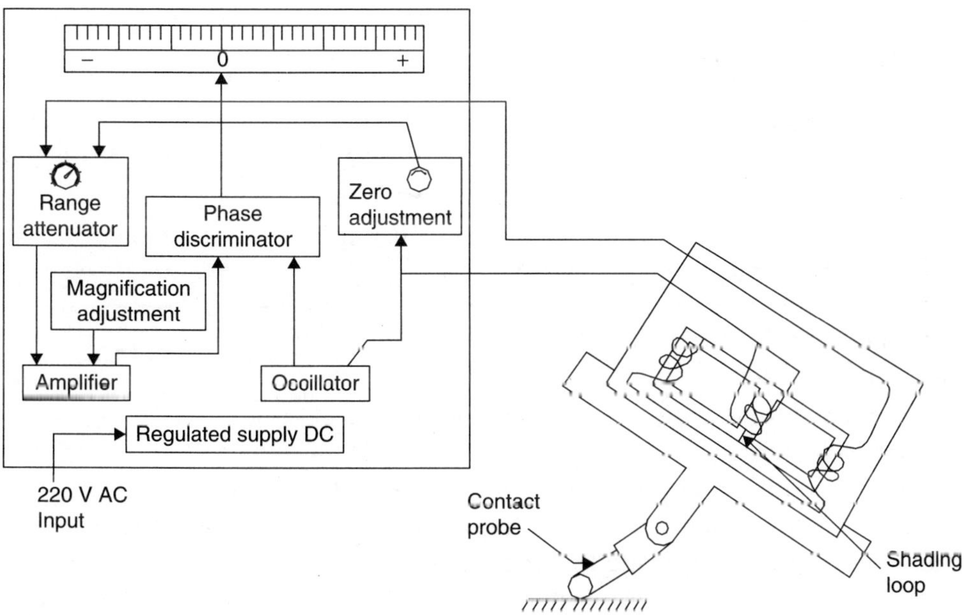

Figure 2.16 Electronic comparator

demodulator, in which the carrier waves are cut off. Finally, the demodulated signal is passed to the electric meter to convert the probe tip movement into the corresponding linear measurement of the plunger.

2.4 SLIP GAUGES

These gauges are otherwise called gauge blocks or block gauges and are universally accepted as end standards of length in industry. Slip gauges are rectangular blocks of high-grade steel (or tungsten carbide) with a small coefficient of thermal expansion. These blocks are highly hardened (more than 800 HV) throughout the material to ensure maximum resistance to wear and are then stabilised by heating and cooling successively in stages so that the hardening stresses are removed. After hardening, they are subjected to lapping to a high degree of finish, flatness and accuracy. The cross-section of these gauges is 9 × 30 mm for sizes up to 10 mm and 9 × 35 mm for larger sizes. The dimension (height) is marked on one of the measuring faces of the gauge blocks.

Wringing: Slip gauges are wrung together by hand through a combined sliding and twisting motion. The air gap between the gauge faces is expelled and adhesion is caused partly by molecular attraction and partly by atmospheric pressure. The gap between the two wrung slip gauges is only of the order of 0.00635 μm, which is negligible.

Note: When taking the slip gauges apart, they should be slid apart in the same manner as used for their attachment.

Procedure for wringing

1. The slip gauges are first cleaned using a lint-free cloth or chamois leather or a cleansing tissue.
2. One slip gauge is placed at 90° to the other using light pressure, and it is then rotated until the blocks are brought in one line.

Indian standards for slip gauges

- Grade II: Workshop grade for rough checks
- Grade I: Used for setting up sine bars, checking gap gauges and setting dial test indicators to zero
- Grade 0: Used in tool rooms and in inspection departments
- Grade 00: Kept in standard rooms and used for high-precision work such as checking Grade I and Grade II slip gauges

Applications

- Direct precise measurement

- For calibration of Vernier calipers, micrometers and other measuring instruments
- For setting up a comparator to a specific dimension
- For angular measurement in conjunction with a sine bar
- For checking the gap between parallel locations such as in gap gauges or between two mating parts

Selection of slip gauges for the required dimension: The last decimal place is deducted from the required dimension. The next smallest figure is selected in the same way and the remainder is found. This process is continued till the required dimension is completed. The minimum number of slip gauges should be selected to build the given dimension. At the ends of a stack of more than 4 mm size, two 2-mm protector slip gauges are wrung to reduce wear on the remaining part of the set.

Details of slip gauges are given in Tables 2.1 and 2.2.

Table 2.1 Normal set of slip gauges

Size (in mm)	Increment (in mm)	Number of pieces
1.001–1.009	0.001	9
1.01–1.09	0.01	9
1.1–1.9	0.1	9
1–9	1	9
10–90	10	9
Total		45

Table 2.2 Special set of slip gauges

Size (in mm)	Increment (in mm)	Number of pieces
1.001–1.009	0.001	9
1.01–1.49	0.01	49
0.5–9.5	0.5	19
10–90	10	9
Total		86

A typical set of slip gauges of 86 pieces is shown in Fig. 2.17.

Figure 2.17 Slip gauge set

Example Problem 1: A length of 52.215 mm is to be built using a normal set of slip gauges. Suggest a suitable arrangement of slip gauges.

Solution:

Desired dimension = 52.215 mm

Protector gauges = 4 mm (for both ends, two protector gauges of 2 mm are added), and hence, remainder = 48.215 mm

For the thousandth place, use one gauge of 1.005 mm, and hence, remainder = 47.210

For the hundredth place, use one gauge of 1.010 mm, and hence, remainder = 46.200

For the tenth place, use two gauges of 1.1 mm, and hence, remainder = 44.000

For the unit place, use four gauges of 1 mm, and hence, remainder = 40.000

For the tenth place, use four gauges of 10 mm, and hence, remainder = 0.000

Summary of slip gauges set:

10 mm gauge – 4 numbers; 1 mm gauge – 4 numbers

1.01 mm gauge – 2 numbers; mm gauge – 1 number

1.005 mm gauge – 1 number; 2 mm protector gauge – 2 numbers

2.5 ROLLER GAUGES

Cylindrical rollers with their lengths equal to their diameters may be used as gauges, secondary to block gauges (slip gauges). These are produced to fine tolerances. A roller gauge is shown in Fig. 2.18.

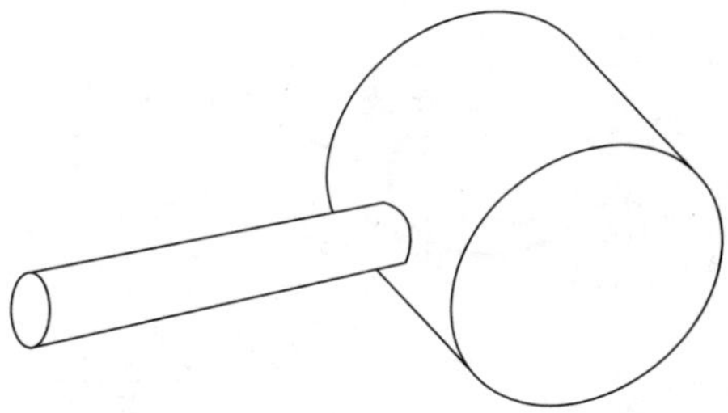

Figure 2.18 Roller gauge

2.6 LIMIT GAUGES

These are inspection tools for rigid design, without a scale, which serve to check the dimensions of manufactured parts. They do not indicate the actual value of the inspected dimension on the work-piece. They can only be used to determine whether the inspection parts are made within the specified limits. These gauges are composed of suitable wear-resisting steel, are normally hardened to not less than 750 HV, suitably stabilised and ground and lapped.

The GO and NO GO gauges may be in the form of separate single-ended gauges or may be combined on one handle to form a double-ended gauge. A progressive gauge is the single-ended gauge with one gauging member having two diameters to the GO and NO GO limits, respectively.

A plug gauge is used for checking the dimensions of holes, whereas a snap gauge is used for checking the external dimensions of work-pieces. A typical limit plug gauge and limit snap gauge are shown in Figs 2.19 and 2.20, respectively.

Ring gauges: Ring gauges are used to check the diameter of shafts. These gauges have a central hole that is hardened and ground to a fine finish. The periphery of a ring gauge is usually knurled to provide grip while handling it. The hole of the GO ring gauge is made to the upper limit size of the shaft. The hole of the NO GO ring gauge is made to the lower limit size of the shaft. The NO GO ring gauge is identified by a groove cut on its periphery by the red mark on it. Typical ring gauges are shown in Fig. 2.21.

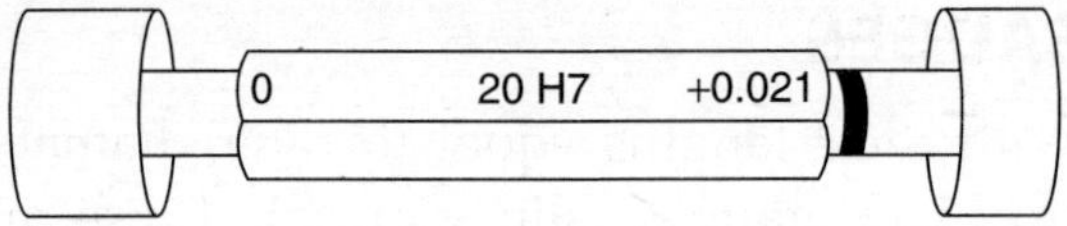

Figure 2.19 Limit plug gauge

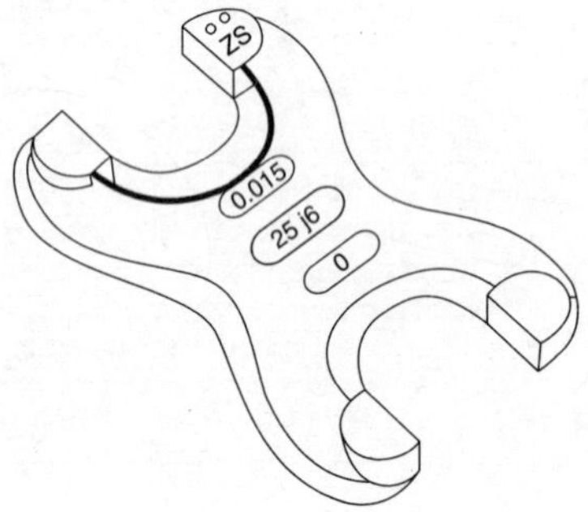

Figure 2.20 Limit snap gauge

Figure 2.21 Ring gauges

2.7 LIMITS, FITS AND TOLERANCES

It is not practically possible to produce two components with identical dimensions through any manufacturing process. Regardless of how well the production process is designed or maintained, a certain level of natural variation will always exist in the dimensions of the components produced. These natural variations are random and are the collective effect of many small and essentially uncontrollable causes. Hence, it is believed that no component can be manufactured precisely to a given dimension and but can only be made to lie between two limits. Namely, the upper (maximum) and lower (minimum) limits. The designer of a component must specify the acceptable tolerance limits for each of the dimensions used to define the shape and form of the component.

The following are the terms and definitions relating to the design of components to be manufactured, with the dimensions falling within tolerance limits. These terms are shown in Fig. 2.22.

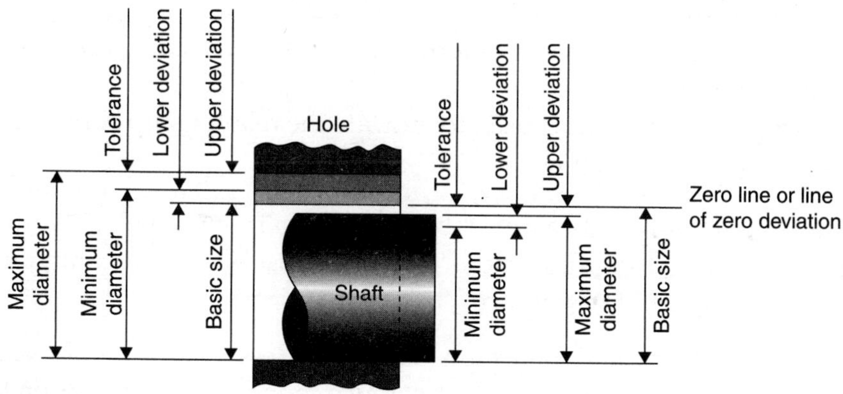

Figure 2.22 General terminology of fit

Shaft: A member that fits into another member.

Hole: A member that houses or fits the shaft.

Basic size: The size with reference to which the upper and lower limits of size are defined. The basic size and nominal size of a component are the same – also termed the zero line.

Upper deviation: The algebraic difference between the maximum limit of size (of the hole or shaft) and the corresponding basic size.

Lower deviation: The algebraic difference between the minimum limit of size (of the hole or shaft) and the corresponding basic size.

Fundamental deviation: One of the two deviations, chosen to define the position of the tolerance zone.

Basic shaft and basic hole: The shaft and hole that have zero fundamental deviation. Basic hole has zero lower deviation, whereas basic shaft has zero upper deviation.

Tolerance: The difference between the maximum and minimum limits of size for a hole or shaft. For example, if a shaft is required to be manufactured to a diameter of 40 $^{\pm 0.02}$ mm, the permissive tolerance will be equal to 0.04 mm.

Tolerance can be classified as

- **Unilateral tolerance**: The tolerance limits lie wholly on one side of the basic size, either above or below it. For example, $40^{+0.03,\,+0.01}$, $40^{-0.03,\,-0.02}$, $40^{+0.03,\,-0.00}$, $40^{+0.00,\,-0.03}$.
- **Bilateral tolerance**: Tolerance is allowed to vary on both sides of the basic size. For example, $40^{\pm 0.02}$.
- **Compound tolerance**: This is determined by established tolerances in more than one dimension.
- **Geometric tolerance**: This is used to indicate the relationship between one part of a component and another.

Fit: The relative motion that can exist between a shaft and a hole, resulting from the final sizes achieved in production. The three classes of fit in common use are clearance, transition and interference.

- **Clearance fit**: It results when the shaft size is less than the hole size for all possible combinations within their tolerance ranges. Relative motion between the shaft and hole is always possible with clearance fit. Minimum clearance occurs at maximum shaft size and minimum hole size. Similarly, maximum clearance occurs at minimum shaft size and maximum hole size. Clearance fit is shown in Fig. 2.23.
- **Interference fit**: This results when the shaft size is larger than the hole size for all possible combinations within their tolerance ranges. Minimum interference occurs at minimum shaft size and maximum hole size. Similarly, maximum interference occurs at maximum shaft size and minimum hole size. Interference fit is shown in Fig. 2.24.
- **Transition fit**: It results when the shaft and hole are exactly the same size. It is theoretically the boundary between clearance and interference. Practical transition fit results when the maximum hole size is larger than the minimum shaft size and the maximum shaft size is larger than the minimum hole size. Transition fit is shown in Fig. 2.25.

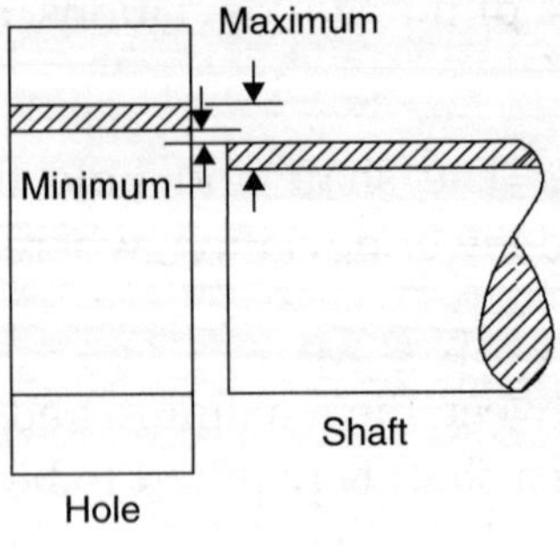

Figure 2.23 Clearance fit

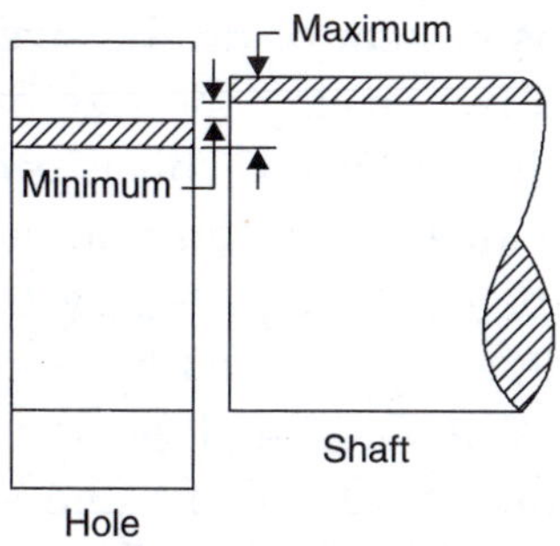

Figure 2.24 Interference fit

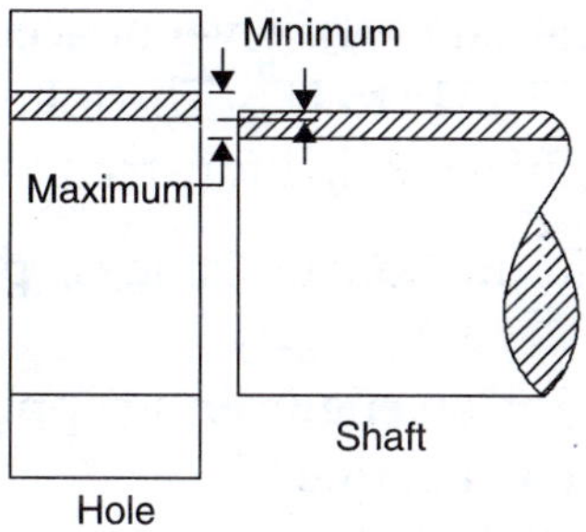

Figure 2.25 Transition fit

Allowance: The term representing the minimum clearance (positive allowance) or the maximum interference (negative allowance) that exists between the mating components, i.e., between the shaft and hole.

Systems of fit: The two systems by which a fit can be accomplished are the hole basis system and the shaft basis system, shown in Fig. 2.26.

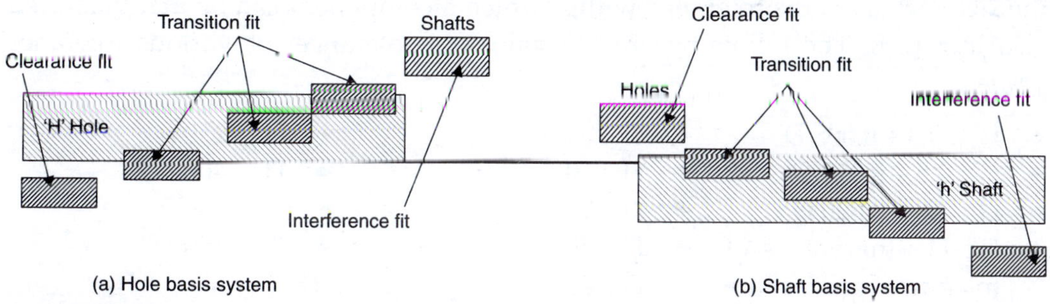

Figure 2.26 Systems of fit

- **Hole basis system**: In this system of fit, the size of the hole is constant and different fits are obtained by varying the size of the shaft. It may be noted that this system is generally preferred. This is because holes are produced by standard-size drills and reamers whose size cannot be adjusted easily, whereas the size of the shaft can be easily adjusted by turning and grinding. As per Indian standards, the hole is represented by uppercase letters – A, B, C.....Z (excluding I, L, O, Q and W and adding JS) and ZA, ZB and ZC.
- **Shaft basis system**: In this system of fit, the size of the shaft is constant and different fits are obtained by varying the size of the hole. As per Indian standards, the shaft is represented by lowercase letters – *a*, *b*, *c*...*z* (excluding *i*, *l*, *o*, *q*, *w* and adding *js*) and *za*, *zb* and *zc*.

Grade of tolerance: An indication of the level of accuracy. There are 18 grades of tolerance, ranging from IT_{01}, IT_0, IT_1 to IT_{16}. The following are the applications of the various grades in the manufacturing industry:

- $IT0_1$–IT_4, used for the production of gauges, plug gauges and measuring instruments
- IT_5–IT_7, used for fits in precision engineering applications
- IT_8–IT_{11}, used for general engineering
- IT_{12}–IT_{14}, used for sheet metal working or press working
- IT_{15}–IT_{16}, used for processes like casting and general cutting work

Various grades of tolerance are defined using the following standard tolerance unit (i) in μm, which is a function of basic size.

$$i \text{ (micron)} = 0.45\,(D)^{1/3} + 0.001\,D$$

where D (mm) is the basic size (the geometric mean of the lower and upper limits of a particular diameter within which the chosen diameter D lies).

Table 2.3 gives the steps to fix the value for D.

The tolerances have a parabolic relationship with the size of the component. As the size increases, the tolerance within which a component can be manufactured also increases. The following are the values of tolerance for various tolerance grades.

- IT_{01}: $0.3 + 0.008\,D$
- IT_0: $0.5 + 0.012\,D$
- IT_1: $0.8 + 0.020\,D$
- IT_2–IT_4: the values of tolerance grades are placed geometrically between the tolerance grades of IT_1 and IT_5
- IT_5: $7\,i$
- IT_6: $10\,i$
- IT_7: $16\,i$
- IT_8: $25\,i$
- IT_9: $40\,i$
- IT_{10}: $64\,i$
- IT_{11}: $100\,i$
- IT_{12}: $160\,i$
- IT_{13}: $250\,i$
- IT_{14}: $400\,i$
- IT_{15}: $640\,i$
- IT_{16}: $1000\,i$

Table 2.3 Basic sizes up to 500 mm

Values in millmetres

a) Basic sizes up to 500 mm (incl.)			
Main steps		**Intermediate steps (1)**	
Above	**Up to and Including**	**Above**	**Up to and Including**
–	3		
3	6	**No subdivision**	
6	10		
10	18	10	14
		14	18
18	30	18	24
		24	30
30	50	30	40
		40	50
50	80	50	65
		65	80
80	120	80	100
		100	120
120	180	120	140
		140	160
		160	180
180	250	180	200
		200	225
		225	250
250	315	250	280
		280	315
315	400	315	355
		355	400
400	500	400	450
		450	500

Table 2.4 gives the formulae for calculating the fundamental deviations of shafts up to a size of 500 mm.

Table 2.4 Formulae for fundamental deviation of the shaft (up to 500 mm)

Upper deviation (es)		**Lower deviation (ei)**	
Shaft deviation	**In microns (for *D* in mm)**	**Shaft deviation**	**In microns (for *D* in mm)**
a	$= -(285 + 13D)$ for $D \leq 120$	*j*	No formula
	asd $= -3.5\ D$	*js*	IT1/2
	for $D > 120$	k4 to k7	$= +0.6\sqrt[3]{D}$
b	$= -(140 + 0.85D)$ for $D \leq 160$	k for grade k3 and k7	
	$= -18\ D$ for $D > 160$		
c	$= -52\ D^{12}$ for $D \leq 40$	*ps*	= + (IT1-IT6)
	$= -(96 + 0.8D)$	*d*	$= +5D^{0.34}$
	for $D > 40$	*p*	= +IT7+0 to 5
cd	GM of values for *c* and *d*	*g*	= geometric mean of values for *p* and *s*
d	$= -16\ D^{0.46}$	*o*	–IT8 + 1 to 4 for $D \leq 50$
e	$= -11\ D^{0.41}$		= +IT7 τo + 0.4*D* for $D > 50$
ef	GM of values for *e* and *f*		
f	$= -5.5\ D^{0.41}$	*i*	= +IT7 + 0.43*D*
fg	GM of values for *f* and *g*	*w*	= +IT7 + *D*
g	$= -2.5\ D^{0.34}$	*o*	= +IT7 + 1.25*D*
h	= 0	*x*	= +IT7 + 1.6*D*
	y		= +IT7 + 2*D*
	z		= +IT7 + 2.5*D*
	za		= IT8 + 3 + 3.15*D*
	zb		= +IT9 + 4*D*
	zc		= +IT10 + 5*D*

Example Problem 2: Evaluate the limits and fits for a pair of mating components of diameter 4 H_8/g_7.

Solution: The size 4 mm lies in the diametral step of 3–6, and hence, the basic size is given by $D = \sqrt{3 \times 6} = 4.24$ mm

The fundamental tolerance is given by

$$i \text{ (micron)} = 0.45\,(D)^{1/3} + 0.001\,D$$

$$= 0.45\,(4.24)^{1/3} + 0.001\,(4.24) = 0.7327 \text{ mm}$$

Limits of tolerance for hole H_8:

Standard tolerance is $25\,i = 25 \times 0.7327 = 18.3175 = 18\,\mu$m

Fundamental deviation for H hole is 0.

Limits of tolerance for g_7:

Standard tolerance is $16\,i = 16 \times 0.7327 = 11.7232 = 12\,\mu$m

Fundamental deviation for g shaft: $-2.5\,D^{0.34} = -2.5\,(4.24)^{0.34} = -4.085 = -4\,\mu$m

Disposition of tolerance zone around the zero line is marked in Fig. 2.27.

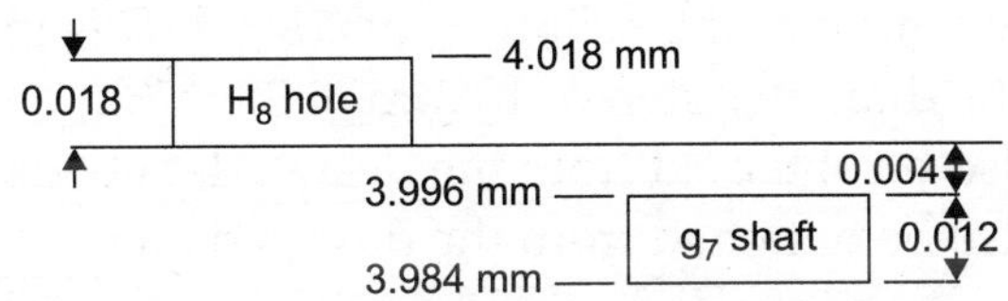

Figure 2.27 Disposition of tolerance

Fit

$$\text{Maximum clearance} = \text{Maximum size of hole} - \text{Minimum size of shaft}$$

$$= 4.018 - 3.984 = 0.114 \text{ mm}$$

$$\text{Minimum clearance} = \text{Minimum size of hole} - \text{Maximum size of shaft}$$

$$= 4.000 - 3.996 = 0.004 \text{ mm}$$

This type of fit is a clearance fit.

2.7.1 Design of Limit Gauges

Every gauge is a copy of the part that mates with the counterpart for which the gauge is designed. For example, a bush is made to mate with a shaft; in this case, the shaft is the mating part. The bush is checked using a plug gauge and its size should be a copy of the mating part (shaft).

Taylor's principle: According to Taylor, GO and NO GO gauges should be designed to check the maximum and minimum material limits.

- **GO limit**: This is applied to the upper limit of a shaft and the lower limit of a hole.
- **NO GO limit**: This is applied to the lower limit of a shaft and the upper limit of a hole.

Taylor's principle states that the GO gauges should check all the possible elements of dimensions at a time (roundness, size, location, etc.) while the NO GO gauge should check only one element of the dimension at a time.

Based on Taylor's principle, GO gauges are designed for maximum material condition while NO GO gauges are designed for minimum material condition.

Example Problem 3: Design general GO and NO GO gauges for components having $20H_7f_8$ fit. Given:

(i) i (micron) = $0.45\,(D)^{1/3} + 0.001\,D$ (can also be obtained from the design data book)
(ii) Fundamental deviation of the f shaft = $-5.5\,(D)^{0.41}$ (D is in mm)
(iii) 20 mm falls in the diameter step of 18 mm to 30 mm.
(iv) $IT_7 = 16\,i$ (can also be obtained from the design data book)
(v) $IT_8 = 25\,i$ (can also be obtained from the design data book)

Solution: From the standard table, it is found that 20 lies between 18 and 30. Hence, $D = \sqrt{D_1 \times D_2} = \sqrt{18 \times 30} = 23.2379$ mm

$$i = 0.45\,(D)^{1/3} + 0.001\,D \text{ microns}$$

$$= 0.45\,(23.2379)^{1/3} + 0.001\,(23.2379) = 1.3074 \text{ microns}$$

$$IT_7 = 16\,i \text{ microns}$$

$$= 16 \times 1.3074 \text{ microns} = 20.918 \text{ microns} = 0.021 \text{ mm}$$

$$IT_8 = 25\,i \text{ microns}$$

$$= 25 \times 1.3074 \text{ microns} = 33 \text{ microns} = 0.033 \text{ mm}$$

Limits for hole 20 H_7:

Fundamental deviation of shaft = $-5.5\,D^{0.41}$ microns

$= -5.5 \times 23.2379^{0.41}$

$= -20$ microns $= -0.02$ mm

Lower deviation of H hole = zero (hole basis system)

Hence, the lower limit of hole = $20^{-0.000}$

Higher limit of hole = $20 + IT_7 = 20^{+0.021}$

Hence, limits for hole 20 $H_7 = 20^{+0.021, -0.000}$ mm

Limits for shaft 20 f_8:

Higher deviation of shaft = zero

Hence, higher limit of shaft = 20 + Fundamental deviation + 0

$= 20^{-0.02}$ mm

Lower limit of shaft = 20 + Fundamental deviation – IT_8

$= 20^{-0.02-0.033} = 20^{-0.053}$ mm

Hence, limits for shaft 20 $f_8 = 20^{-0.02, -0.053}$ mm

Plug gauge design:

Work tolerance for hole = Higher limit of hole – Lower limit of hole

$= 20.021 - 20 = 0.021$ mm

Gauge tolerance for plug gauge = 10% of work tolerance (hole)

$= 0.1 \times 0.021 = 0.0021$ mm

Wear allowance for plug gauge = 10% of gauge tolerance (hole)

$= 0.1 \times 0.0021 = 0.00021$ mm

Higher limit of GO plug gauge = Lower limit of hole + Gauge + Wear tolerance

$= 20^{+0.0021 + 0.00021} = 20^{+0.00231}$ mm

Lower limit of GO plug gauge = Lower limit of hole + Wear tolerance

$= 20 + 0.00021$ mm $= 20^{+0.00021 \text{ mm}}$

Higher limit of NO GO plug gauge = Higher limit of hole + Gauge tolerance

$= 20.021 + 0.0021 = 20.0231$ mm

Lower limit of NO GO plug gauge = Higher limit of hole

= 20.021 mm

Snap gauge design:

Work tolerance for shaft = Higher limit of shaft – Lower limit of shaft

= 19.98 – 19.947 = 0.033 mm

Gauge tolerance for snap gauge = 10% of work tolerance (shaft)

= 0.1 × 0.033 = 0.0033 mm

Wear allowance on snap gauge = 10% of gauge tolerance (shaft)

= 0.1 × 0.0033 = 0.00033 mm

Higher limit of GO snap gauge = Higher limit of shaft – Gauge tolerance

= 19.98 – 0.00033 = $20^{-0.02033}$ mm

Lower limit of GO snap gauge = Higher limit of shaft – Wear tolerance – Gauge tolerance

= 19.98 – 0.0033 – 0.00033 = $20.00^{-0.02363}$ mm

Now, higher limit of NO GO snap gauge = Lower limit of shaft

= 19.947 mm = $20^{-0.0530}$ mm

Lower limit of NO GO snap gauge = Lower limit of shaft – Gauge tolerance

= 19.947 – 0.0033 mm = $20^{-0.0563}$ mm

The limits for plug gauge and snap gauge are shown in Fig. 2.28.

2.7.2 Types of Limit Gauges

The types of limit gauges used for measuring the internal diameters of holes are given below.

- **Full form cylindrical plug gauge**: The gauging surface is in the form of an external cylinder (Fig. 2.29). Generally, a small circumferential groove is cut near the leading end of the gauge and the remaining short cylindrical surface is slightly reduced in order to act as a pilot.
- **Full form spherical plug or disc gauge**: The gauging surface is in the form of a sphere, from which two equal segments are cut off by planes normal to the axis of the handle (Fig. 2.30).

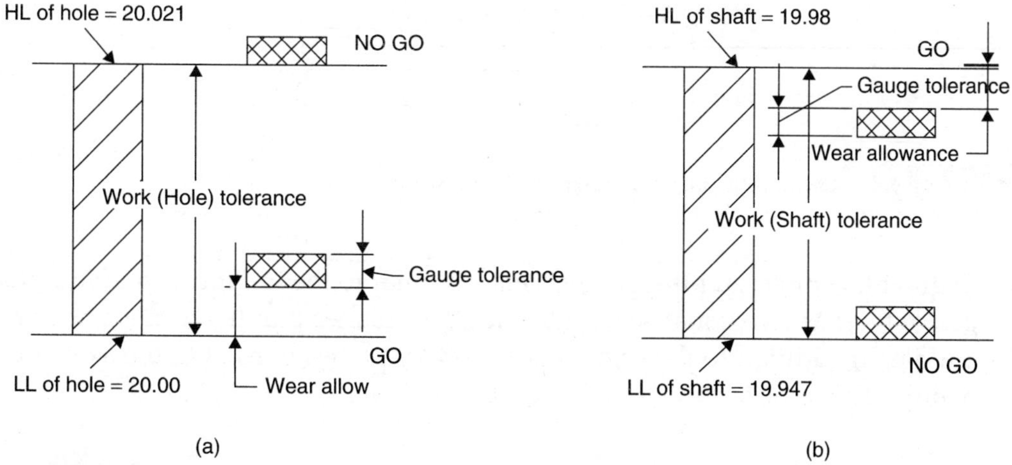

Figure 2.28 Limits for (a) plug gauge and (b) snap gauge

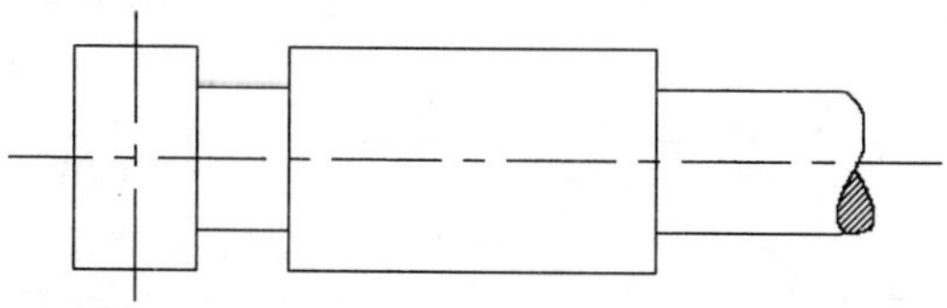

Figure 2.29 Full form cylindrical plug gauge

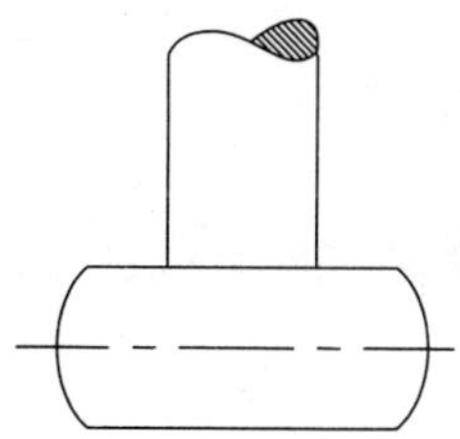

Figure 2.30 Full form spherical plug gauge

- **Segmental cylindrical bar gauge**: The gauging surface is in one of two forms – an external cylindrical form from which two axial segments are made by lowering the surface at other places (Fig. 2.31), and an external cylindrical form in which segments are formed by removing the remaining material (Fig. 2.32).

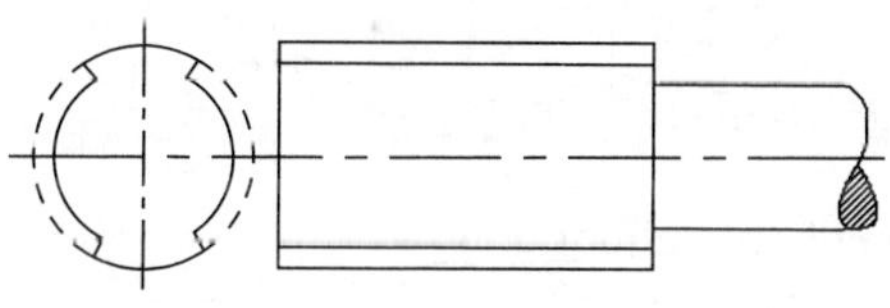

Figure 2.31 Segmental cylindrical bar gauge (type 1)

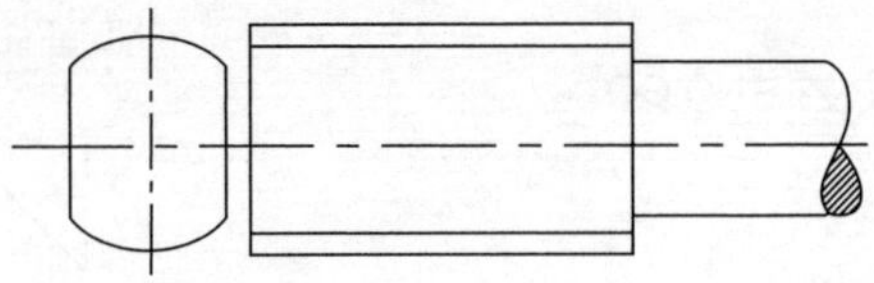

Figure 2.32 Segmental cylindrical bar gauge (type 2)

- **Segmental spherical plug gauge**: This is similar to the full form spherical plug gauge but has two equal segments cut off by planes parallel to the axis of the handle, in addition to the segments cut off by planes normal to the axis of the handle. This type is shown in Fig. 2.33.

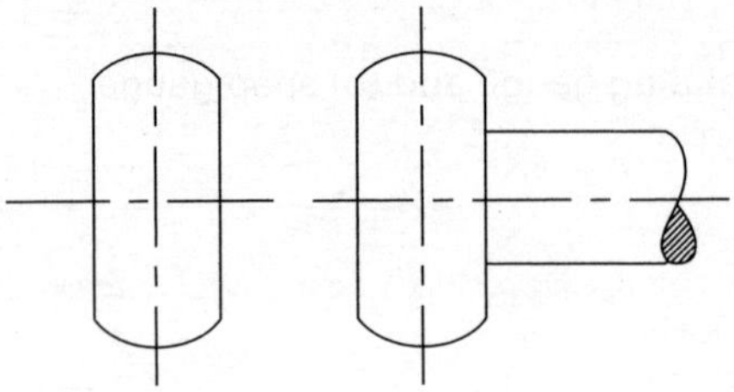

Figure 2.33 Segmental spherical plug gauge

- **Segmental cylindrical bar gauge with reduced measuring faces**: It is similar to the segmental cylindrical bar gauge, but has reduced measuring faces in a plane parallel to the axis of the handle. This type is shown in Fig. 2.34.
- **Rod gauge with spherical ends**: It has spherical end surfaces which form part of one single sphere. This type is shown in Fig. 2.35.

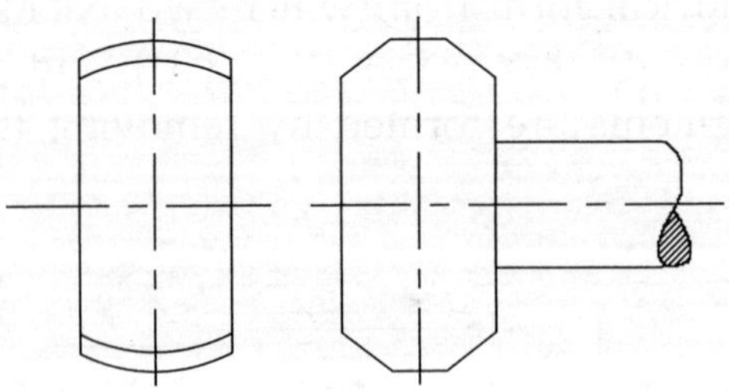

Figure 2.34 Segmental cylindrical bar gauge

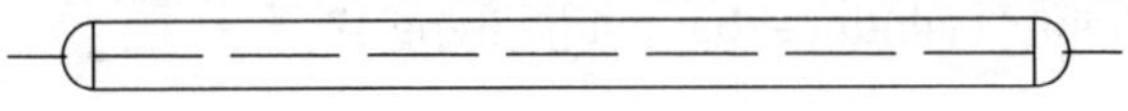

Figure 2.35 Rod gauge with spherical ends

The types of limit gauges used for gauging external diameters of shaft are given below.

- **Full form cylindrical ring gauge**: The gauging surface is in the form of an internal cylinder, whose wall is thick enough to avoid deformation under normal conditions of use.
- **Gap gauge**: This has one flat surface and one cylindrical surface, the axis of the two surfaces being parallel to the axis of the shaft being checked. The surfaces constituting the working size may both be flat or cylindrical.
- **Limit gauges for tapers**: The limit gauges for measuring the internal taper of a hole and the external taper of a shaft are shown in Figs 2.36 and 2.37, repectively.

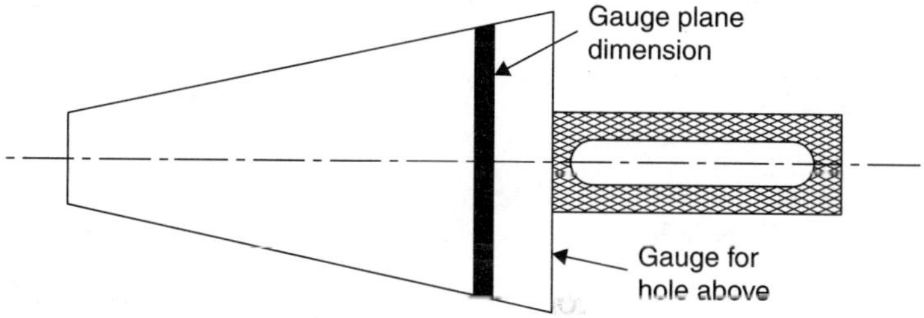

Figure 2.36 Limit plug gauge for internal taper of a hole

The gauges (for internal taper) are marked with a ring on the gauge planes, while another ring indicates the minimum depth of the internal taper of the hole. The distance between the two rings marked Z corresponds to the permissible deviation of the gauge plane for a particular taper. For testing the

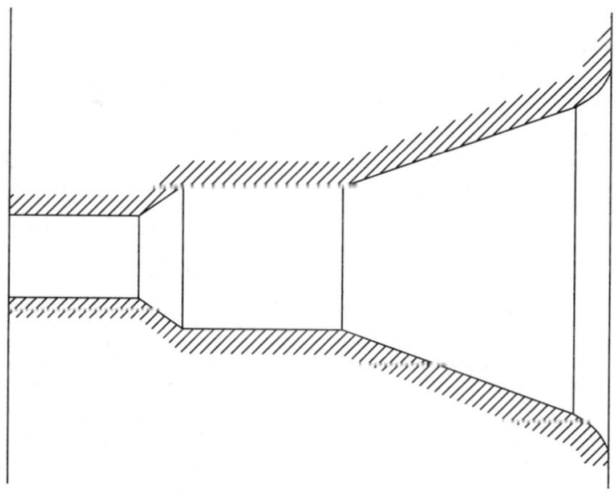

Figure 2.37 Limit snap gauge for external taper of a shaft

external taper of the shaft, the ring gauge is inserted as far as it can go with light pressure.

SUMMARY

- The Vernier calipers is used to measure the width, depth and diameter of large cylindrical components (internal and external), whereas a screw gauge is used to measure the thickness of a wire, the diameter of small balls, etc.
- A comparator compares the dimensions of a given component with those of the working standard. It does not measure the actual dimensions but indicates by how much it differs from the basic dimensions (working standard).
- Amplification of Johansson Mikrokator is $\frac{d\theta}{d\ell} = \text{Amplification} = -\frac{9.1\ell}{W^2 n}$, where ℓ is the length of the twisted strip, measured along its neutral axis; W is the width of the twisted strip; n is the number of turns and θ is the twist of the midpoint of the strip with respect to end.
- The amplification of a Sigma comparator is $LR/(xr)$, where the effective length of the Y arm is L, the distance from the hinge pivot to the knife edge is x, the radius of the drum is r and the pointer length is R.
- The amplification of an optical comparator is $(L_2 / L_1) \times (L_4 / L_3) \times 2$ units, where L_1 is the distance between the plunger and the pivot of the lever, L_2 is the distance between the pivot of the lever and the mirror, L_3 is the distance between the lever end and the pivot of the mirror, and L_4 is the distance between the mirror and the scale.
- Principle of pneumatic comparator: When the measuring probe is directed towards measurement, the back pressure varies when the gap or distance between the work and head varies. Back pressure variation indirectly measures the dimension.
- Wringing of slip gauges means the process of attaching slip gauges together by hand through a combined sliding and twisting motion. The air gap between the gauge faces is expelled and adhesion is caused partly by molecular attraction and partly by atmospheric pressure.
- Limit gauges serve to check the dimensions of manufactured parts without a scale. Limit gauges do not indicate the actual value of the inspected dimension on the work-piece. They can only be used to determine whether the inspection parts are made within the specified limits. A plug gauge is used to check the dimensions of holes, whereas a snap gauge is used to check the exertnal

dimensions of a work-piece. Ring gauges are used to check the diameter of shafts.

- Taylor's principle states that the GO gauge should check all the possible elements of dimensions at a time (roundness, size, location, etc.) while the NO GO gauge should check only one element of the dimension at a time.
- Limit gauges include full form cylindrical plug gauges; full form spherical plug gauges; segmental cylindrical bar gauges; segmental spherical plug gauges; segmental cylindrical bar gauges with reduced measuring faces; and rod gauges with spherical ends.

EXERCISES

Multiple Choice Questions

1. Measurement of length and diameter is referred to as ____________ measurement.
   a) Angular b) Linear
   c) Non-linear d) Concurrent

2. A steel rule is an example of ______________ measurement.
   a) End b) Line
   c) Angular d) Non-linear

3. A block gauge is an example of ____________ measurement.
   a) Line b) End
   c) Angular d) Non-Linear

4. Least count in the Vernier scale is given by
   a) 1 MSR – 1 VSD b) 1 MSR × 1 VSD
   c) 1 MSR/1 VSD d) 2MSR – 2VSD

5. ____________ is used to measure the width, depth and diameter of components, both internal and external.
   a) Micrometer b) Vernier height gauge
   c) Vernier calipers d) Screw gauge

6. The working principle of ____________ is that the distance moved by the nut along the screw is proportional to the number of revolutions made by the nut.
   a) Vernier height gauge b) Vernier calipers
   c) Vernier depth gauge d) Screw gauge

7. ______________ is the reading indicated when there is no gap between the anvil and spindle faces.
   a) Zero error
   b) Screw pitch error
   c) Location error
   d) Dirt error

8. ______________ is the error that is due to constant pitches.
   a) Zero error
   b) Screw pitch error
   c) Location error
   d) Dirt error

9. Range of the instrument is ______________
   a) Length of the instrument
   b) Variety of the instrument
   c) Minimum and maximum values of measurement
   d) Period of measurement

10. ______________ is an instrument employed to compare the dimensions of a given component with the working standard.
    a) Comparator
    b) Vernier calipers
    c) Vernier height gauge
    d) Vernier depth gauge

11. The following is not an essential characteristic of a good comparator:
    a) Robust design and construction
    b) Versatility
    c) High magnification
    d) Maximum wear of contact point

12. ______________ consists of a thin metal strip which carries at the centre of its length a very light glass tube pointer.
    a) Comparator
    b) Vernier calipers
    c) Johansson Mikrokator
    d) Brooke's level comparator

13. Deadbeat readings can be obtained by
    a) Sigma mechanical comparator
    b) Brooke's level comparator
    c) Johansson Mikrokator
    d) Vernier calipers

14. Total amplification in optical comparator is given by
    a) $(L_2/L_1) \times (L_4/L_3) \times 2$ units
    b) $(L_2/L_1) + (L_4/L_3) \times 2$units
    c) $(L_2/L_1) \times (L_4/L_3) \times 4$ units
    d) $(L_2/L_1) + (L_4/L_3) \times 4$ units

15. ______________ uses the principle of varying back pressure when the gap between the work and head varies.
    a) Sigma mechanical comparator
    b) Brooke's level comparator
    c) Pneumatic comparator
    d) Johansson Mikrokator

16. Plug gauge is used for checking______________ and snap gauge is used for checking ______________.
    a) Outside dimensions, dimensions of holes
    b) Dimensions of holes, outside dimensions

c) Dimensions of holes, outside dimensions
d) Diameter of holes, inner diameter

17. Based on Taylor's principle, ______________ is designed for maximum material condition.
a) GO
b) NO GO
c) Both 'a' and 'b'
d) Neither 'a' nor 'b'

18. Based on Taylor's principle, ___________ is designed for minimum material condition.
a) GO
b) NO GO
c) Both 'a' and 'b'
d) Neither 'a' nor 'b'

19. The amplification of Johansson Mirokator is given by
a) $d\theta/dl = -9.1l/w^2n$
b) $d\theta/dl = 9.1l/w^2n$
c) $d\theta/dl = w^2/9.1l$
d) $d\theta/dl = w^2n/9.1l$

20. Ring gauges are used to check the
a) Thickness of rings
b) Diameter of spheres
c) Outer diameter of shafts
d) Inner diameter of hollow shafts

21. Wringing is due to
a) Atmospheric pressure
b) Molecular attraction
c) Both a and b
d) None of the above

22. The following is the theorectical size which is common to both the parts of a mating pair:
a) Normal size
b) Actual size
c) Base size
d) All of the above

23. The following is equal to the differences between the two limits of size of the part:
a) Tolerance
b) Low limit
c) High limit
d) Design size

23. The following is used to check the diameters of holes:
a) Plug gauge
b) Ring gauge
c) Slip gauge
d) Screw gauge

24. The following is not a type of comparator:
a) Electrical
b) Pneumatic
c) Optical
d) Hydraulic

25. The following is not a type of direct measuring instrument:
a) Micrometer
b) Vernier calipers
c) Divider
d) All of the above

26. The following is not used to measure angles:
  a) Bevel protractor b) Calibrated level
  c) Clinometer d) Optical flat

Answers

1. b	2. b	3. b	4. a	5. b
6. d	7. a	8. b	9. c	10. a
11. d	12. c	13. a	14. a	15. c
16. b	17. a	18. b	19. a	20. c
21. c	22. c	23. a	24. d	25. c
26. d				

Review Questions

1. Define the following with respect to an instrument:
   i) Least count ii) Range
2. What is the difference between measurement and gauging?
3. Distinguish between line measurement and end measurement.
4. Name the types of standard (end or line) used for the following measurements:
   i) Micrometer ii) Slip gauge iii) Vernier height gauge
5. How are micrometer and Vernier calipers specified?
6. Describe the relative advantages of micrometer and Vernier calipers.
7. Explain the instrument suitable for measuring the following:
   i) A bar of 150 mm diameter to within 0.06 mm
   ii) The diameter of a bar which is 60 mm to within 0.1 mm
   iii) Length of a bar 220 mm long to within 0.5 mm
   iv) A bar 75 mm in diameter to within 0.8 mm
8. Describe the relative advantages of micrometer and Vernier calipers.
9. Describe the Vernier height gauge. Why it is considered a superior marking tool?
10. With a neat diagram, explain the construction and working principle of the depth micrometer.
11. What is the principle of a dial indicator?
12. How can you calibrate an end standard from a line standard?
13. What is the basic requirement for any calibration?
14. What is a comparator? Classify comparators.
15. What are the essential characteristics of a good comparator?
16. Explain the working principle of Johansson Mikrokator with a neat sketch.

17. Explain with a sketch the working principle of a Sigma comparator. Write the salient features of the Sigma comparator.
18. With a neat sketch, explain the principle of operation of the reed type mechanical comparator.
19. Explain how magnification of readings is obtained in an optic-mechanical comparator.
20. Why are front reflecting mirrors used in optical comparators?
21. What is the principle of the working of a pneumatic comparator?
22. With a neat sketch, describe the principle of Solex pneumatic comparator.
23. Explain the working principle of flow–velocity type pneumatic comparator.
24. Differentiate electrical comparator from electronic comparator.
25. What is a block (slip) gauge? What are the Indian standards for it?
26. What is wringing of slip gauges? How should you remove slip gauges from one another?
27. Discuss the characteristics and applications of slip gauges.
28. How are gauge blocks selected to build lengths of 45.525 mm and 53.971 mm?
29. What is a limit gauge? Write Taylor's principle of gauge design.
30. Explain the following in connection with gauge design:
    i) Gauge maker's tolerance ii) Wear allowance
31. Evaluate the limits and fits for a pair of mating components of diameter 60 mm with tolerance grade H_8/f_7. Also show the limits graphically.
    Ans: For shaft, 59.97 mm and 59.94 mm; for hole, 60 mm and 60.046 mm.
32. A limit gauge is to be designed for checking a hole of 40 $^{+0.050,\ +0.000}$ mm. Sketch the GO and NO GO gauges with dimensions.
33. The limits of size for a hole are 25.033 mm and 25 mm. The limits of size for a shaft are 24.980 mm and 24.959 mm. Design the plug gauge and gap gauge to check the fit.
34. Shafts of 75 $^{\pm 0.02}$ mm diameter are to be checked with the help of GO and NO GO snap gauges. Design the gauge, sketch it and show its GO size and NO GO size dimensions. Assume normal wear allowance and gauge maker's tolerance.
35. Design the plug and ring gauges to monitor the production of a shaft and hole pair of specification φ 50 H_7 d_8 (dimensions in mm).
36. What is meant by hole basis system and shaft basis system? Which is better?
37. What is the difference between measurement and gauging?
38. What is the difference between a comparator and a measuring instrument?
39. What are roller gauges and limit gauges?
40. What is Taylor's principle of gauge design?
41. Describe the types of limit gauges with neat sketches.
42. Explain with neat sketches the suitable methods of measuring or checking the a) diameter of hole up to 50 mm, and b) diameter of a hole greater than 50 mm.

3

Angular Measurement

3.1 INTRODUCTION

Angular measurement is another important measurement in metrology. It involves the measurement of angles of tapers, bent surfaces, and so on. In angular measurement, two types of angular measuring instruments are used: one is the angle gauge corresponding to the slip gauge (end standard), while the other is a divided scale that corresponds to the line standard. For angular measurement, no absolute standard is required. This is the main difference between linear measurement and angular measurement. An angle is defined as the gap between two lines which meet at a point.

The common measuring instruments used for angular measurement are as follows:

- Autocollimator
- Alignment telescope
- Vernier bevel protractor
- Angle dekkor
- Sine bar

3.2 AUTOCOLLIMATOR

It is essentially an infinity telescope and a collimator combined in one instrument. This optical instrument is used for non-contact measurement of small angular differences. It is used to align components and measure deflections in optical systems. It works by projecting an image onto a target mirror and measuring the deflection of the returned image against a scale.

3.2.1 Principle of Autocollimation

If a point source of light O is placed at the principal focus of a collimating lens, it will be projected as a parallel beam of light, as in Fig. 3.1. If this parallel beam strikes a plane reflector which is normal to the optical axis, it will be reflected back along its own path and refocused at the source O. If the plane reflector is now tilted through some small angle θ, the reflected parallel beam will turn through 2θ, and will be brought to a focus at O_1, in the focal plane, a distance x from O. This effect is shown in Fig. 3.2. If the ray passing through the geometric centre of the lens is considered, as it is, unaffected by refraction, it can be seen that $x = 2\theta f$ mm, where f is the focal length of the lens.

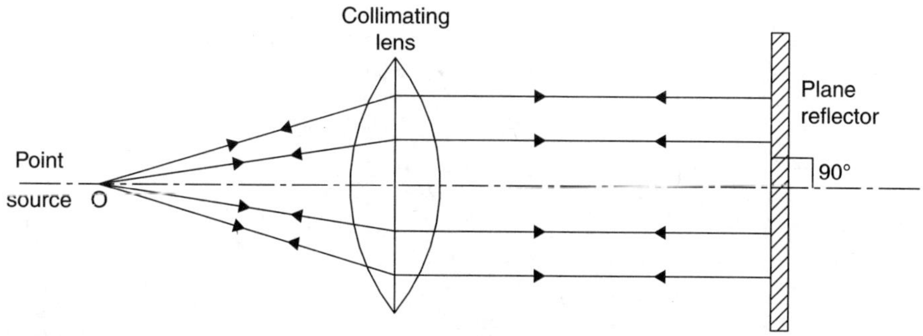

Figure 3.1 Point source of light in the focal plane of a collimating lens

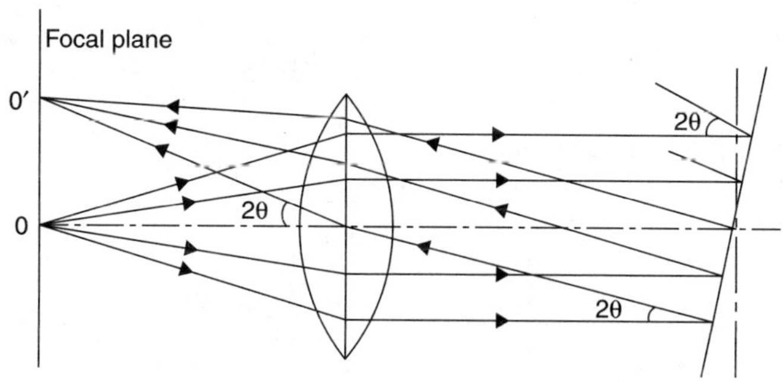

Figure 3.2 Projection of a point source being reflected from an inclined reflector

The points to note about collimation of a beam of light are:

- The distance between the reflector and the lens has no effect on the separation x between the source and the image.
- For high sensitivity, i.e., a large value of x for a small angular deviation θ, a long focal length is required.
- Although the distance of the reflector does not affect the reading x, if, at a given value of θ, it is moved too far back, all of the reflected rays will miss the lens completely, and no image will be formed. Thus, for a wide range of readings, a minimum distance between the lens and the reflector is essential.

3.2.2 Micro-optic Autocollimator

Figure 3.3 shows a typical micro-optic autocollimator. It consists of three main parts, namely, a micrometer microscope, a lighting unit and a collimating lens.

The concept of projecting the image of a point source of light is not practical. So, in this instrument, a pair of target wires in the focal plane of the collimating lens is illuminated from behind and their images are projected. Figure 3.4 shows

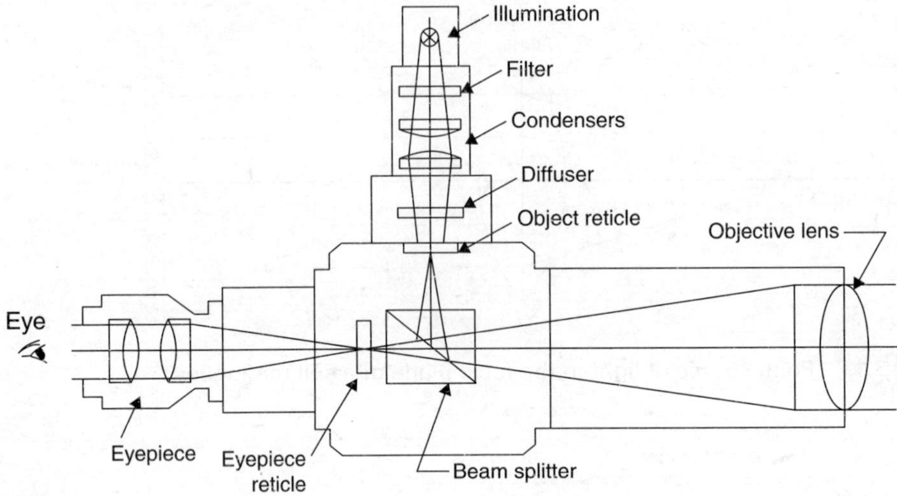

Figure 3.3 Micro-optic autocollimator

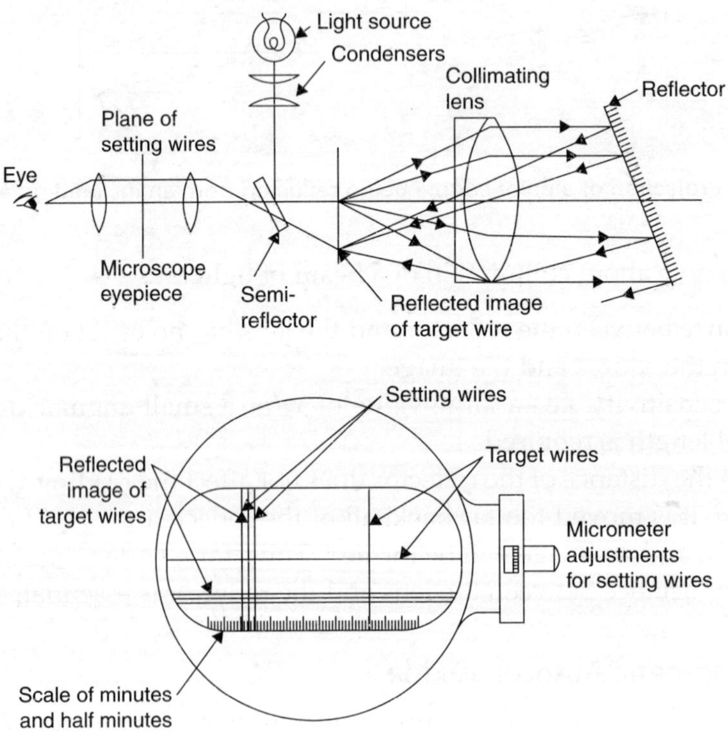

Figure 3.4 Optical arrangement of an autocollimator

the optical arrangement of the instrument, the projected image striking a plane reflector and the reflection of the image being brought to focus in the plane of the target wires.

The target wires and their reflected images are viewed simultaneously in an eyepiece, which also contains a pair of adjustable setting wires and a scale. The setting wires are adjusted by a micrometer until they straddle the reflected image (not the target wire). The scale is read to the nearest ½ min, and the micrometer drum, which moves the wires ½ min per revolution, is divided into 60 equal parts. Thus, 1 division of the micrometer drum represents an angular deflection of the reflector of ½ sec of arc. With care, and given a rigid mounting for the instrument, repeat readings of 0.2 sec are possible. The instrument normally has a range of readings of 10 min of arc up to 10 m.

The longer the focal length of the autocollimator, the greater the angular resolution and the smaller the field of view. A bright halogen light source makes it possible for the autocollimator to measure the angle of target surfaces as small as a 1-mm diameter. A TV camera can be installed on the eyepiece for ease of viewing.

3.3 ANGLE DEKKOR

This is also a type of autocollimator. In this system, an illuminated scale is set in the focal plane of the collimating lens outside the field of view of a microscope eyepiece. It is then projected as a parallel beam and strikes a plane reflector below the instrument. It is reflected and refocused by the lens so that its image is in the field of view of the eyepiece. The image falls, not across a simple datum line, but across a similar fixed scale at right angles to the illuminated image.

Thus, the reading on the illuminated scale measures angular deviations from one axis at 90° to the optical axis, and the reading on the fixed scale gives the deviation about an axis mutually at right angles to the other two. This feature enables angular errors in two planes to be dealt with or, more importantly, to ensure that the reading on the setting master and on the work is the same in one plane, the error being read in the other. Thus, induced compound angle errors are avoided.

The set-up consists of a lapped flat and reflective base above which the optical details are mounted in a tube on an adjustable bracket. Figure 3.5 shows a typical angle dekkor. The optical system and a view of the illuminated scale in the eyepiece are shown in Fig. 3.6.

In use, a master, either a sine bar or a group of combination angle gauges, is set up on the base plate and the instrument is adjusted until a reading on both sides is obtained. It is then replaced by the work-piece and a gauge block to provide a good reflective surface for the face to be checked. The gauge block can be held in place with elastic bands. The work-piece is then slowly rotated until the illuminated scale moves across the fixed scale and is adjusted until the fixed scale reading is the

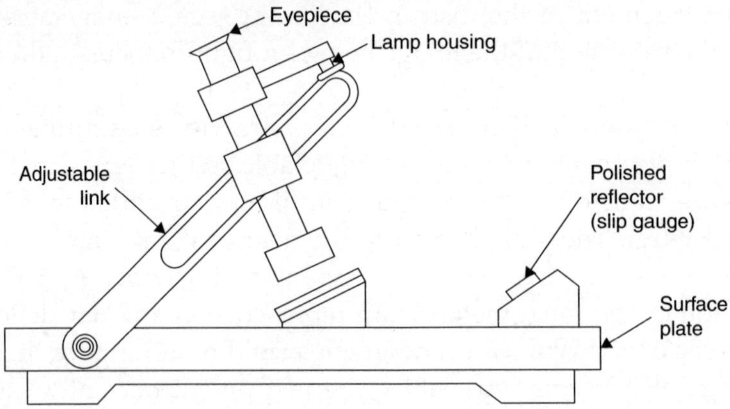

Figure 3.5 Angle dekkor

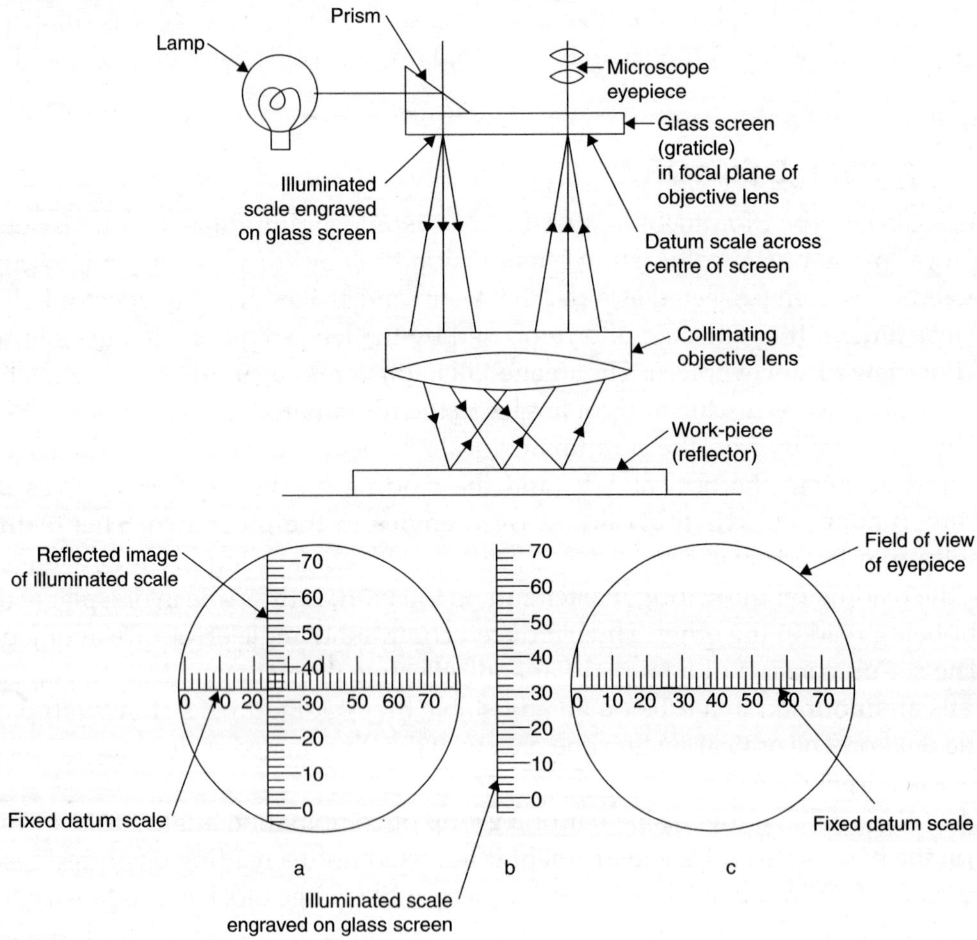

Figure 3.6 Optical system of angle dekkor

same as on the setting gauge. The error in the work-piece angle is the difference in the two readings on the illuminated scale.

To check the squareness of the work-piece, no master is necessary. If gauge blocks are held against both faces and the angle dekkor is adjusted to give a reading, it will be found to consist of two mirror images (due to double reflection) of the illuminated scale, superimposed on each other. The misalignment of the readings of these images will be double the error in the right angle. The type of view obtained in the eyepiece when the angle dekkor is used in this manner is shown in Fig. 3.6. Initially, it is difficult to read, but with practice, the method provides a quick and accurate means of testing squareness.

Although this instrument is not as sensitive as the autocollimator, it is extremely useful for a wide range of angular measurements at short distances. It is used in tool room inspection. Readings direct to 1 min over a range of 50 min may be taken, and by estimation, readings down to about 0.2 min are possible.

3.4 ANGLE GAUGES

Angle gauges are used as master in angle dekkor. They are simply blocks, hardened and lapped to precise angles so that they can be wrung together. The angle gauges, which are manufactured to the same high standards as gauge blocks, are stabilised, have 'wringing' characteristics, and are calibrated to a high degree of precision. The reflective properties of their lapped surfaces make them particularly suitable for use with collimating instruments.

Unlike gauge blocks (slip gauges), angular blocks can be added or subtracted, as in Fig. 3.7.

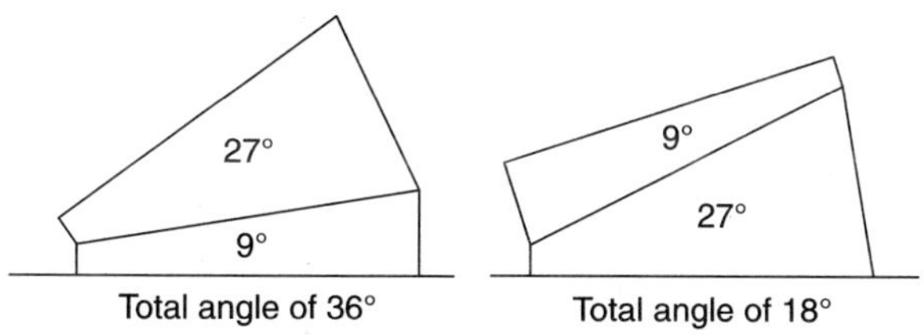

Figure 3.7 Addition and subtraction of combination angle gauges

The values of the angles used are arranged in a modified geometric progression, with a common ratio of 3 for a 13-piece set of angle gauges, as shown in Table 3.1. These 13 gauges, along with a square block, each of whose angles is calibrated, enable any angle between 0° and 90° to be realised in increments of 3 sec. The gauges are used either in combination with each other or may be subtracted from the square block.

Table 3.1 Nominal values of angle gauges (set of 13 pieces)

Degrees	Minutes	Seconds
1	1	3
3	3	6
9	9	18
27	27	30
41		

The gauges are available in a 16-piece set as well, as given in Table 3.2.

Table 3.2 Nominal values of angle gauges (set of 16 pieces)

Degrees	Minutes	Seconds
1	1	1
3	3	3
5	5	5
15	20	20
30	30	30
45		

For example, to measure 43° 24′ 12″ using the angle dekkor, angle gauges (from a 13-piece set) form the standard angle set required to obtain the angle, as shown below:

$$\text{Degrees}: \quad 41 + 3 - 1 = 43^\circ$$

$$\text{Minutes}: \quad 27 - 3 = 24'$$

$$\text{Seconds}: \quad 30 - 18 \text{ (or) } 18 - 6 = 12''$$

The arrangement of angle gauges for 43° 24′ 12″ is shown in Fig. 3.8.

Thus, a total of seven angle gauges are required to set the angle 43° 24' 12". These angle gauges are placed on the table of the angle dekkor and aligned parallel to its edge. The instrument is then adjusted to obtain a reflection from the top surface of the gauge combination, and the reading obtained is recorded. This angle gauge set is then replaced by the work-piece whose angle is to be measured. A reflector is used to ensure a clear reading. Any angular variation between the angle gauges and the work-piece is indicated by vertical and horizontal displacements of the

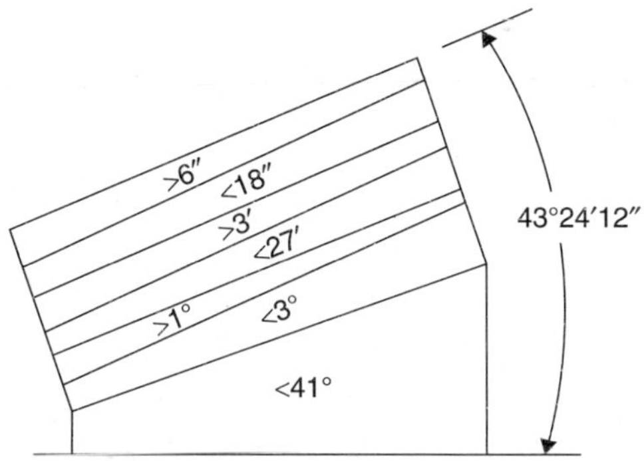

Figure 3.8 Combination of angle gauges for 43° 24' 12"

reflected vertical scale. This angular deviation is then added or subtracted with the angle gauges set to obtain the true taper angle of the given specimen.

3.5 ALIGNMENT TELESCOPE

The alignment telescope is a precision instrument used for the alignment of objects on a reference line defined by the line of sight or optical axis of the telescope. The optical axis runs concentrically with respect to the tubular barrel to a very high degree of accuracy. The mounting of the alignment telescope is accurately tied to its optical axis and is specially designed for aligning boreholes, bearings and optical set-ups. The alignment telescope has a wide range of focus, from a few inches to infinity. The sturdy, stainless steel construction will maintain specifications even in rough working conditions. Optional autocollimation using the alignment telescope is accomplished by specifying an autocollimation eyepiece in place of the standard eyepiece.

It consists of two units: a collimating unit and a focusing telescope, the body of each of which is ground truly cylindrical and to a precise external diameter. Further, the optical axis and the mechanical axis are coincident. Thus, each unit may be fitted directly or by precision bushing into two bearings, a considerable distance apart, and sightings taken from the telescope unit to the collimating unit. The collimating unit contains a light source and condensers, in front of which is placed an angular graticule in the focal plane of the collimating lens. This graticule scale is thus projected as a parallel beam of light. If the telescope is focused at infinity, it will bring to focus the parallel rays, and the angular scale is seen against the datum lines in the telescope. Thus, angular misalignment in both planes is determined.

The collimating unit also contains, in front of the collimating lens, a second graticule also having two scales at right angles to each other. If the focus of the telescope is shortened, this graticule is seen against the datum lines of the telescope and linear displacements are measured directly. In this case,

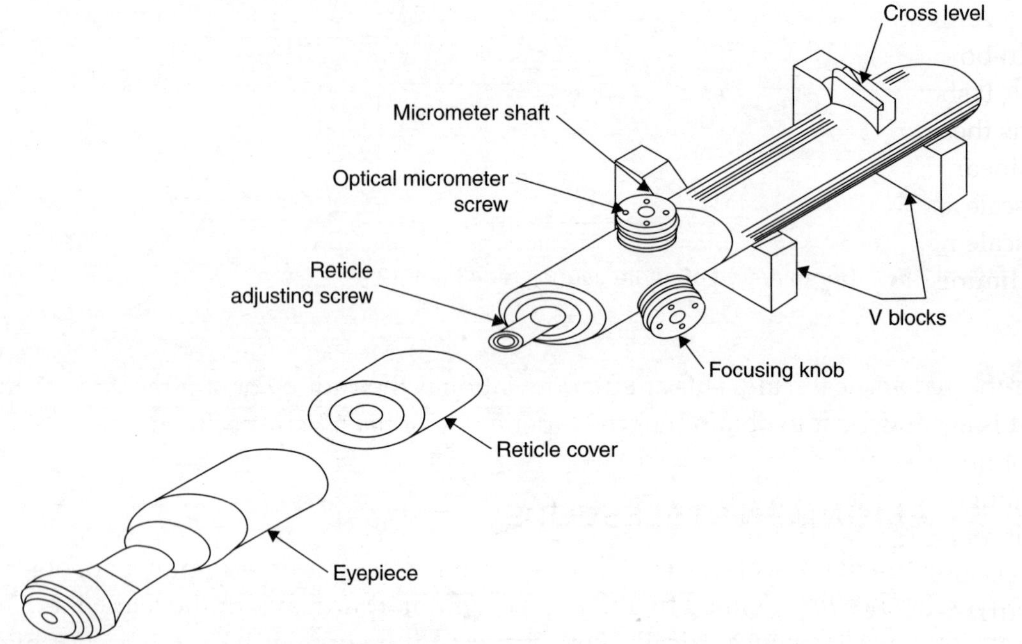

Figure 3.9 Taylor–Hobson alignment telescope

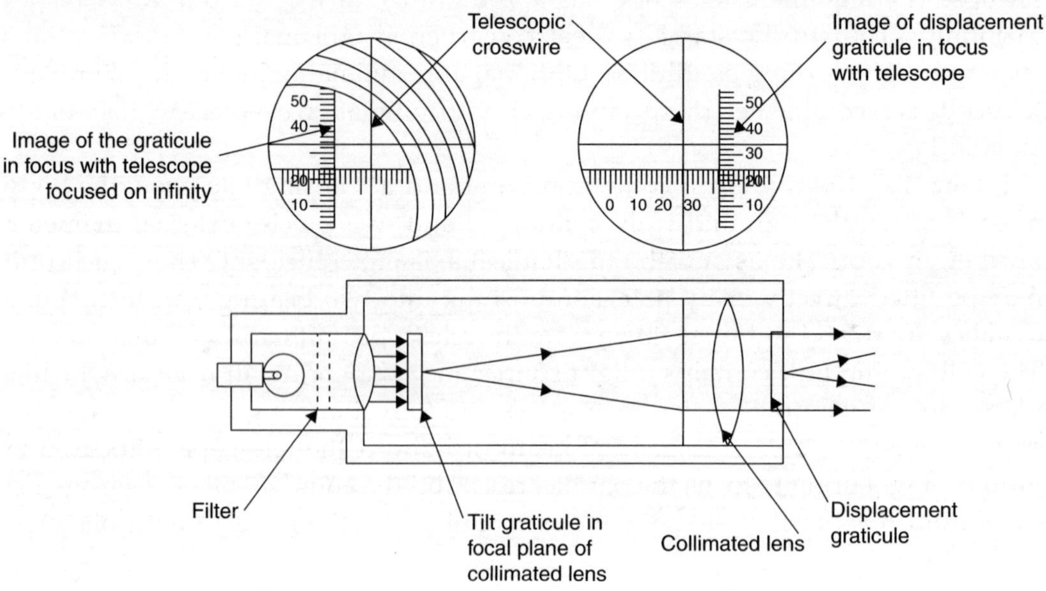

Figure 3.10 Optical arrangement of alignment telescope

the collimating lens simply provides even illumination for the displacement graticule, and the angular misalignment graticule cannot be seen because it is so far out of focus. Figure 3.9 shows a typical alignment telescope.

A line diagram of the collimating unit and the view in the telescope eyepiece in both conditions of focus is shown in Fig. 3.10.

It should be noted that distance has no effect on angular misalignment readings, as these are taken using the collimating principle. However, this is not so with the linear displacement scale. The telescope only magnifies the apparent size of the scale as seen by the eye. Thus, as the distance increases, the size of the displacement scale reduces and its 'readability' also reduces. Thus, the accuracy of this reading diminishes with distance.

3.6 SINE BAR

It is a precision measuring instrument and an excellent example of a combination of linear measurement and angular measurement when used in conjunction with gauge blocks (slip gauges). It consists of a bar carrying a suitable pair of rollers set at a known central distance. It is made of high carbon, high chromium corrosion-resistant steel, suitably hardened, precision ground and stabilised. The working surfaces of the rollers are finished to a 0.2 μm R_a value. Relief holes are provided

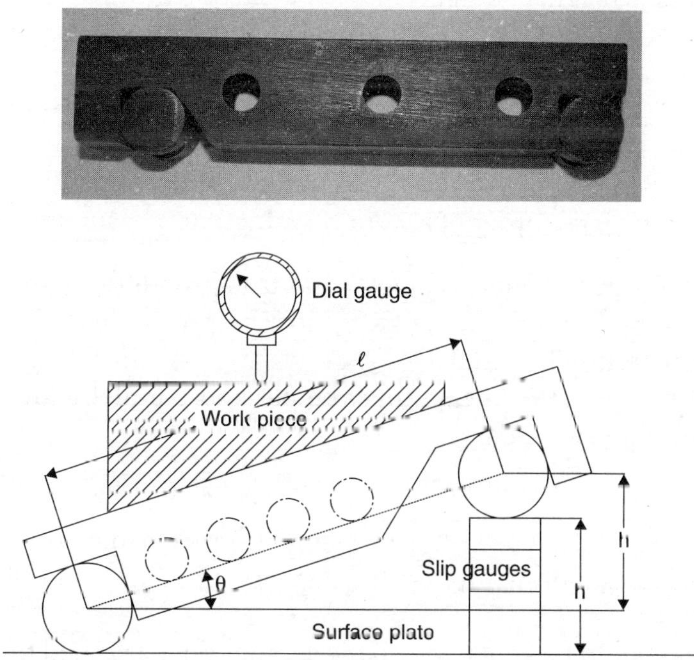

Figure 3.11 Sine bar

for easy handling of the sine bar and to reduce its weight. It should be used on a grade A surface plate. Figure 3.11 shows the arrangement of a sine bar with slip gauges and the work-piece whose taper angle needs to be measured.

Working principle of sine bar: The working of the sine bar is based on the principle of trigonometry. First, the sine bar should be placed on the surface plate. Then, the given specimen should be placed on the working surface of the sine bar such that the tapered surface, whose taper angle is to be measured, faces upwards. To measure the angle of a given specimen, the roller near the taper provided at the bottom of the sine bar should be placed over the surface of the slip gauges. The slip gauges should be built till the tapered surface of the given specimen becomes flat, i.e., parallel to the surface plate. This can be ensured by moving the plunger end of the dial gauge on the tapered surface to check whether there is any deflection in the dial gauge. No deflection of the hand in the dial gauge indicates that the tapered surface of the specimen is parallel to the surface plate.

Take readings with the dial gauge at both ends and note their difference, specifying which end is low. Assuming that the end nearest the high end of the sine bar is low, the height of the slip gauges must be increased by an amount equal to the difference in the dial gauge readings multiplied by the proportion of sine bar length to work length. For example, assuming that the end of a work-piece is 0.01 mm low, the sine bar being 250 mm long and the work-piece 100 mm long, the required increase in height of the slip gauge set will be $0.01 \times 250/100$, i.e., 0.025 mm. This will not provide an immediately correct setting from first approximation, but it is much quicker than a trial-and-error method.

If ℓ is the linear distance between the axes of the rollers and h is the height of the slip gauges, then $\sin\theta = h/\ell$.

The design requirements of a sine bar are as follows, and unless these are carefully maintained, the order of accuracy of angular measurement will fall:

- The rollers must be of equal diameter and true geometric cylinders.
- The distance between the roller axes must be precise and known, and these axes must be mutually parallel.
- The upper surface of the beam must be flat and parallel with the roller axes, and equidistant from each.

Note:

- No sine bar should be used to set up angles greater than 45°, as beyond this angle, the errors due to the centre distance of rollers and slip gauges are magnified.
- Slip gauges should be kept beneath the setting roller attached to the end which is with the taper shape but not beneath the hinge roller. This is to ensure that the slip gauges do not hit the bottom surface of the sine bar.

- If the work-piece is large, the sine bar must be kept upside down on the surface of the work–piece, as shown in Fig. 3.12. The height over the rollers can be measured using a Vernier height gauge, with a dial test gauge mounted on the anvil of the height gauge to ensure constant measuring pressure. The difference in roller heights is equal to the height of the slip gauges set used in the previous case.

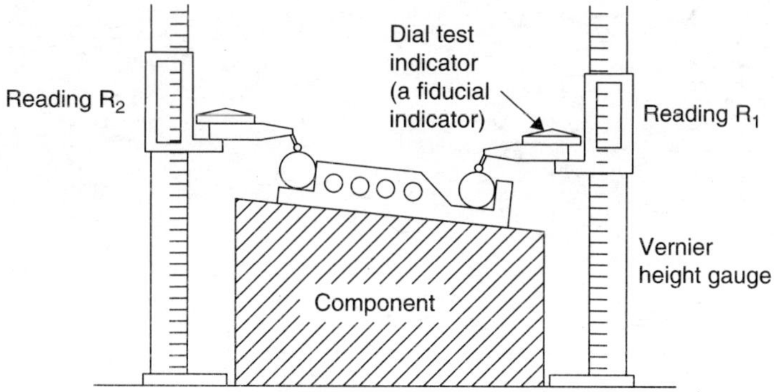

Figure 3.12 Use of sine bar with large work-piece

3.7 SINE CENTRE

A sine centre is a sine bar that has two dead centres to hold a conical work-piece. A typical sine centre set-up is shown in Fig. 3.13. It is difficult to mount the conical work-piece on the sine bars. To overcome this problem, a sine centre is used. In the sine centre, two blocks are mounted on the top surface of the sine bar at each end. These blocks have centres and can be clamped in any position. The centres can be adjusted for the required length of the work-piece. The sine centre cannot be used to measure angles greater than 45°.

The principle of setting is the same as in the sine bar, although a hazard to be avoided is that of the work and centres not being co-axial. To overcome this, the work-piece should be rotated on the centres until the maximum dial gauge reading is at the top. The angle is calculated from the slip gauges set in this condition, the work-piece is turned through 180° and the process is repeated. The mean of the two angles determined will be the semi-angle of the work-piece, although it must be pointed out that any work that runs out to a measurable extent would probably be considered sub-standard in quality and be rejected on this account.

3.7.1 Sine Table

It is the most convenient and accurate design for heavy work-pieces. A sine table consists of a self-contained sine bar hinged at one roller and mounted on its datum surface (Fig. 3.14). The table is quite rigid and the weight of the unit and work-piece

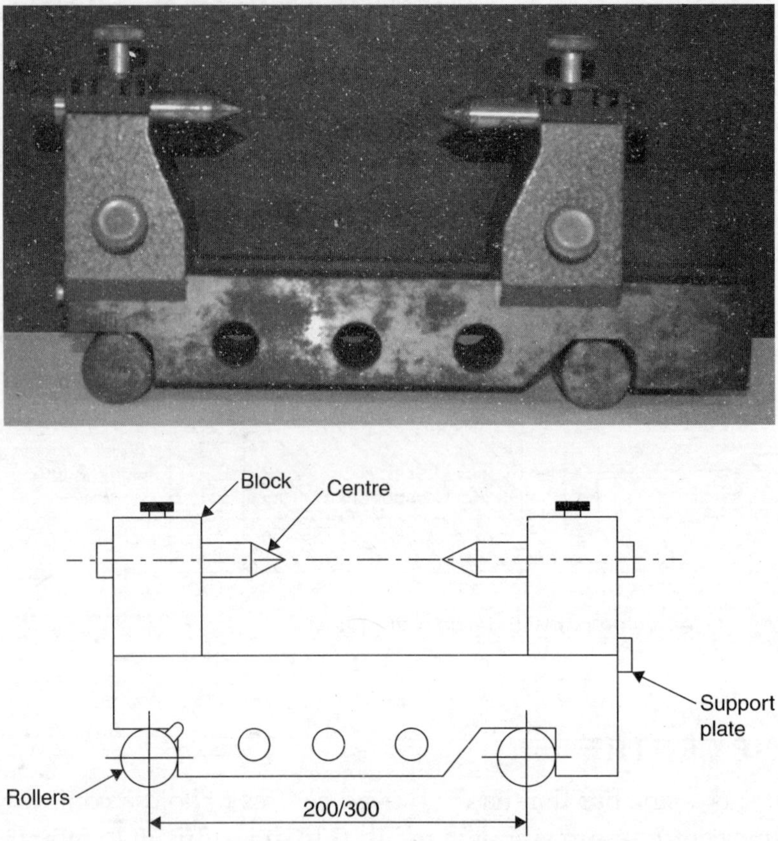

Figure 3.13 Sine centre

is given fuller and safer support. The sine bar may be safely swung to any angle from 0° to 90° by pivoting it about its hinged end. The set-up consisting of two

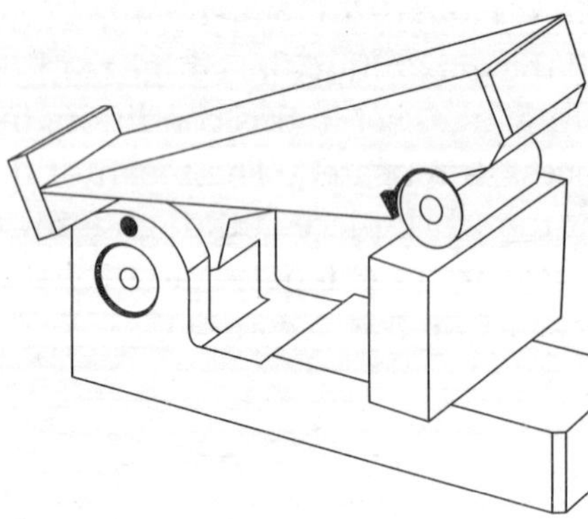

Figure 3.14 Sine table

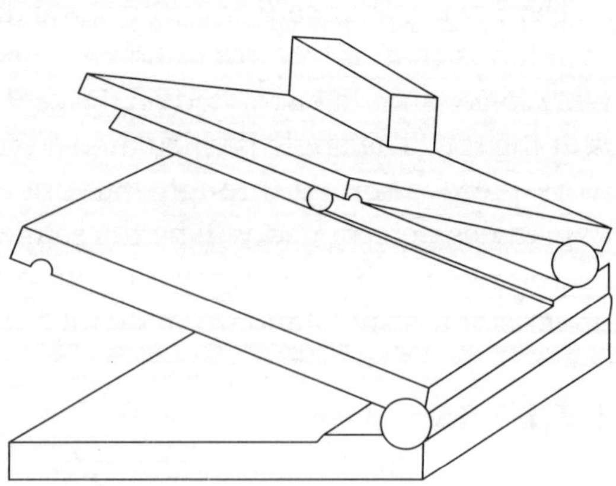

Figure 3.15 Compound sine table

sine tables mounted on a common base, with axes set at right angles, is called a compound sine table. A typical compound sine table is shown in Fig. 3.15.

3.7.2 Sources of Error in Sine Bar

The different sources of errors pertaining to sine bars are listed below:

- Error in distance between roller centres
- Error in slip gauge combination
- Error in checking parallelism
- Error in equality of roller size and cylindricity of rollers
- Error in parallelism of roller axes with each other
- Error in flatness of the working surface of the sine bar

3.8 VERNIER BEVEL PROTRACTOR

This is the simplest angle-measuring instrument. A simple Vernier bevel protractor with its various elements is shown in Fig. 3.16. The body of the bevel protractor is

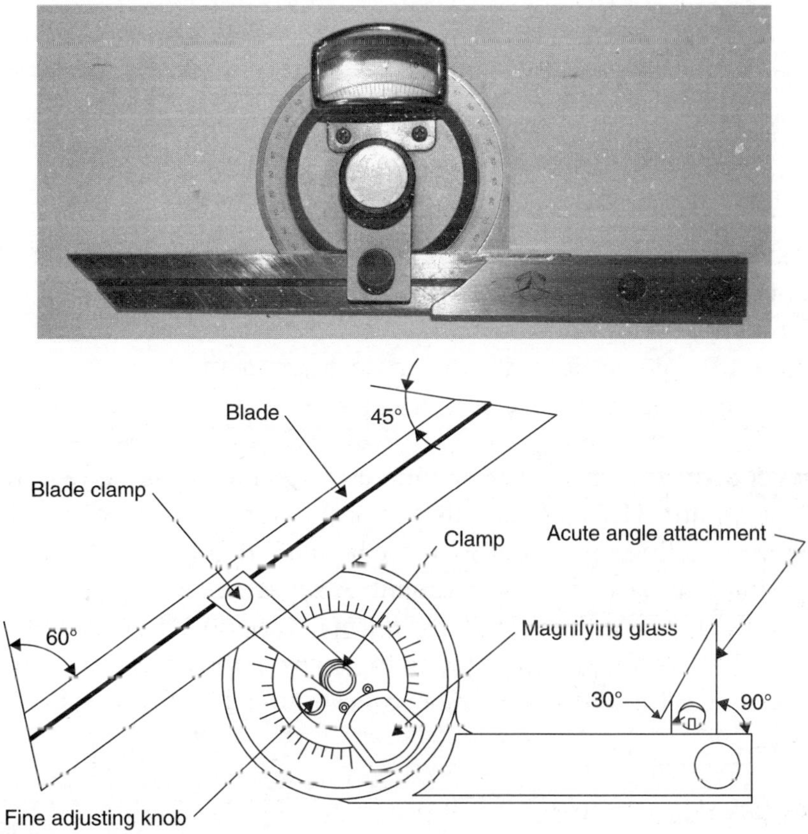

Figure 3.16 Vernier bevel protractor

designed in such a way that its back is flat and there are no projections beyond its back. The base plate is attached to the main body and an adjustable blade is attached to a circular plate containing the Vernier scale. The main body carries a main scale graduated in degrees from 0° to 90° in both directions. The adjustable blade, which is capable of rotating freely about the centre of the main scale engraved on the body of the instrument, can be locked in any position. An acute angle attachment is provided at the top. The base of the base plate is made flat so that it can be laid flat on the work-piece. The blade can be moved throughout its length and can also be reversed. The ends of the blade are bevelled at angles of 45° and 60°. The acute angle attachment can be readily fitted into the body and clamped in any position.

Universal bevel protractor: The protractor dial is slotted to hold a blade that can be rotated using the dial to the required angle. It can also be adjusted independently to any desired length. The blade can be locked in any position. This instrument is capable of measuring readings precisely within 5 minutes.

The main scale is graduated, as one main scale division is 1°, and the Vernier scale is graduated into 12 divisions on each side of zero. Therefore, the least count is calculated as

$$\text{Least count} = \frac{\text{One main scale division}}{\text{Number of divisions on the Vernier scale}}$$

$$= \frac{60 \text{ minutes}}{12} = 5 \text{ minutes}$$

Types of bevel protractors: As per IS practice, there are two bevel protractors:

- Mechanical bevel protractor
  - With Vernier scale and acute angle attachment
  - Without Vernier scale and acute angle attachment
- Optical bevel protractor

Optical bevel protractor: This instrument is capable of taking readings within 2 minutes of an arc. The circular interval scale is graduated in divisions of 10 minutes of arc. Readings are taken with the help of an optical magnifying system, which is an integral part of the instrument. A separate arrangement is provided for adjusting the focus of the system for normal variations in eyesight. The main and Vernier scales are always arranged in the focus of the optical system. A typical optical bevel protractor with the Vernier scale engraved on the magnifying glass is shown in Fig. 3.17.

Uses of the bevel protractor: Bevel protractors can be used to test the flatness, squareness, parallelism, straightness and angular measurements of objects. One

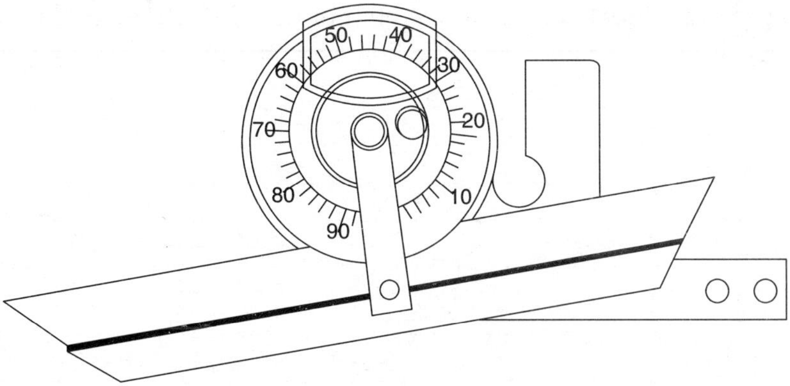

Figure 3.17 Optical bevel protractor

use of the bevel protractor, i.e., checking inside the bevelled face of a ground surface, is shown in Fig. 3.18.

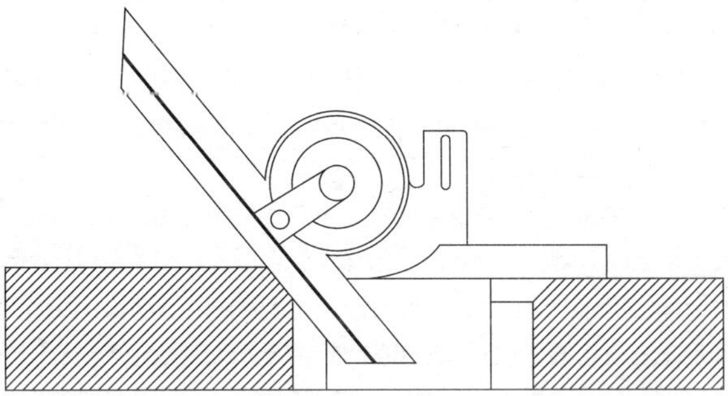

Figure 3.18 Checking inside the bevelled face of a ground surface

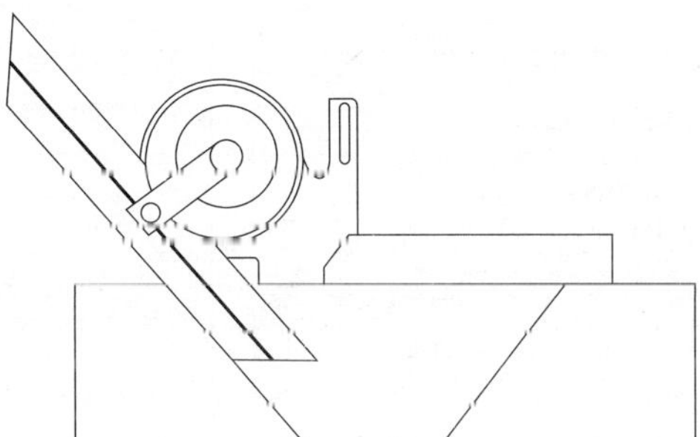

Figure 3.19 Checking the V block

Checking of the V block and measurement of the acute angle of the surface of a work-piece are shown in Figs 3.19 and 3.20, respectively.

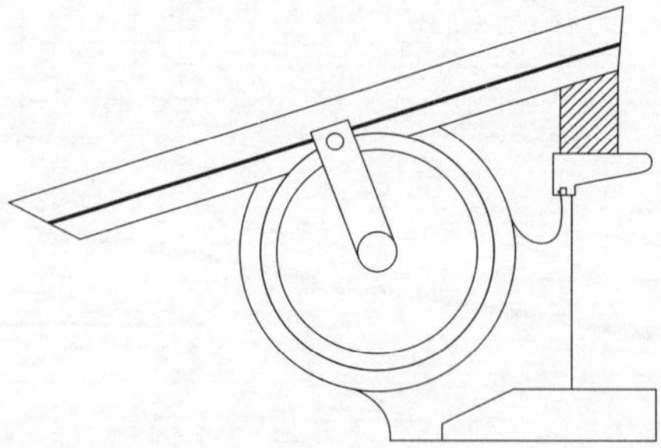

Figure 3.20 Measuring the acute angle

3.9 VERNIER CLINOMETER

This is a special case of an application of the spirit level. In this instrument, the spirit level is mounted on a rotatable member carried in a housing, one face of which forms the base of the instrument. The instrument is mainly used to measure the included angle between two adjacent faces of the work-piece. A typical Vernier clinometer is shown in Fig. 3.21.

In use, the instrument base is placed on one of the surfaces of the work-piece and the rotatable member is adjusted until a zero reading of the bubble is obtained. The angle of rotation is then noted on the circular scale against an index. The instrument base is then placed on the other surface of the work-piece and the reading is taken in the similar manner. If θ and ϕ are the readings of the instrument, the included angle between the surfaces of the work-piece is $\alpha = 180 - (\theta + \phi)$, which is shown in Fig. 3.22.

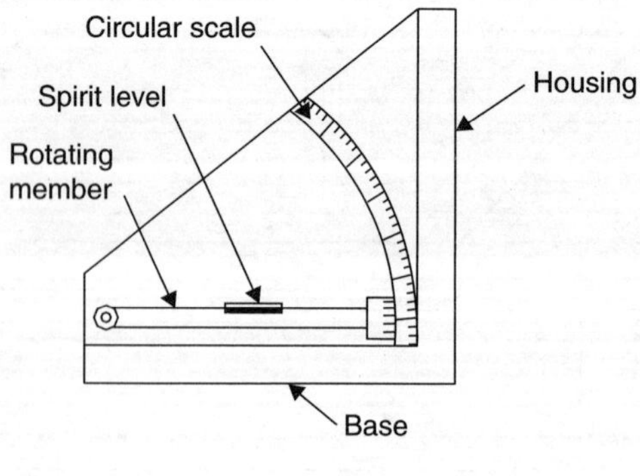

Figure 3.21 Vernier clinometer

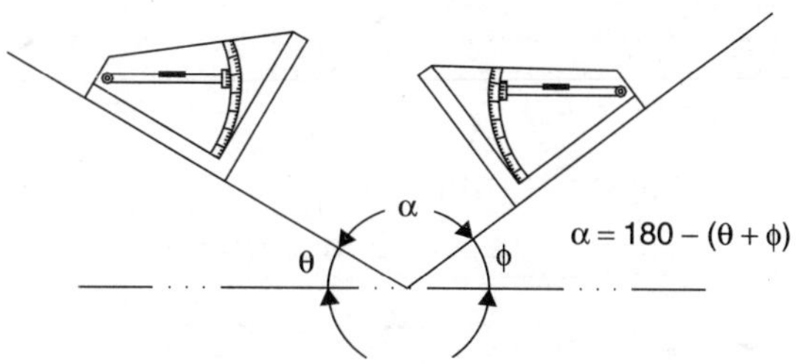

Figure 3.22 Measuring the included angle of adjacent faces

3.10 MEASUREMENT OF V-ANGLE USING ROLLERS

If a roller of diameter d_1 is placed in the V groove, it will touch its flank at a point A, as shown in Fig. 3.23. Similarly, if a roller of diameter d_2 is placed in the V groove, it will touch its flank at point B. Then, the line joining the centres of the two rollers and the line drawn parallel to one of the flanks will form a triangle O_1O_2C.

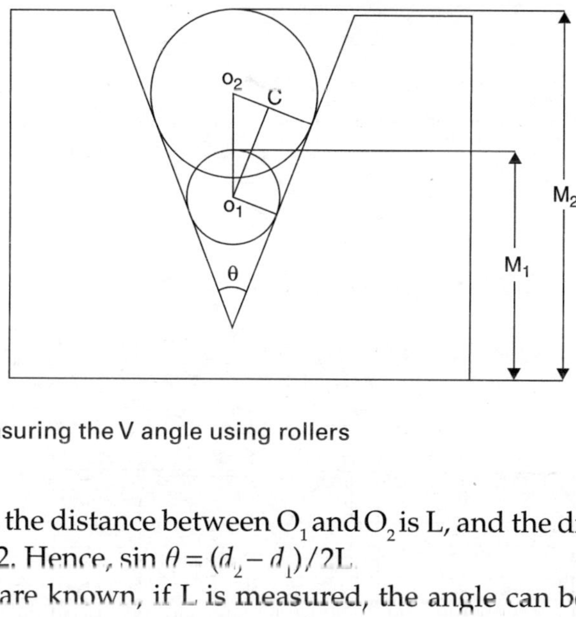

Figure 3.23 Measuring the V angle using rollers

In triangle O_1O_2C, the distance between O_1 and O_2 is L, and the distance between O_2 and C is $(d_2 - d_1)/2$. Hence, $\sin\theta = (d_2 - d_1)/2L$.

Since d_1 and d_2 are known, if L is measured, the angle can be calculated easily. This distance L can be obtained by measuring the height of the tops of the smaller and larger rollers with reference to a base, using the Vernier height gauge.

Let the height of the smaller roller $= M_1$

Height of the larger roller $= M_2$

Then, $$L = \left(\frac{M_2 - d_2}{2}\right) - \left(\frac{M_1 - d_1}{2}\right) = \frac{(M_2 - M_1) - (d_2 - d_1)}{2}$$

Then, $\sin\theta = \dfrac{(d_2 - d_1)/2}{(M_2 - M_1) - (d_2 - d_1)/2}$

Hence, by measuring the heights of the two precision rollers of known diameter, the V-groove angle can be calculated using the equation.

3.11 MEASUREMENT OF THE TAPER OF A BORE USING BALLS

The concept is the same as that used for measuring the V angle of a groove. Two precision balls of different diameters are placed inside the bore instead of rollers, one at a time, and the depth of the top point of each ball from the top edge of the taper bore is measured using a depth micrometer (Fig. 3.24).

As the top surface of the balls can be reached only from the top side of the bore, the depth micrometer is placed on two slip gauge block supports of equal height and the depth to the top of the precision ball is measured.

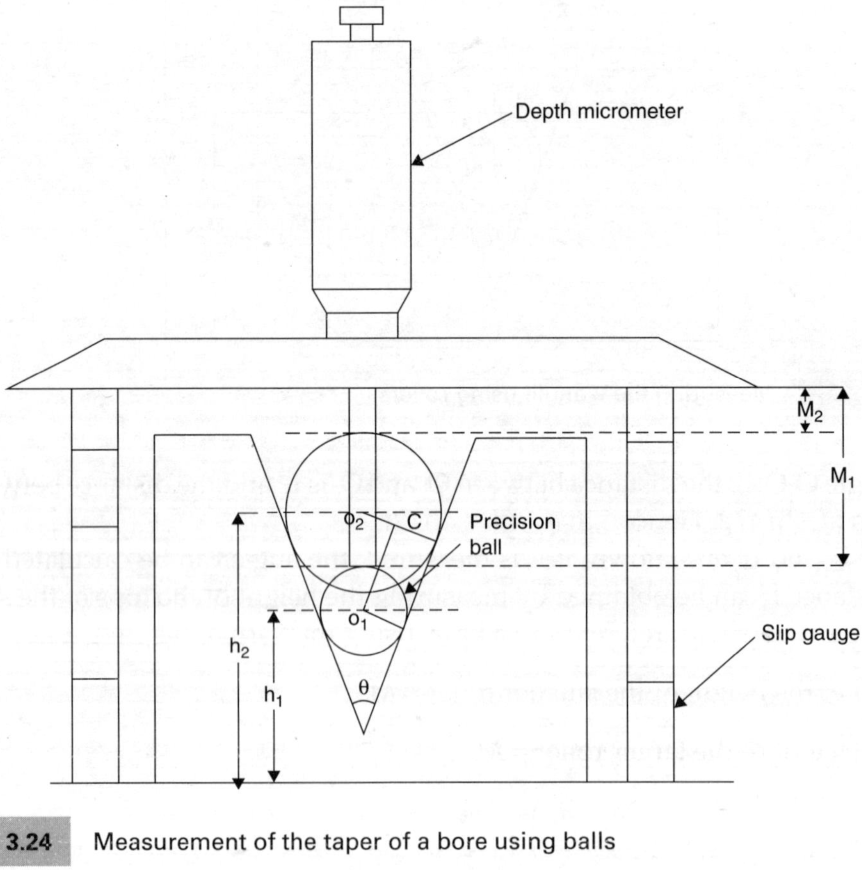

Figure 3.24 Measurement of the taper of a bore using balls

Let the diameter of the smaller ball = d_1

Diameter of the larger ball = d_2

Height of the slip gauge support = x

Depth to the top of the smaller ball = m_1

Depth to the top of the larger ball = m_2

Then, the height of the centre of the smaller ball $h_1 = \dfrac{x - m_1 - d_1}{2}$

Height of the centre of the larger ball $h_2 = \dfrac{x - m_2 - d_2}{2}$

Then, the distance between the centres of the two balls, L is as follows

$$L = \left(\frac{x - m_2 - d_2}{2} \right) - (x - m_1 - d_2)$$

$$= \frac{x - m_2 - d_2}{2} - \frac{x + m_1 + d_2}{2} = (m_1 - m_2) - \frac{1}{2}(d_2 - d_1)$$

Then, $\sin\theta = \dfrac{(d_2 - d_1)/2}{(m_1 - m_2) - (d_2 - d_1)/2}$

By this method, the maximum and minimum diameters of the bore can also be calculated (3.25).

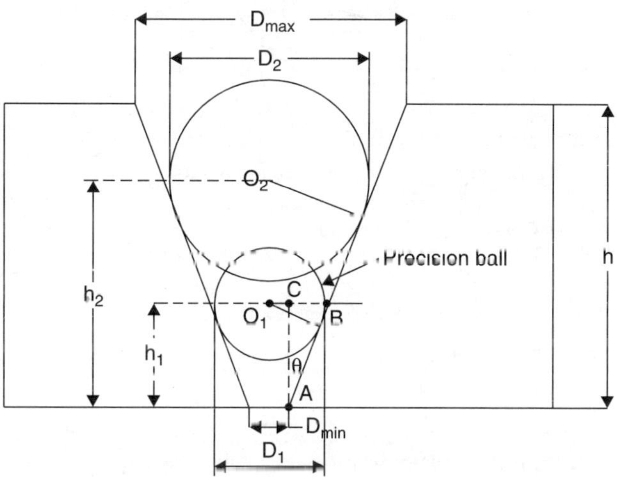

Figure 3.25 Measuring the diameter of a bore using balls

Let the diameter of the bore at the section that passes through the centre of the smaller precision ball be $D_1 = d_1 \sec \theta$.

Similarly, the diameter of the bore at the section that passes through the centre of the larger precision ball is $D_2 = d_2 \sec \theta$.

From the ΔABC $\tan\theta = \dfrac{BC}{h_1}$ or $BC = h_1 \tan\theta$

$$\therefore \quad D_{min} = D_1 - 2h_1 \tan\theta.$$

Similarly, $$D_{max} = D_2 + 2(h - h_2)\tan\theta.$$

The value h is calculated by detecting the depth measurement made on the top of the specimen surface from the slip gauge length; i.e., $h = x - m_3$, where m_3 is the depth of the top of the specimen surface and x is the height of the slip block.

3.12 MEASUREMENT OF THE EXTERNAL CIRCULAR TAPER USING ROLLERS

In this method, two precision rollers of equal diameter are placed at a certain height on either side of the circular tapered work-piece, and the distance between the micrometer anvils holding the rollers is measured using a micrometer (Fig. 3.26). Similarly, the distance is measured at a lower height using the same precision rollers and micrometer. The height can be set accurately using slip gauge blocks.

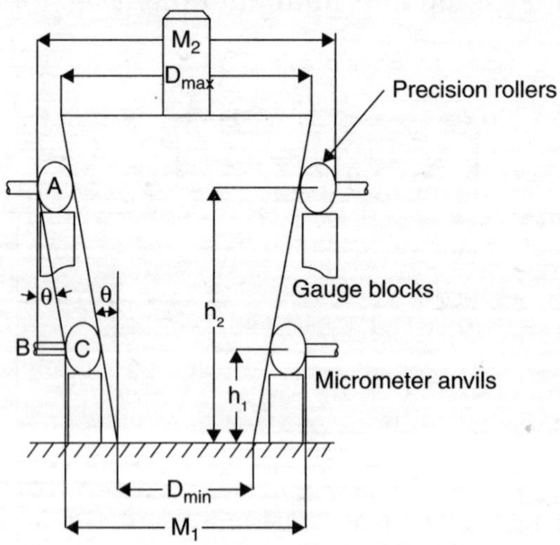

Figure 3.26 Measuring the external taper using rollers

Let the distance between the micrometer anvils at height $h_1 = M_1$ and the distance between the micrometer anvils at height $h_2 = M_2$.

To find the taper angle (θ) of one side of the work-piece:
In triangle ABC

$$\text{Distance AB} = h = (h_2 - h_1)$$

$$\text{Distance BC} = \frac{(M_2 - M_1)}{2}$$

$$\therefore \quad \tan\theta = \frac{\text{BC}}{\text{AB}} = \frac{(M_2 - M_1)}{2(h_2 - h_1)}$$

To find different diameters of the circular tapered work-piece:
In triangle ODE (Fig. 3.27)

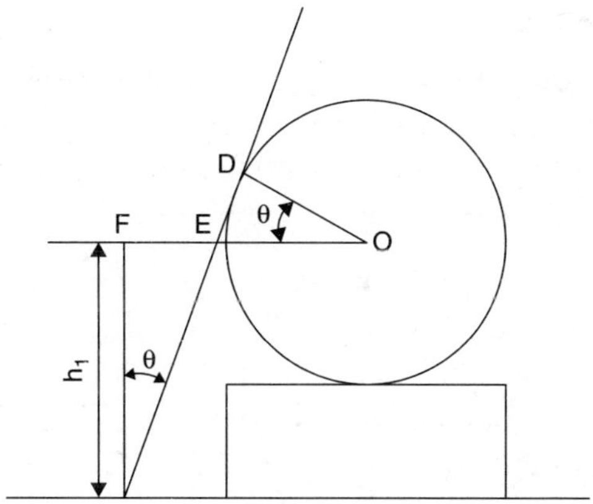

Figure 3.27 Measurement of maximum and minimum diameter

$$\text{Angle DOE} = \theta = \text{Taper angle}$$

Then, $\text{OE} = \dfrac{\text{OD}}{\cos\theta} = \dfrac{d}{2} \div \cos\theta = \dfrac{d}{2\cos\theta}$, where d is the diameter of the roller

Therefore, the diameter at the section joining the centres of the rollers is given by

$D_1 = M_1 - d(1 + \sec\theta)$, at lower height h_1

$D_2 = M_2 - d(1 + \sec\theta)$, at higher height h_2

Using these values, D_{min} and D_{max} can be computed as follows:

$$D_{min} = D_1 - 2h_1 \tan\theta$$

$D_{max} = D_2 + (h - h_2)\tan\theta$, where h is the height of the circular taper.

SUMMARY

- Principle of autocollimation: If a point source of light is placed at the principal focus of a collimating lens, it will be projected as a parallel beam of light. If this parallel beam strikes a plane reflector which is normal to the optical axis, it will be reflected back along its own path and refocused at the source O. If the plane reflector is then tilted through some small angle θ, the reflected parallel beam will turn through 2θ, and will be brought to a focus at O_1, in the focal plane, at a distance x from O. If the ray passing through the geometric centre of the lens is considered, as it is, unaffected by refraction, it can be seen that $x = 2\,\theta\,f$ mm, where f is the focal length of the lens.
- The concept of projecting the image of a point source of light is not practical. So, in the micro-optic autocollimator, a pair of target wires in the focal plane of the collimating lens are illuminated from behind and their images are projected. The projected image strikes a plane reflector and the reflection of the image is brought to a focus in the plane of the target wires.
- In an angle dekkor, an illuminated scale is set in the focal plane of the collimating lens, outside the field of view of a microscope eyepiece. It is then projected as a parallel beam and strikes a plane reflector below the instrument. This is reflected, and refocused by the lens so that its image is in the field of view of the eyepiece. The image falls, not across a simple datum line, but across a similar fixed scale at right angles to the illuminated image.
- The alignment telescope is a precision instrument used for the alignment of objects on a reference line defined by the line of sight or optical axis of the telescope. The optical axis runs concentrically with respect to the tubular barrel to a very high degree of accuracy. The mounting of the alignment telescope is accurately tied to its optical axis and is specially designed for aligning bore holes, bearings and optical set-ups.
- A sine bar is used along with slip gauges to measure the taper of a surface of a solid work-piece. It works on the concept of sine angle of a right-angled triangle. It is also available in the form of sine centre and sine table.
- A bevel protractor is used to measure the angle of a plate, taper surface of a solid work-piece, without using slip gauges. It is available with Vernier scale, acute angle attachments and optical viewing arrangements.
- A Vernier clinometer is used to measure the V angle of adjacent surfaces of a solid work-piece.

❖ Using only rollers and a Vernier height gauge, the taper angle, minimum and maximum diameters of a tapered bore and tapered solid shafts can be easily measured without typical measuring instruments, though the measurement is not more accurate.

EXERCISES

Multiple Choice Questions

1. ______________ standard is not required for angular measurement.
   a) Equal b) Higher
   c) Absolute d) Physical

2. ______________ works by projecting an image onto a target mirror and reading the deflection of the returned image against a scale.
   a) Sine bar b) Autocollimator
   c) Angle dekkor d) Vernier calipers

3. In an autocollimator, the greater the resolution, the ______________ the field of view
   a) Smaller b) Larger
   c) Equal d) Very large

4. Readings down to 0.2 mm are positive in
   a) Angle dekkor b) Sine bar
   c) Vernier calipers d) Vernier bevel protractor

5. Micrometer, micro scale, lighting unit and collimating lens are parts of
   a) Micro-optic autocollimator b) Sine bar
   c) Vernier bevel protractor d) Vernier calipers

6. ______________ is a precision instrument for alignment of objects on a reference line defined by the line of sight.
   a) Alignment telescope b) Sine bar
   c) Vernier calipers d) Vernier bevel protractor

7. ______________ consists of a bar carrying a suitable pair of rollers set at a known centre distance.
   a) Alignment telescope b) Sine bar
   c) Autocollimator d) Vernier bevel protractor

8. Relief holes in the sine bar are provided for easy handling and __________
   a) Increasing weight b) Equalising weight
   c) Reducing weight d) Reducing mass

9. Sine bar works on the principle of __________
   a) Pascal's law b) Trigonometry
   c) Newton d) Gaussian

10. The maximum set-off angle of a sine bar is __________
    a) 45° b) 90°
    c) 60° d) 30°

11. The sine bar is kept upside down on a work-piece whose size is __________
    a) Large b) Small
    c) Sphere d) Circular

12. __________ is a sine bar carrying two dead centres to hold a conical work-piece.
    a) Sine centre b) Slip gauge
    c) Autocollimator d) Vernier calipers

13. The sine centre cannot be used to measure a taper angle that is greater than
    a) 30° b) 60°
    c) 90° d) 45°

14. __________ is most convenient and accurate design for heavy components.
    a) Sine bar b) Sine centre
    c) Sine table d) Vernier scale

15. __________ is not a source of error in a sine bar.
    a) Slip gauge combination b) Parallelism
    c) Distance between roller centres d) Misalignment of scale

16. Checking a V block can be done easily using a __________
    a) Slip gauge b) Bevel protractor
    c) Sine bar d) Sine centre

17. __________ instrument measures the included angle of two adjacent faces of a work–piece.
    a) Vernier calipers b) Bevel protractor
    c) Vernier clinometer d) Sine bar

18. The angle between two surfaces of a work-piece using a Vernier clinometer is given by
    a) $\alpha = 180 - (\theta + \varphi)$ b) $\varphi = 180 - (\theta + \alpha)$
    c) $\theta = 180 - (\varphi + \alpha)$ d) $180 + (\theta + \alpha) = \varphi$

19. Calculate the taper angle if 8 mm and 12 mm are the diameters of balls A and B, respectively. Ball A is at a height of 25 mm from the horizontal ground surface and ball B is at 15 mm.
    a) 14.47°
    b) 28.95°
    c) 7.23°
    d) None of the above

20. Which among the following is an optical instrument?
    a) Angle dekkor
    b) Autocollimator
    c) Both a and b
    d) None of the above

Answers				
1. c	2. b	3. b	4. a	5. a
6. a	7. b	8. c	9. b	10. a
11. a	12. a	13. d	14. c	15. d
16. b	17. c	18. a	19. b	20. c

Review Questions

1. What is the basic difference between linear and angular measurement?
2. What are the common angular measuring instruments?
3. Explain the principle of autocollimation.
4. What is the minimum and maximum angular displacement that can be resolved by the autocollimator?
5. Explain the working principle of the micro-optic autocollimator and briefly explain its applications.
6. What is the advantage of angle dekkor over autocollimator?
7. Explain the working principle of angle dekkor and how a job of angle 43° 24′ 12″ is checked.
8. Explain the working principle of the alignment telescope.
9. What are the differences between autocollimator and angle dekkor?
10. What are the standard angle gauges? Explain how the various combinations of angle gauges give different measurements.
11. The taper angle was measured using a 100-mm sine bar and the height of gauge block set was 1.25 mm. What is the taper angle?
12. With a neat sketch, explain the use of sine bar for measuring the angle of a taper plug gauge.
13. Explain how a sine bar is used in angular measurement of a small job and a heavy job.

14. With the aid of sketches, explain the methods of using sine bar for a) measuring known angles and b) checking unknown angles.
15. Why is a taper provided at the bottom of the sine bar near one of the bottom rollers?
16. What are the sources of error in sine bar? Name the factors affecting the accuracy of a sine bar.
17. What is the use of a sine centre?
18. What is a sine table?
19. Explain why sine bars are not suitable for measuring angles above 45°.
20. Why are holes provided in a sine bar?
21. What are the precautions to be taken while building slip gauges?
22. Why should the minimum number of slip gauges be used for making the required dimension?
23. Explain the constructional features of a bevel protractor, with a neat sketch. What are the various types?
24. How does a universal bevel protractor differ from a bevel protractor?
25. What is the need for acute angle attachment in bevel protractor?
26. Briefly explain the various elements of an optical bevel protractor, with a neat sketch.
27. Explain the uses of bevel protractor with neat sketches.
28. Describe the method of checking the angle of a taper plug gauge using rollers, micrometer and slip gauges.
29. How is the taper angle of a bore and the dovetail angles of a specimen checked using standard balls, gauge blocks and Vernier scale?
30. What is a clinometer? Explain the working principle of the clinometer. Where is it used?

4

Form Measurement

4.1 INTRODUCTION

Form measurement is typically the measurement of contours. It is also defined as the measurement of geometrical tolerances.

The following are a few important terms used in form measurement:

Screw thread: The helical ridge produced by forming a continuous helical groove of uniform section on the external or internal surface of a cylinder or cone. A screw thread formed on a cylinder is known as a straight or parallel screw thread, while one formed on a cone or the frustum of a cone is known as a tapered screw thread.

Form of a thread: The shape of the contour of one complete thread as seen in the axial section.

Crest of a thread: The prominent part of a thread.

Root of a thread (Trough): The bottom of the groove between the flanks of the thread.

Flanks of a thread: The straight edges that connect the crest of a thread with the root.

Angle of a thread: The angle between the flanks of a thread measured in an axial plane.

Pitch: The distance measured parallel to the axis of the thread, between corresponding points on adjacent thread forms in the same axial plane and on the same side of the axis.

Lead: The axial distance moved by the threaded part, when it is given one complete revolution about its axis with respect to a fixed mating thread.

Helix angle: The angle made by the helix of a thread at the pitch line with the axis.

Addendum: For an external thread (Fig. 4.1), the radial distance between the major and pitch cylinders. For an internal thread, the radial distance between the minor and pitch cylinders.

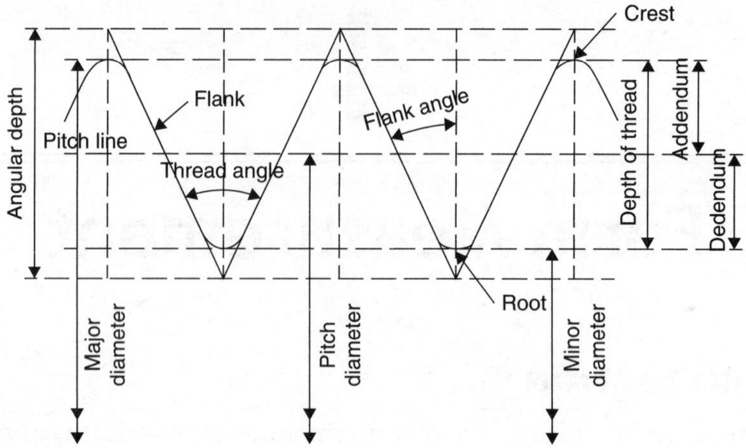

Figure 4.1 Screw thread terms relating to external thread

Dedendum: For an external thread, the radial distance between the pitch and minor cylinders. For an internal thread, the radial distance between the pitch and major cylinders.

Major (outside/crest/full) diameter: The diameter of the major cylinder that is an imaginary cylinder, co-axial with the screw, just touching the crest of an external thread or the root of an internal thread.

Minor (root/cone) diameter: The diameter of the minor cylinder that is an imaginary cylinder, co-axial with the screw, just touching the root of an external thread or the crest of an internal thread.

Effective (Pitch) diameter: The diameter of the pitch cylinder that is an imaginary cylinder which is co-axial with the axis of the screw, intersecting the flank of the threads in such a way as to make the width of the threads and the width of the spaces between the threads equal.

Errors can occur in any one of the following elements of a thread:

- Major diameter
- Minor diameter
- Effective diameter
- Pitch
- Flank angles
- Profile at crest and trough

Pitch errors: The errors in the effective diameter measured parallel to the axis of the screw thread. Threads are usually generated by a point-cutting tool. The ratio of the linear velocity of the tool and the angular velocity of the work must be correct, and this ratio must be constant, otherwise pitch error will occur.

4.2 CLASSIFICATION OF PITCH ERRORS

The following are the various types of pitch errors generally found during measurement of a thread:

- **Progressive error**: If the pitch of a thread is uniform but is longer or shorter than its nominal value, the error is called progressive. The causes of progressive pitch error are:
  - Incorrect ratio of the linear velocity of the tool and the angular velocity of the work
  - Pitch error in the lead screw of the lathe or other machines used for thread cutting
  - Change in length due to hardening
- **Periodic error**: In this case, successive portions of the thread length are either shorter or longer than the mean value. This error repeats itself at equal intervals along the thread. The causes of this error are:
  - When the tool velocity is not uniform
  - The lead screw of the machine lacks squareness, causing the lead screw to move forward or backward once in each revolution
  - Errors in the teeth of the gears of machines used for thread cutting
- **Drunken error**: It is similar to periodic error but repeated once per turn of a thread in a drunken thread. In such a thread, the pitch measured parallel to the thread axis will always be correct; however, this error arises because the thread is not cut to the true helix. Threads with drunken error are called drunken threads.
- **Erratic error**: Errors that vary in an irregular manner along the length of the thread. The possible causes of these errors are:
  - Fault in the machine
  - Disturbances in the machining set–up
  - Irregular cutting action resulting from non-uniformity in the material of the screw

4.3 MEASUREMENT OF THREAD PARAMETERS

A bench micrometer, otherwise called a floating carriage micrometer, designed by the National Physical Laboratory (NPL), is generally used to determine various types of thread parameters, namely, the major diameter, the minor diameter and the effective diameter of a given thread.

The deficiencies inherent in a normal hand micrometer are eliminated in the bench micrometer. The following deficiencies are generally found in normal hand micrometers:

- Variations in measuring pressure
- Pitch errors in the micrometer thread

Though the procedures to determine the various diameters of the thread are similar, the use of prisms and wires determines the parameter to be ascertained.

4.3.1 Measurement of the Major Diameter of an External Thread

The major diameter of an external thread is measured using a bench micrometer, designed by NPL to eliminate the deficiencies inherent in a normal hand micrometer.

The fixed anvil is replaced by a fiducial indicator so that all measurements are made at the same pressure, and this indicator has a positional adjustment which not only increases the range of the instrument but also makes it virtually impossible for a direct reading. Instead, it must be used as a comparator and set to a standard. The setting standard is a calibrated setting cylinder. The set-up is shown in Fig. 4.2.

The procedure consists of simply noting the reading obtained on the setting cylinder, and that obtained on the thread. The difference between these two

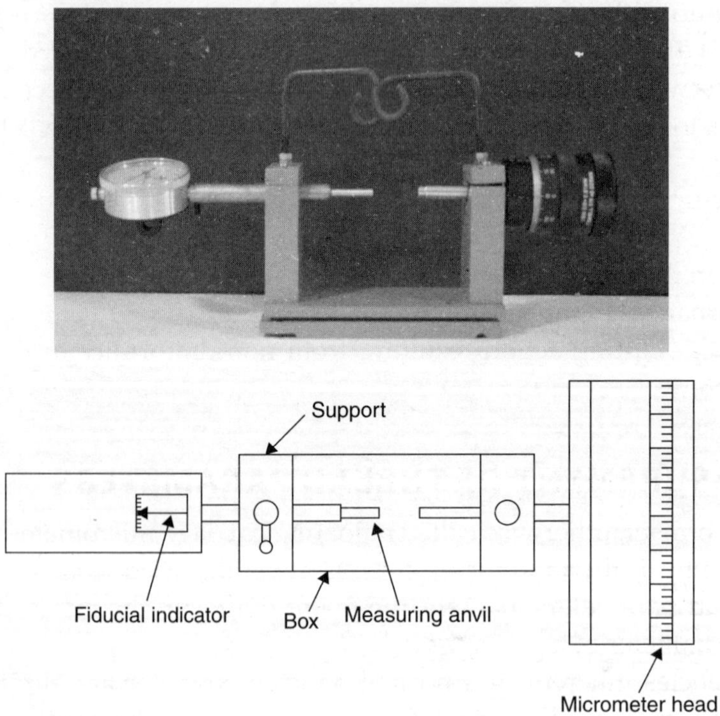

Figure 4.2 Bench micrometer

readings is the size difference between the cylinder and the thread. If this difference is added to the diameter of the setting cylinder, the result is the major diameter of the thread.

Major diameter = $D_c + (R_t - R_c)$, where D_c is the calibrated diameter of the setting cylinder; R_c is the micrometer reading on the setting cylinder and R_t is the micrometer reading on the thread.

4.3.2 Measurement of the Minor Diameter of an External Thread

It is measured by a comparative process, similar to that used in the measurement of the major diameter of an external thread. However, hardened and ground steel prisms are used to probe the root of the thread, as shown in Fig. 4.3.

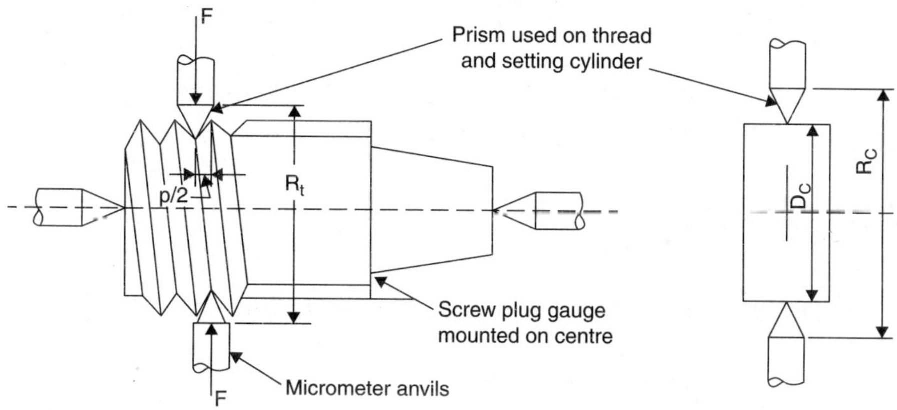

Figure 4.3 Use of prisms to measure the minor diameter

It can be seen that due to the thread helix, a couple of ½ *Fp N-m* is produced, where *F* is the measuring pressure and *p* is the thread pitch. This couple would tend to rotate the thread through an angle depending on the pitch, and an erroneous reading would result. The measurement is, therefore, carried out on a floating carriage diameter measuring machine, in which the thread is mounted between centres and a type of bench micrometer is constrained to move at right angles to the axis of the centres by a V ball slide. The readings are taken on the setting cylinder with the prisms in position, and on the thread. If the prisms are considered as extensions of the micrometer anvils, as shown in Fig. 4.3, their size is unimportant and the minor diameter = $D_c + (R_t - R_c)$, where D_c is the calibrated diameter of the setting cylinder; R_c is the micrometer reading on the setting cylinder and R_t is the micrometer reading on the thread.

4.3.3 Measurement of the Effective Diameter of an External Thread (Two-wire Method)

The prisms used for measuring the minor diameter are replaced by steel wires or cylinders whose size is chosen so that they pitch approximately at the effective diameter of the thread. Such cylinders are known as 'best wires' and enable nominal values of pitch and flank angle to be used in subsequent calculations. The best wire is shown in Fig. 4.4.

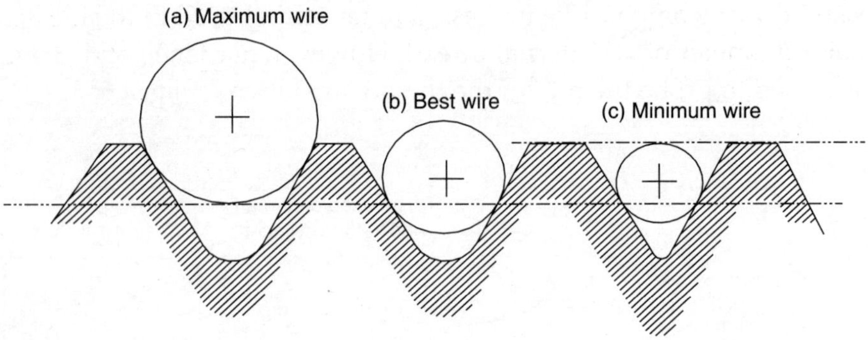

Figure 4.4 Best wire

The formula to find out the 'best' wire sign is given as follows:

$$\text{Best wire diameter} = \frac{P}{2\cos\dfrac{x}{2}}$$

where P is the pitch of the thread and x is the thread angle.

As with the minor diameter, a reading is taken with the cylinders, over a setting cylinder. A reading is then taken with the cylinders engaged in the thread. The difference in the readings is the difference between the diameter of the setting cylinder and the dimension T under the wires when engaged with the thread. The geometry of this arrangement is shown in Fig. 4.5 and is used to derive the expression required to calculate the simple effective diameter.

In Fig. 4.5, θ = Half of the form angle = 30° for a metric thread

= 27.5° for a Whitworth thread

$$E_d = T + 2x$$

In Δ ABC, $\tan\theta = \dfrac{BC}{AB}$; $AB = \dfrac{BC}{\tan\theta}$

∴ $AB = BC \cot\theta$

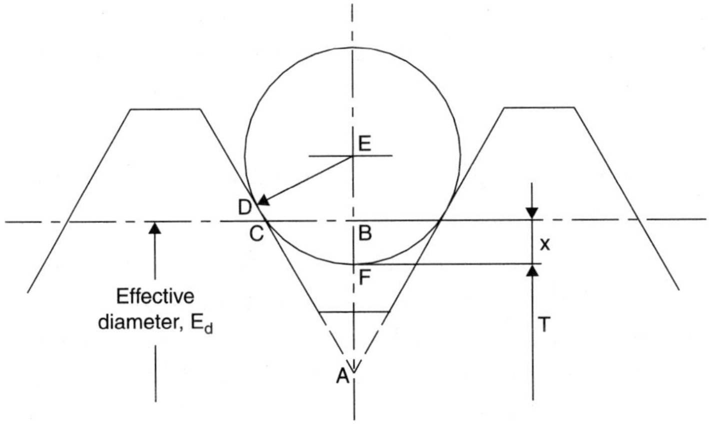

Figure 4.5 Calculation of simple effective diameter

By the definition of the simple effective diameter, BC = ¼ pitch = $p/4$

$$\therefore \quad AB = \frac{p}{4}\cot\theta \qquad (4.1)$$

In Δ ADE, $\sin\theta = \frac{DE}{AE}$; $AE = \frac{DE}{\sin\theta}$

$$\therefore \quad AE = DE\ \text{cosec}\,\theta$$

However, DE = ½ diameter of measuring cylinders = $d/2$

$$\therefore \quad AE = \frac{d}{2}\text{cosec}\,\theta$$

Now, $x = AB - AF$

And, $AF = AE - EF$

$$= \frac{d}{2}\text{cosec}\,\theta - \frac{d}{2} \quad (\text{since } EF = \frac{d}{2}) = \frac{d}{2}(\text{cosec}\,\theta - 1) \qquad (4.2)$$

Subtracting Eq. (4.2) from Eq. (4.1), we get

$$x = \frac{p}{4}\cot\theta - \frac{d}{2}(\text{cosec}\,\theta - 1)$$

However, $E_d = T + 2x = T + 2\left\{\frac{p}{4}\cot\theta - \frac{d}{2}(\text{cosec}\,\theta - 1)\right\}$

$$= T + \frac{p}{2}\cot\theta - (\text{cosec}\,\theta - 1)\,d$$

where E_d is the effective diameter of the thread; T is the measured dimension under the cylinders (wires); p is the nominal pitch; d is the cylinder diameter and θ is the nominal flank angle, equal to the semi-angle of the thread.

4.3.4 Measurement of the Effective Diameter of an External Thread (Three-wire Method)

Three best size wires, as described in the two-wire method, are used to engage in the given thread such that one is the opposite to the other two wires engaged in the subsequent threads. A reading is taken with the cylinders engaged in the thread and noted as M. The geometry of this arrangement is shown in Fig. 4.6 and is used to derive the expression required to calculate the simple effective diameter.

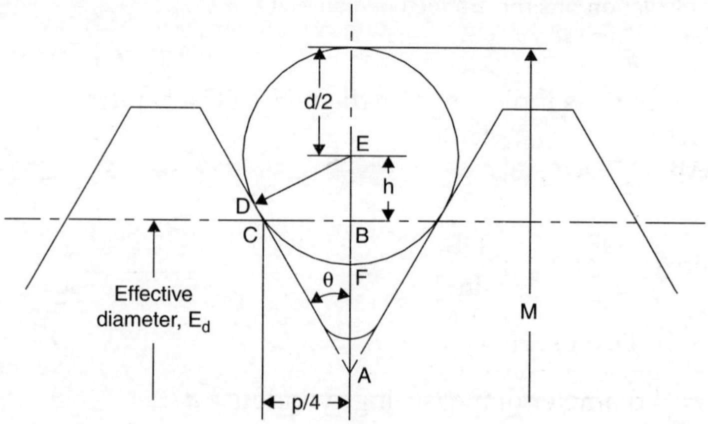

Figure 4.6 Effective diameter using the three-wire method

In Fig. 4.6, θ = Half of the form angle = 30° for a metric thread

= 27.5° for a Whitworth thread

$$M = E_d + 2h + d \tag{4.3}$$

In Δ ABC, $\text{cosec}\,\theta = \dfrac{AE}{DE}$

$$AE = DE\ \text{cosec}\,\theta = \frac{d}{2}\text{cosec}\,\theta$$

Also, $\cot\theta = \dfrac{AB}{BC}$

$\therefore$ $$AB = BC\cot\theta = \frac{p}{4}\cot\theta$$

Hence, $h = \text{AE} - \text{AB}$

$$= \frac{d}{2}\,\text{cosec}\theta - \frac{p}{4}\cot\theta \tag{4.4}$$

Substituting Eq. (4.3) in Eq. (4.4), we get

$$M = E_d + 2h + d = E_d + 2\left[\frac{d}{2}\,\text{cosec}\theta - \frac{p}{4}\cot\theta\right] + d$$

$$= E_d + \left[d\,\text{cosec}\theta - \frac{p}{2}\cot\theta\right] + d = E_d + d\,[1 + \text{cosec}\theta] - \frac{p}{2}\cot\theta$$

Hence, $E_d = M - d\,[1 + \text{cosec}\theta] + \frac{p}{2}\cot\theta$

4.3.5 Measurement of the Diameter of an Internal Thread

As measurement of the elements of an internal thread is more cumbersome, an indirect approach of making a cast of the thread is followed. Once a good cast is available, various elements can be measured for external threads.

The cast may be made using Plaster of Paris or sulphur. The part whose internal thread is to be measured is first cleaned and brushed with fine oil. The part is then mounted between two wooden blocks whose upper surface lies about halfway up the ring. The cast material is then poured to a depth of less than the radius of the part to permit easy removal of the cast without screwing it out. After the plaster is set, it should be taken out without rotating, by pulling the middle portion of the cast. The measurements are then made on this external thread (cast) whose major diameter is equal to the minor diameter of the internal thread, and whose minor diameter is equal to the major diameter of the internal thread of the given part.

4.3.6 Measurement of the Pitch of a Thread

The pitch of an external thread can be measured by using a screw pitch or profile gauge. Such a gauge consists of a series of thread forms with varying pitch. The one that coincides perfectly with the thread under test conditions gives the pitch. The accuracy of measurement depends on the method of sighting used to judge the perfectness.

4.3.7 Measurement of Thread Parameters using Toolmaker's Microscope

This instrument uses an optical head. As shown in Fig. 4.7, it consists of a heavy, hollow cast-iron base that accommodates an illuminating unit underneath. The work table is mounted on the base of the instrument on cross-slides and is equipped with micrometer screws to move it in two mutually perpendicular directions in the horizontal plane. The optical head is mounted on a vertical column and can be moved up and down and focused over the work. A clamping screw is used to fix the optical head in any desired position.

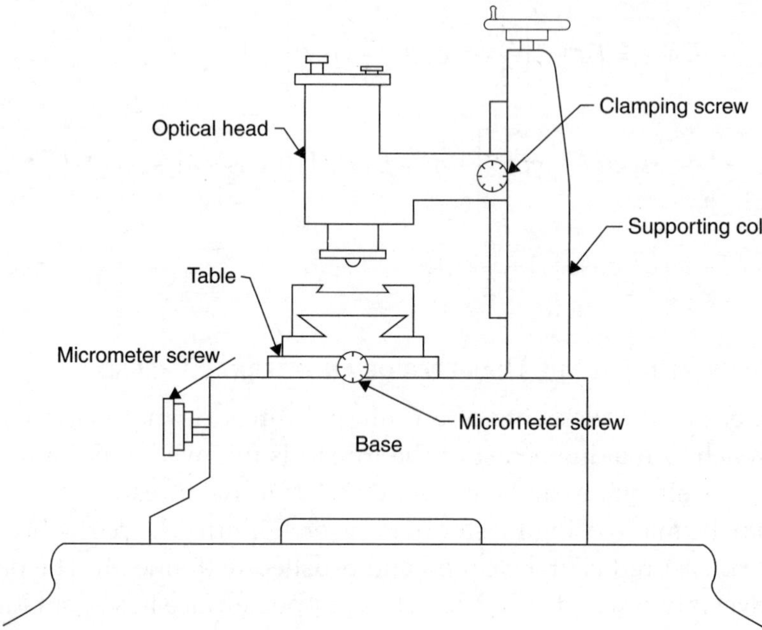

Figure 4.7 Toolmaker's microscope

The top of the table can be swivelled about a vertical axis to align the measurement direction with the direction of longitudinal or transverse displacements of the table. The angle of table rotation is determined using a Vernier scale with a division of 3′ of arc. The screw thread whose dimensions are to be measured is mounted on a glass plate placed on the work table. A light source provides a horizontal beam of light which is reflected from a mirror 90° upwards towards the work table. The illuminating device comprises a lamp, a condenser, a light filter and a diaphragm that can be adjusted by a ring. The beam of light passes through the transparent glass plate on which the work-pieces (screw threads) to be checked are placed. A shadow image of the contour of the screw thread passes through the objective of the optical head and is projected by a system of three prisms to a ground-glass screen. Observations are made through the eyepiece of the optical head.

The protractor goniometric head has a round body with an eyepiece mounted at its centre. A reading protractor microscope is installed from the side of the head with a mirror to illuminate the scale on the underside. A degree scale and a graticule microscope produce an inverted image of the degree-scale marks on the plane of the goniometric graticule, with a minute scale marked. The minute scale and the degree-scale image superimposed on the former are viewed through the eyepiece.

Measurements are made by means of cross lines engraved on the ground-glass screen. The screen can be rotated through 360° and the angle of rotation can be read through an auxiliary eyepiece provided on the optical head. Different types of graduated and engraved screens and corresponding eyepieces are used for measuring different elements.

Linear measurement: To find out the length, width, diameter, etc., one end of the screw thread is made to coincide with the cross line in the microscope and the reading of the micrometer is taken. The table is then moved until the other end of the screw thread coincides with the cross line on the screen, and the final reading is taken. The difference between these two readings gives the desired linear measurement.

Measurement of pitch: The contour is set so that some point on the contour coincides with the cross line on the screen. The reading on the micrometer is noted. The table is then moved using the micrometer screw until a corresponding point on the contour of the next thread coincides with the cross line. The reading of the micrometer is again noted and the difference in the two readings gives the pitch of the screw thread.

Measurement of thread angle: The screen is rotated until a line on the screen coincides with one flank of the thread profile. The angle of screen rotation is noted and the screen is further rotated till the same line coincides with the other flank of the thread. The difference in the two readings gives the angle of the thread.

4.4 MEASUREMENT OF GEAR

A gear is defined as a toothed wheel. The following are the various types of gears:

- **Spur gear:** A cylindrical gear whose tooth traces are straight lines (Fig. 4.8).
- **Helical gear:** A cylindrical gear whose tooth traces are straight helices.
- **Spiral gear:** A gear whose tooth traces are curved lines.
- **Bevel gear:** A gear whose tooth traces are straight line generators of a cone.
- **Worm gear pair:** The worm and mating worm wheels have non-parallel and non-intersecting axes.

The most commonly used forms of gear teeth are the involute and the cycloidal.

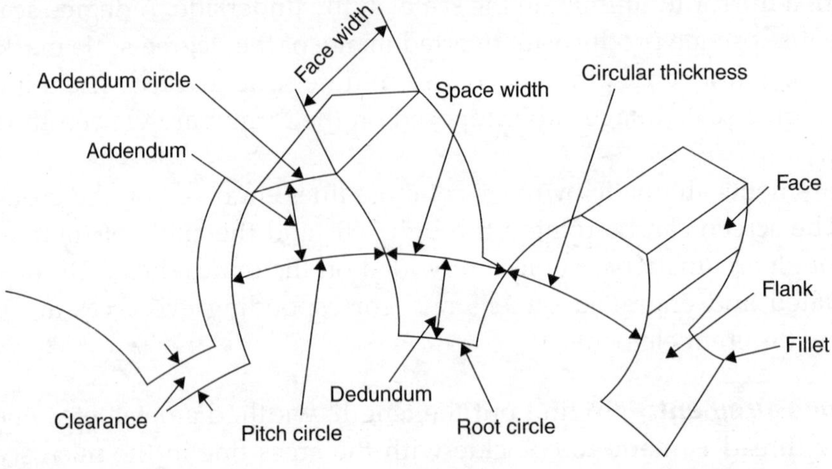

Figure 4.8 Spur gear terminology

Terminology used for gear teeth

Base circle: The circle from which an involute form is generated.

Pitch circle: An imaginary circle useful in calculations. It may be noted that an infinite number of pitch circles can be chosen, each associated with its own pressure angle.

Pitch circle diameter: The diameter of a circle which, by pure rolling action, would produce the same motion as the toothed gear wheel.

Line of action: The common tangent to the base circles of meshing gears at the point of contact.

Pressure angle (Ψ): The angle between the line of action and the common tangent to the pitch circles.

Diametral pitch (P): The number of teeth per inch of pitch circle diameter.

Module: The length of the pitch circle diameter per tooth. It is the reciprocal of diametral pitch.

Circular pitch (CP): The arc distance measured around the pitch circle from the flank of one tooth to a similar flank in the next tooth.

Base pitch (P_b): The arc distance measured around the base circle from the origin of the involute on one tooth to the origin of a similar involute on the next tooth.

Addendum: The radial distance from the pitch circle to the tip of the tooth. The nominal value is equal to the module.

Clearance: The radial distance from the tip of a tooth to the bottom of a mating tooth space when the teeth are symmetrically engaged.

Dedendum: The radial distance from the pitch circle to the bottom of the tooth space.

Tooth thickness: The arc distance measured along the pitch circle, from its intercept with one flank to its intercept with the other flank of the same tooth.

Backlash: The circumferential movement of one gear of a mating pair, the other gear being fixed, measured at the pitch circle, bearing clearances being eliminated.

4.5 PROFILE PROJECTOR

It is an optical instrument that can be used for measuring forms and contour images. It is also known as an optical comparator or shadowgraph. A typical profile projector is shown in Fig. 4.9.

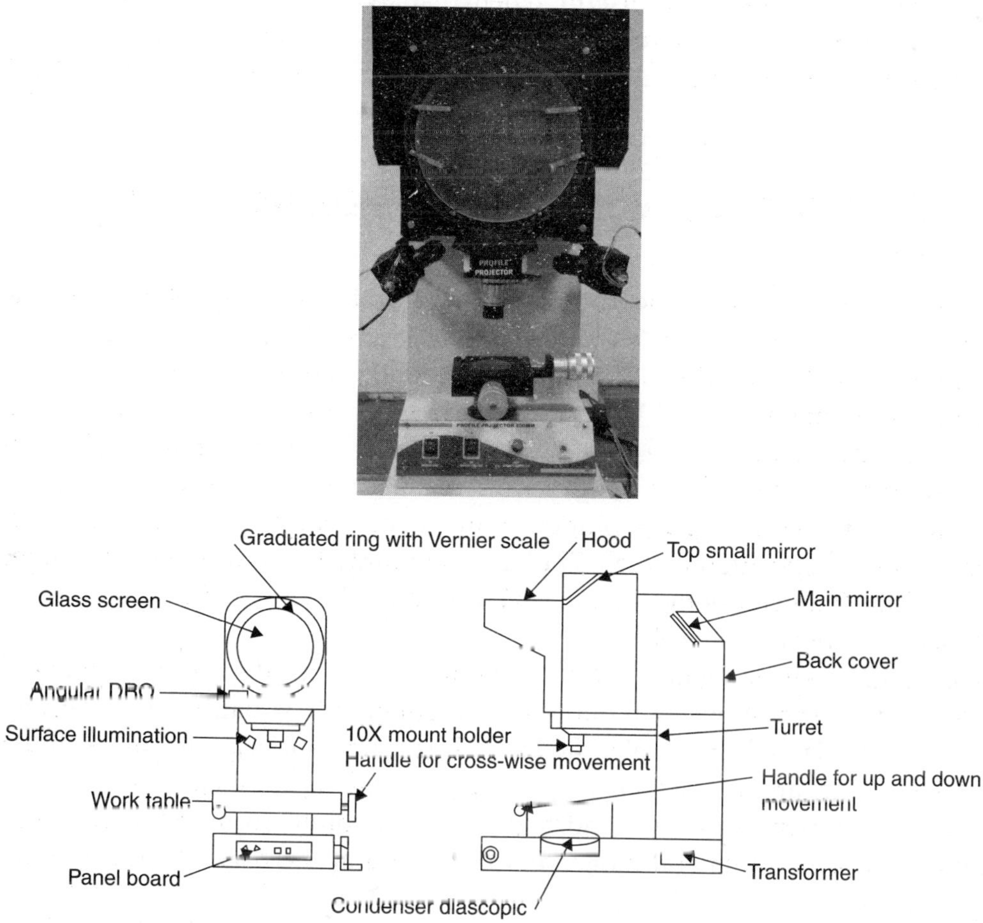

Figure 4.9 Profile projector

The profile projector magnifies the profile of the specimen kept on the work table of the projector. The magnified image is displayed on the built-in projection screen. On this screen, there is a grid that can be rotated 360° so that the X–Y axis of the screen can be aligned with the straight edge of the machined part to measure the required parameter. An edge of the specimen to be measured may be lined up with the grid on the projection screen. From there, simple measurements may be taken for distances to other points. This is generally done on a magnified profile of the specimen.

The typical method of lighting is by diascopic illumination, which is lighting from behind the screen. When the specimen is translucent, light from behind the screen can pass through the specimen, and hence, the light can be transmitted. If the specimen is opaque, the light will not go through the specimen, but will form a profile of it. A profile projector may also have episcopic illumination, which is light shining from above the specimen. This type of illumination is useful when displaying bores and internal areas that may need to be measured. Manual or digital micrometers attached to the specimen table are used for measuring the linear movement of the table in the required direction.

4.6 TESTING THE INVOLUTE FORM – DAVID BROWN TESTER

If a measuring instrument incorporating a dial indicator gauge is made to trace the path of an involute and its stylus is set against the involute, variations in the involute form would show up as variations in the dial gauge reading. The instrument outlined in Fig. 4.10 suggests itself as a possible design. If the straight edge is rolled around the base circle without slipping, the plunger of the dial gauge traverses the involute curve generated from that base circle. If the tooth is of correct involute form, the dial gauge plunger will traverse the tooth and be displaced from its original position of zero. The gear is mounted on a disc whose diameter is exactly the diameter of the theoretical base circle, D_b. The David Brown involute form tester is shown in Fig. 4.10.

The stylus, bearing against the involute form, is accurately located above the straight edge. The stylus is mounted on a small V/flat ball slide constrained to move parallel to the straight edge. Thus, slide movements measured by the dial gauge are in fact errors in the involute form measured along the line of the straight edge, i.e., normal to the involute form. Motion without slipping is caused by lateral movement of the rod at the back of the instrument (Fig. 4.11).

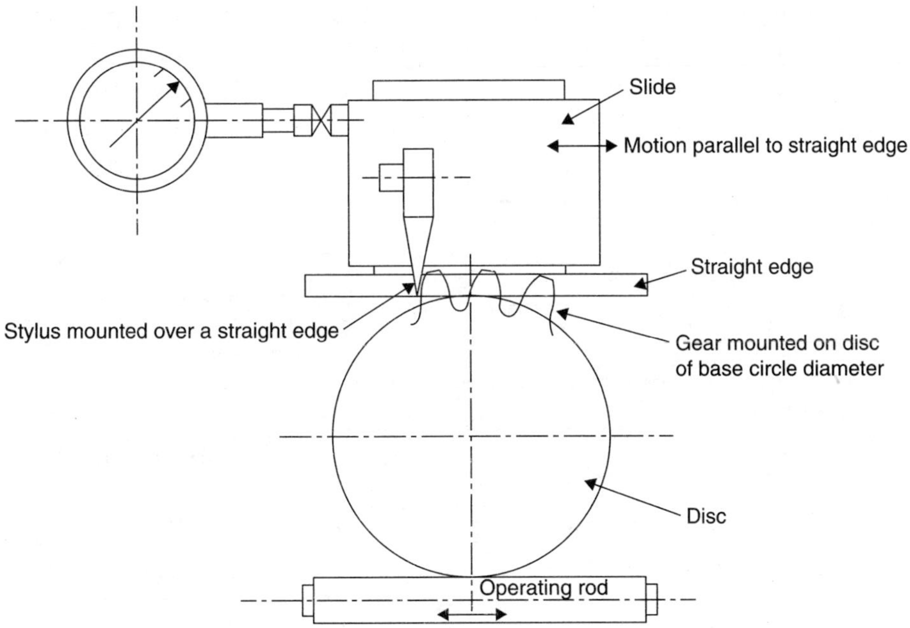

Figure 4.10 David Brown involute form tester

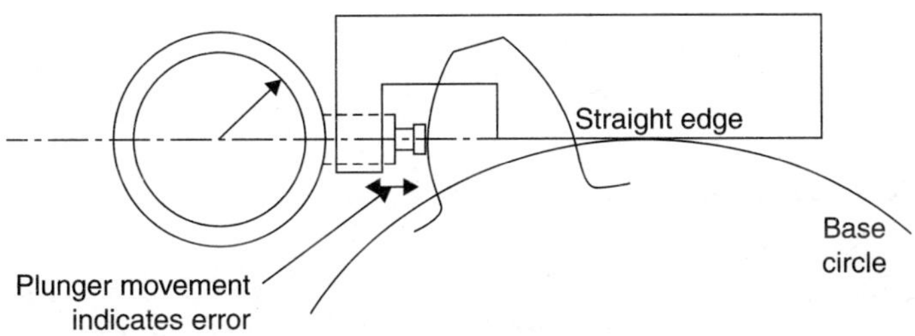

Figure 4.11 Principle of testing the involute form

4.7 PARKINSON GEAR TESTER

The principle behind this instrument is to mount a standard gear on a fixed vertical spindle and the gear to be tested on another similar spindle mounted on a sliding carriage, maintaining the gears in mesh using spring pressure. Movements of the sliding carriage as the gears are rotated are indicated by a dial indicator and these variations are a measure of any irregularities in the gear under test; alternatively, a recorder can be fitted in the form of a waxed circular chart and records made of gear variation in accuracy of the mesh. This set-up is shown in Fig. 4.12. The gears are mounted on the two mandrels so that they are free to rotate without measurable clearance. The left spindle can be moved along the table and clamped in any desired position. The right mandrel slide is free to move, running on steel balls against spring pressure and it has limited movement. The two mandrels can be adjusted so that their axial distance is equal to the designed gear centre distance.

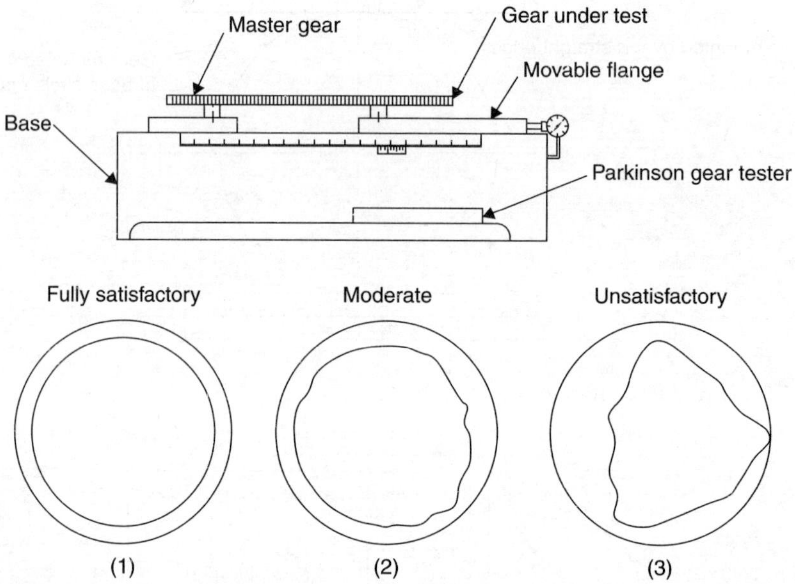

Figure 4.12 Parkinson gear tester

The spring pressure can be regulated. There are also screws for limiting the movement of the sliding carriage. A scale is attached to one carriage and a Vernier scale to the other; this enables centre distances to be measured to within 0.025 mm. The dial indicator on the right is in contact with the right end of the sliding carriage and, therefore, indicates any radial variations of the gear under test as the gears are rotated.

When the waxed paper recorder is fitted, the chart makes a revolution for each of the gears mounted on the sliding carriage. As the chart moves or rotates, the line traced records the movements of the floating carriage; a circle is drawn at the same time as the record.

The method discussed thus far is the dual-flank method, i.e., both tooth flanks come in contact with each other. However, this is seldom the case in actual practice. This method is suitable for checking whether the gear is to be accepted or rejected, and not for finding out detailed causes for rejection.

The correct method of mesh testing is the single-flank method, in which angular variation and not centre distance variation is measured. This instrument is complex and more expensive. The simplest machine of this type consists of two shafts, each carrying a gear and a plain disc having a diameter equal to the nominal pitch diameter of the gear. One shaft has a rotary joint between the gear and its associated pitch disc. An indicator is used to measure the angular variation between the gear and the disc on this shaft. In use, two discs are brought into frictional contact so that one can drive the other without slip.

This single-flank method is not popular because it requires the manufacture of two very accurate pitch discs for every gear pair of different sizes. Today's single-flank mesh testers do not require different pitch discs. The two shafts carrying the gears are fitted with radial gratings that have an angular band of accurately spaced clear radial lines. When two such gratings rotate in close proximity, interference bands known as moire fringes are formed, moving in the radial direction that generates electric pulses. These pulse trains are continuously phase-compared to provide a detailed chart record of gear transmission errors.

4.8 GEAR TOOTH VERNIER CALIPERS

This instrument is used to measure the thickness of the gear tooth at the pitch line. A typical gear tooth Vernier calipers is shown in Fig. 4.13. It consists of two perpendicular Vernier arms with a Vernier scale on each arm. One of the arms is used to measure the thickness of the gear tooth while the other measures the depth. The calipers is set so that it slides on top of the gear tooth being tested, while the lower ends of the calipers jaws touch the sides of the tooth at the pitch line. The reading on the horizontal Vernier scale gives the value of chordal thickness (width) and the reading on the vertical Vernier scale gives the value of chordal addendum. These measured values are then compared with the calculated values.

This method is not very satisfactory for the following reasons:

- Vernier itself is not reliable for less than 0.05 mm or perhaps 0.025 mm with practice.
- Measurements depend on two Vernier readings, each of which is a function of the other.
- Measurements are made with an edge of the measuring jaw, not its face, which again does not lend itself to accurate measurement.

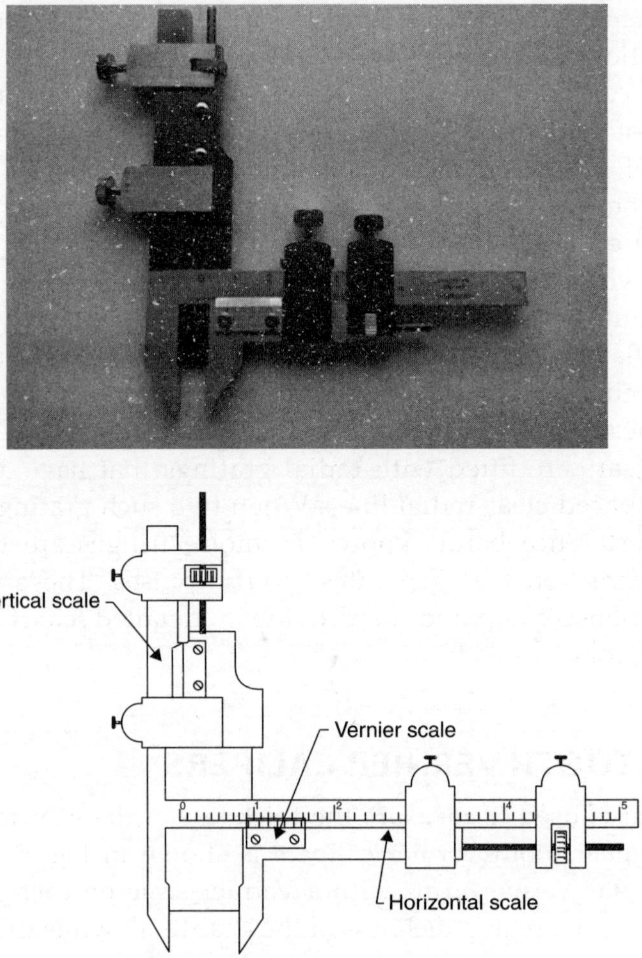

Figure 4.13 Gear tooth Vernier calipers

The chordal width W and chordal addendum can be determined theoretically using the following formulae and the experimental value can be validated with the theoretical values.

$$\text{Chordal width } W = 2R \sin \frac{360}{4n}$$

$$\text{Chordal addendum } d = \frac{nm}{2}\left[1 + \frac{2}{n} - \cos\left(\frac{90}{n}\right)\right]$$

where R = pitch circle radius; n = number of teeth and m = module = $2R/n$.

4.9 RADIUS MEASUREMENT

The radius of curvature of a blunt surface, concave surface or convex surface can be determined using various methods involving mathematical calculations. A few such methods are described here.

4.9.1 Radius of Curvature of Any Job Having a Portion of a Circle

The set-up requires a surface plate, Vernier calipers, C-clamp and two balls of equal size. The job is first clamped to the surface plate using the C–clamp. It should be clamped in such a way that the central position of the circular part is in contact with the surface plate. The two balls of equal diameter are placed on each side of the job, as shown in Fig. 4.14. Let the reading over the balls taken using the Vernier calipers be l, the diameter of balls be d and the radius of the job be R.

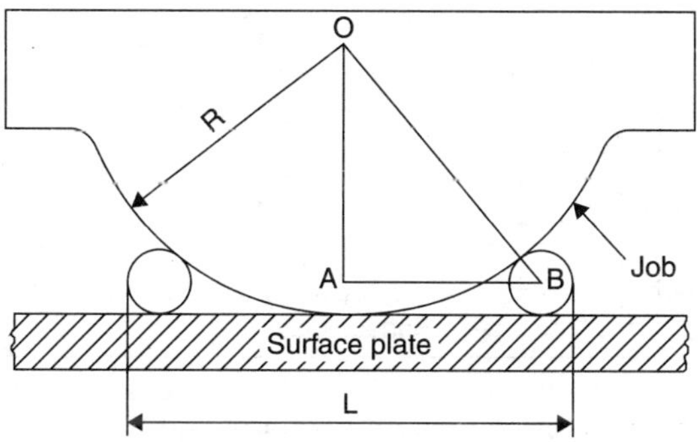

Figure 4.14 Radius of curvature of a job having a portion of a circle

Let O be the assumed centre of the arc of radius of the job. Then, for ΔOAB

$$OB^2 = OA^2 + AB^2$$

$$\left(R+\frac{d}{2}\right)^2=\left(R-\frac{d}{2}\right)^2+\left(\frac{l-d}{2}\right)^2$$

$$R^2+\frac{d^2}{4}+Rd=R^2+\frac{d^2}{4}-Rd+\frac{(l-d)^2}{4}$$

$$2Rd=\frac{(l-d)^2}{4};\ R=\frac{(l-d)^2}{8d}$$

4.9.2 Radius of Curvature of a Blunt Corner of a Job

The template whose blunt corner radius is to be determined is placed on a surface plate with an adjacent surface touching the straight block. The height of the template, H, is found using slip gauges/height gauge. A roller of known diameter D is placed and allowed to rest between the straight block and the surface whose radius (R) is to be determined; h is measured over the roller using the Vernier height gauge/slip gauges. The set-up is shown in Fig. 4.15. Let O be the assumed centre of arc of radius of the job.

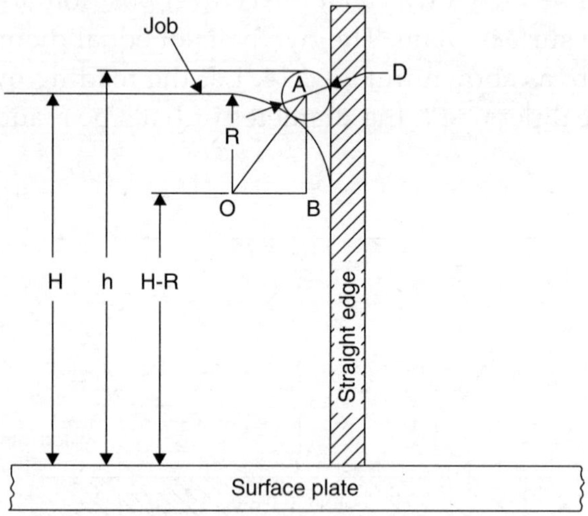

Figure 4.15 Radius of curvature of a blunt corner of a job

For ΔOAB, $OA^2 = OB^2 + AB^2$

$$\left(R+\frac{D}{2}\right)^2=\left(R-\frac{D}{2}\right)^2+\left[\left(h-\frac{D}{2}\right)-(H-R)\right]^2$$

From the above expression, R can be calculated.

4.9.3 Radius of a Concave Surface (Cavity)

When the edges are well defined (sharp corners): The part to be tested is kept on a surface plate and, with the help of a depth micrometer, the maximum depth of the cavity is determined. Let it be h. Next, the part is placed in such a way that the cavity rests against an angle plate and the part is then clamped in this position. The hole is measured from edge to edge with a height gauge having a sharp scribing arm. Let this maximum reading (diameter of the cavity) be d. Let O be the assumed centre of the cavity and R be the radius of curvature (Fig. 4.16).

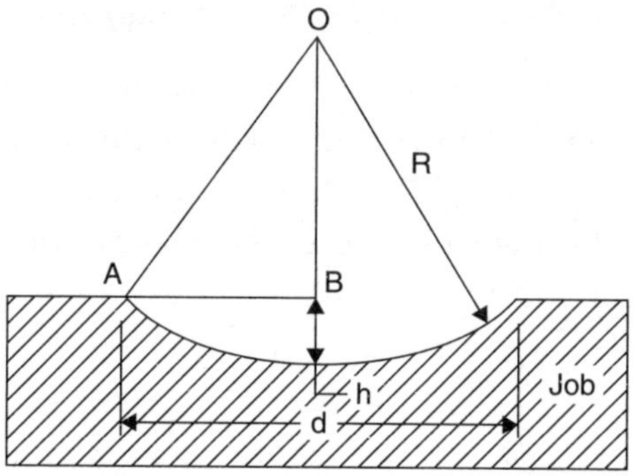

Figure 4.16 Radius of curvature of a concave surface

From Fig. 4.16, for ΔOAB, $OA^2 = AB^2 + OB^2$

$$R^2 = \left(\frac{d}{2}\right)^2 + (R-h)^2 = \frac{d^2}{4} + R^2 + h^2 - 2Rh$$

$$2Rh = \frac{d^2}{4} + h^2; \; R = \frac{d^2}{8h} + \frac{h}{2}$$

When the edges are rounded (blunt corners): A depth micrometer is placed inside the cavity such that it fully rests in the cavity and its frame touches all the sides of the cavity; the depth of the cavity from its base is measured. Let it be h. The width of the depth micrometer base is measured with the help of slip gauges. Let it be d. The set-up is shown in Fig. 4.17.

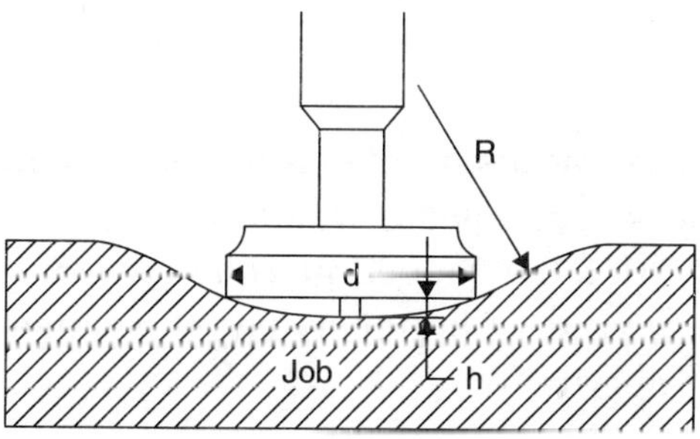

Figure 4.17 Radius of curvature of a concave surface

Now, this case is similar to the previous one, and the radius of curvature R can be found using the same formula, $R = \frac{d^2}{8h} + \frac{h}{2}$.

4.9.4 Measurement of Smaller Radius of Curvature

When two surfaces meet, a radius is generally given to avoid stress concentration. To measure this radius of curvature, radius gauges are used. If the curved surface is external, a male radius (or radius) is used. If the curved surface is internal, a female radius (or fillet) is used. Radius gauges consist of sets of blades, on each of which is stamped the corresponding radius. Figure 4.18 shows a single radius gauge.

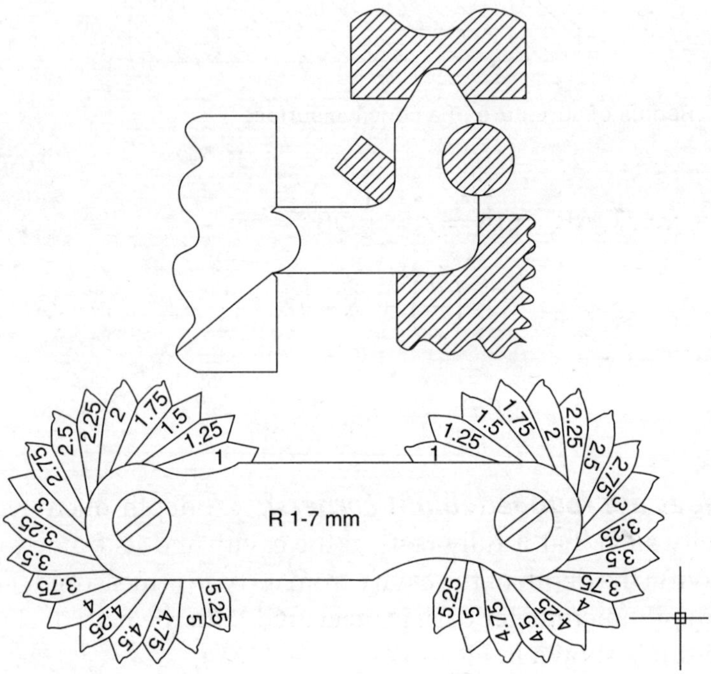

Figure 4.18 Radius gauges

On one side is the external radius, while on the other side is the internal radius. So this is suitable for measuring fillet as well as radius. Radius gauges are used in proper light, with the work to be gauged placed between the light source and the eye. It is only with the passage of light between the gauge and the work that the radius can be gauged properly.

4.10 SURFACE FINISH

The main factors affecting the significance of the control of surface finish are as follows:

- Fatigue life

- Bearing properties
- Wear resistance
- Corrosion resistance
- Lubrication
- Load carrying capacity
- Noise reduction in case of gears
- Appearance
- Ability to hold pressure.

The processes used to achieve good surface finish are given below:

- Turning
- Grinding
- Honing
- Lapping
- Shaping
- Lapping

The factors affecting surface roughness are as follows:

- Vibrations
- Metal cutting conditions, i.e., feed, speed and depth of cut
- Type, form, material and sharpness of cutting speed
- Type of coolant used
- Material of the work-piece
- Type of machining
- Rigidity of the system consisting of the machine tool, a fixture used for clamping the cutting tool and the work-piece

Primary texture (Roughness): This is the surface irregularities of small wavelength, caused by direct action of the cutting elements on the material, i.e., cutting tool shape, tool feed rate, etc.

Secondary texture (Waviness): This is the surface irregularities of considerable wavelength of a periodic character, caused by inaccuracies of slides, wear of guides, misalignment of centres, non-linear feed motion, deformation of work under the action of cutting forces, vibrations of any kind, etc.

4.10.1 Elements of Surface Texture

Profile of the surface: The contour of any section through a surface.

Roughness of the surface: Relatively finely spaced micro geometrical irregularities.

Waviness of the surface: Macro geometrical irregularities. It consists of surface irregularities that are of greater spacing than roughness and it occurs in the form of waves.

Lay: The direction of predominant surface pattern by tool marks or scratches.

Sampling length: The length of the profile necessary for evaluation of irregularities. It is also known as cut-off length.

Surface roughness is expressed as Ra in microns (μm), with minimum and maximum limiting values. The conventional method of expressing the surface finish is shown in Fig. 4.19.

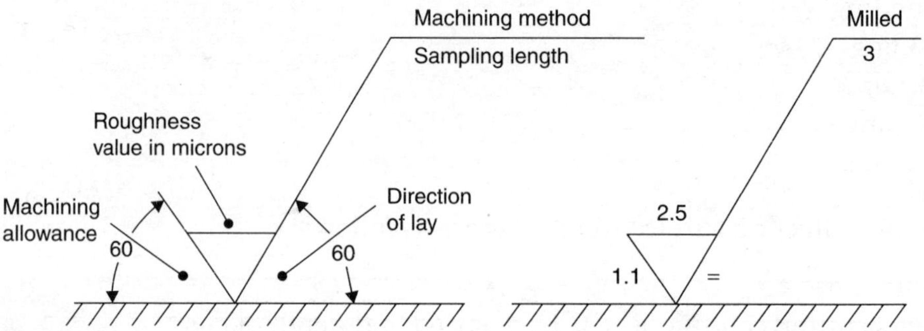

Figure 4.19 Conventional method of measuring surface finish

4.10.2 Measurement of Surface Finish

Surface finish is a measure of the direction of the predominant machining pattern on a part. A lay pattern is a repetitive impression produced on the surface of a part. It is often representative of a specific manufacturing operation. The equipment used to measure the surface finish are discussed in this section.

Tomlinson surface meter: This instrument uses mechanical–optical magnification methods and was designed by Dr. Tomlinson of the National Physical Laboratory (NPL). The set-up is shown in Fig. 4.20.

The shoe is attached to the body of the instrument, its height being adjustable to enable the diamond to be positioned conveniently. The stylus is restrained from all motion except a vertical one by a leaf spring and a coil spring; the tension in the coil spring P causes a similar tension in the leaf spring. These forces hold a cross roller in position between the stylus and a pair of parallel fixed rollers, as shown in the top view. Attached to the cross roller is a light spring steel arm carrying at its tip a diamond which bears against a smoked glass screen.

In operation, the body of the instrument is drawn slowly across the surface by a screw turned at 1 rev/min by a synchronous motor, the glass remaining stationary. Irregularities in the surface cause a vertical movement of the stylus, which causes the cross roller to pivot about point A and thus produce a magnified motion of the diamond on the arm. This motion, coupled with the horizontal movement, produces a trace on the glass magnified in the vertical direction at 100x, there being no horizontal magnification.

The smoked glass is transferred to an optical projector and magnified further at 50x, giving an overall vertical magnification of 5000x and a horizontal magnification of 50x. The trace may be taken off by hand or by photographic methods and analysed.

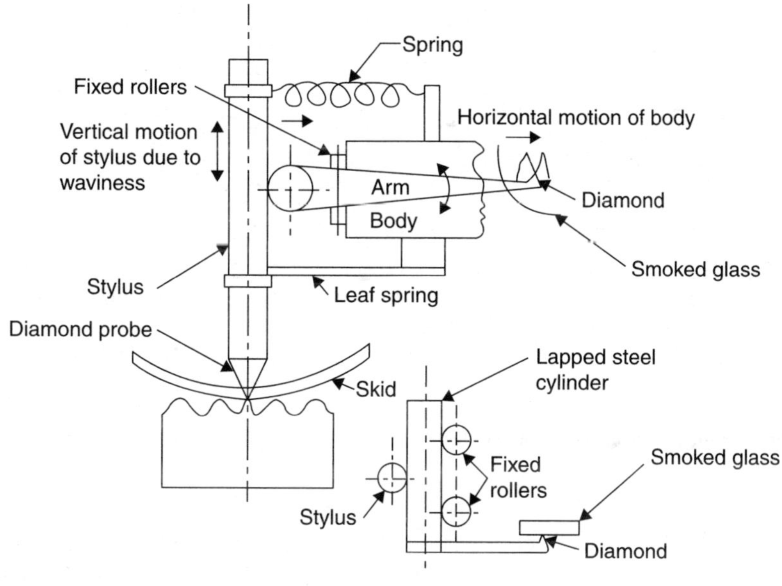

Figure 4.20 Tomlinson surface meter

Taylor–Hobson Talysurf: It is an electronic instrument whose function can be broken down in a similar manner to the Tomlinson surface meter. However, it differs mainly in the method of magnification. It gives the same information much more quickly and probably more accurately. It is used on the factory floor and in laboratories.

The measuring head, as on the Tomlinson surface meter, consists of a stylus and a shoe which are drawn across the surface under test by an electric motor and gear box. In this case, the arm carrying the stylus forms an armature which pivots about the centre piece of a stack of E-shaped stampings around each of the outer pole

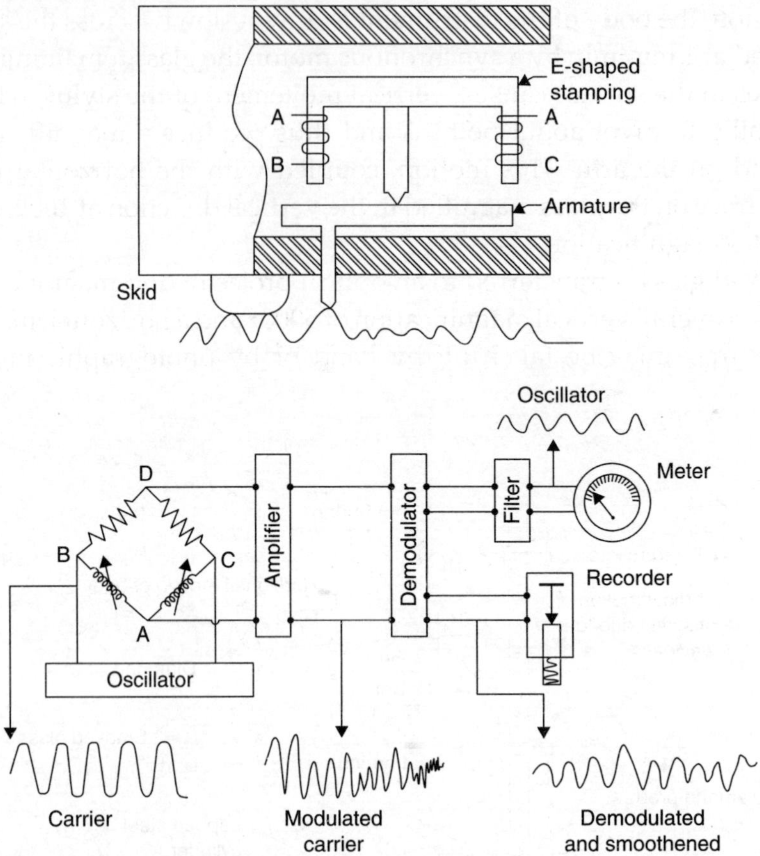

Figure 4.21 Taylor–Hobson Talysurf

pieces, one of which is a coil carrying an AC current, as shown in Fig. 4.21. The Wheatstone bridge circuit is connected to a pen recorder and a deflection meter through an amplifier, a demodulator and a filter.

As the armature pivots about point M, it causes the air gaps to vary and thus modulates the amplitude of the original AC current flowing in the coils. As these form part of a Wheatstone bridge circuit, the output consists of the modulation of AC current only. This modulated AC current is fed into an amplifier and made to operate a pen recorder to produce a permanent record of surface roughness.

The movement of the pen recorder depends upon the strength of amplitude of the AC current flowing through the coils, which in turn depends upon the varying air gap between the armature and the E stampings. Hence, the AC signal varies according to the surface irregularities and causes the pen recorder to move up and down and produce a permanent record of the surface texture on a magnified scale.

The pen recorder is of interest in that the trace is produced on carbon-backed paper by a pointer whose tip arcs electrically across the paper. This produces a much finer line than ink pen recorders, with no distortion due to drag. This AC signal is also fed into a deflection meter to directly obtain a numerical assessment.

Profilometer: This is an electrical instrument. The principle of the instrument is similar to that of the gramophone pick-up. It consists of two principal units: a tracer and an amplifier. The tracer is a finely pointed stylus mounted in the pick-up unit, which consists of an induction coil located in the field of a permanent magnet. The set-up is shown in Fig. 4.22.

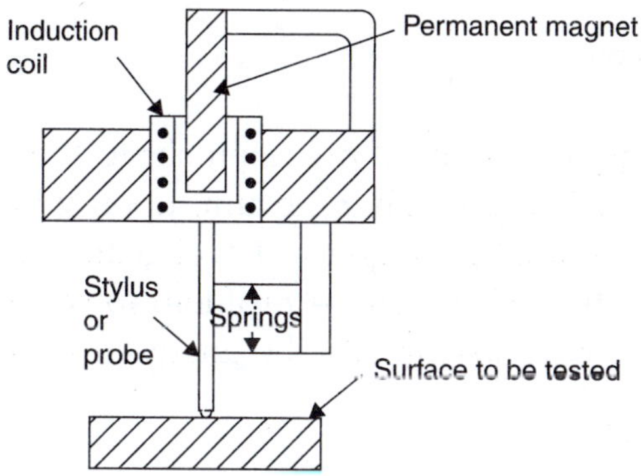

Figure 4.22 Profilometer

When the tracer is moved across the surface to be tested, it is displaced vertically up and down because of surface irregularities. This causes the induction coil to move in the field of the permanent magnet and induce a voltage, the strength of which depends on the air gap between the induction coil and the permanent magnet. The induction coil carrying the stylus moves up and down with the help of a spring attached to the body. The voltage produced is then amplified and recorded. If the potentiometer measuring the voltage is already calibrated, the irregularities can be determined according to the voltage produced during the movement of the stylus over the surface under test for its surface finish.

Profilograph: The principle of a tracer type profilograph based on an optical system is shown in Fig. 4.23.

The work-piece whose surface is to be tested is placed on the table of the instrument. The work-piece and the table are traversed with the help of a lead screw. The stylus, which is pivoted to a mirror, moves over the tested surface. Oscillations of the tracer point are transmitted to the mirror. A light source sends

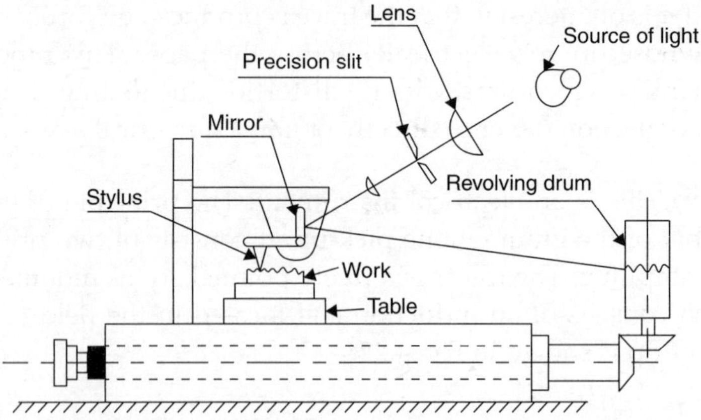

Figure 4.23 Profilograph

a beam of light through a lens and a precision slit to the oscillating mirror. The reflected beam is directed to a revolving drum, upon which a sensitised film is arranged. This drum is rotated through two bevel gears from the same lead screw that moves the table of the instrument. A profilograph will be obtained from the sensitised film, which may subsequently be analysed to determine the value of surface roughness.

4.10.3 Assessment of Surface Roughness as per Indian Standards

Various methods are used to evaluate the surface roughness. Among them, the most commonly used ones are the mean line (M) or centre line average (CLA) method and the envelope (E) method.

CLA method: This method expresses the arithmetical average departure of the actual surface both above and below a mean line, within a specified sampling length, and in a plane substantially normal to the direction of the surface. The CLA value is defined as the average height from a mean line of all the ordinates of the surface regardless of the sign.

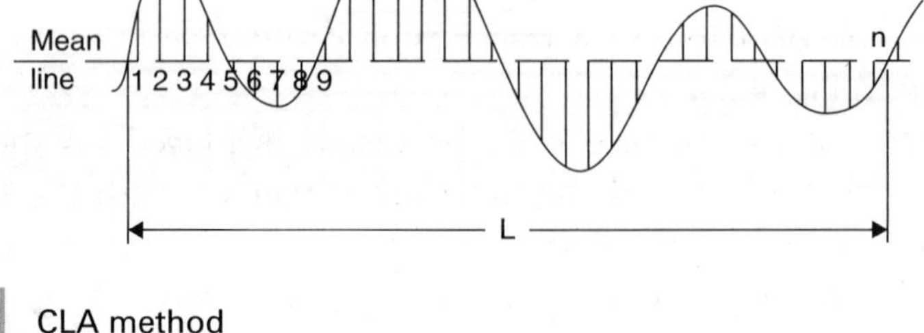

Figure 4.24(a) CLA method

Referring to Fig. 4.24 (a), $\text{CLA} = \frac{h_1 + h_2 + \cdots + h_n}{n}$.

As it is a laborious job to find CLA using the above formula, a simplified formula is given below in which the area of any curve can be found using a planimeter. The revised CLA method is shown in Fig. 4.24(b).

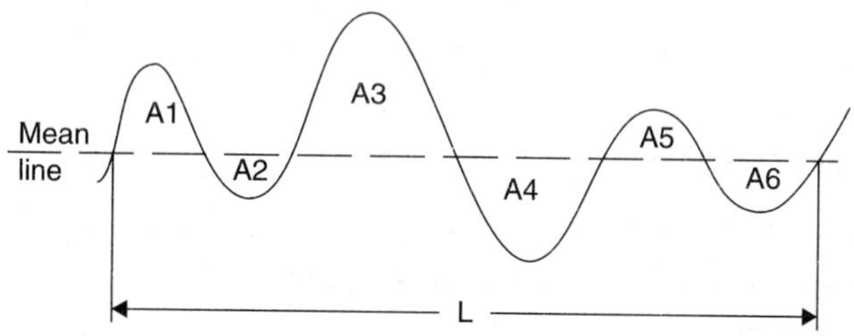

Figure 4.24(b) CLA method (revised)

$$\text{CLA} = \frac{A_1 + A_2 + \ldots +}{L} = \frac{\Sigma A}{L}, \text{ where } L \text{ is the sampling length.}$$

Envelope method: This method expresses the arithmetical departure of a surface whose roughness needs to be measured, both above and below a mean curve. A contacting envelope is first developed by rolling a sphere of small radius – of around 25 mm – across the surface and by shifting the locus of centre of the circle towards the surface by an amount equivalent to the radius of the circle. The mean curve is then developed from the contacting envelope by displacing it to a position where the areas enclosed by the profile above and below the mean curve are equal. The mean curve developed is shown in Fig. 4.25.

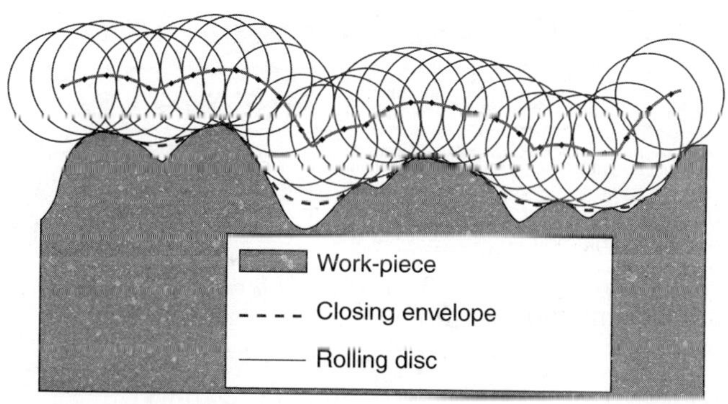

Figure 4.25 Envelope method

4.11 STRAIGHTNESS MEASUREMENT

Straightness is the condition where one line element of an axis or a surface is in a straight line. A line is said to be straight over a given length if the variation in the perpendicular distance of two points on the line from two planes perpendicular to each other and parallel to the line to be tested remains within the specified tolerance limits; the two points being so located on the line that they are closer to the ends of the line to be tested. The following sections discuss various methods used for testing straightness.

4.11.1 Spirit Level Method

The straightness of any surface can be determined by measuring the relative angular positions of a number of adjacent sections of the surface to be tested. So, a straight line is first drawn on the surface whose straightness is to be tested. Then, the line is divided into a number of sections so that the length of each section is equal to the length of the sprit level base.

A sensitive spirit level, fitted with two feet a convenient distance apart, is moved along the line drawn on the surface to be tested in steps equal to the pitch distance between the centre lines of the feet. For each position, the reading is noted. Variations in the bubble position represent angular variations in the surface and these angular variations can be converted into differences in height of the feet above or below the starting point on the line drawn on the surface to be tested. Angular variation can be correlated in terms of the difference in height between two points by knowing the least count of the level and length of the base. The height difference between the two feet can also be obtained by keeping the slip gauges beneath any of the feet such that the bubble comes to the neutral position. The total height of the slip gauges gives the difference in level of the two feet. Thus, the differences in height (the error) with respect to the starting point of the line for various sections of the line are noted and plotted as shown in Fig. 4.26.

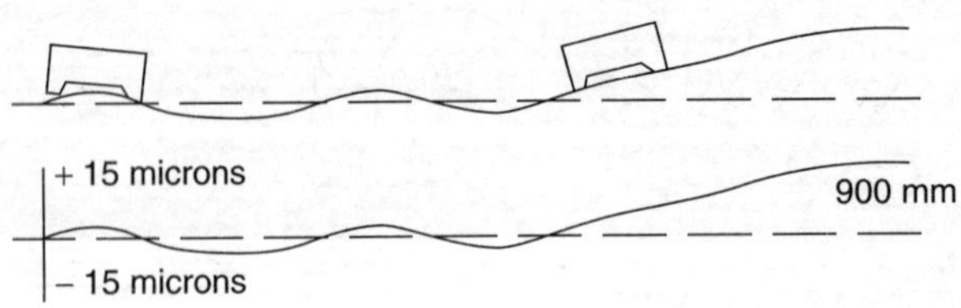

Figure 4.26 Spirit level method

This method is suitable only for horizontal surfaces whereas an autocollimator can be used to check the straightness of both horizontal and vertical surfaces.

4.11.2 Autocollimator Method

The principle of this method is the same as that of the spirit level method. The schematic set-up is shown in Fig. 4.27. The autocollimator is placed at a distance of 0.5–0.75 m from the surface to be tested on any rigid support which is independent of the surface to be tested. A parallel beam from the instrument is projected along the length of the surface to be tested. A block fitted with two feet and with a plane vertical reflector is placed on the surface and the reflector face is shown towards the autocollimator. The reflector and the autocollimator are arranged such that an image of the cross wires of the autocollimator appears closer to the centre of the field. For the complete movement of the reflector along the surface straight line, the image of cross wires will appear in the field of the eyepiece. The reflector is then moved to the other end of the surface in steps equal to the centre distance between the feet and the tilt of the reflector is noted in seconds from the eyepiece. The tilt (angular variation) can be correlated in terms of the difference in height of the feet as

$$1 \text{ min of arc} = \frac{2\pi}{360 \times 60}$$

Then, on a base length l, 1 min of arc $= \dfrac{2\pi}{360 \times 60} \times l$

where, l is the distance between the centres of the feet in mm, which in general is equal to 103.5 mm.

When testing a vertical surface, an optical square may be interposed so that the autocollimator can be used in the horizontal position. This is usually more convenient than arranging the autocollimator on its vertical axis.

4.11.3 Straight Edge Method

Parallel straight edges are made from suitable high-quality non-distorting and non-shrinking steel and are hardened on the working surface (not less than 54

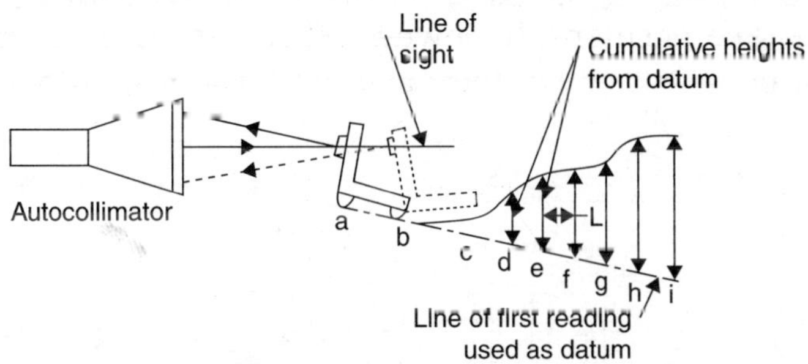

Figure 4.27 Autocollimator checking the straightness of lathe bed guide-ways

HRC or 600 HV). The cross-sectional view of the straight edges (I section and rectangular section) are shown in Fig. 4.28.

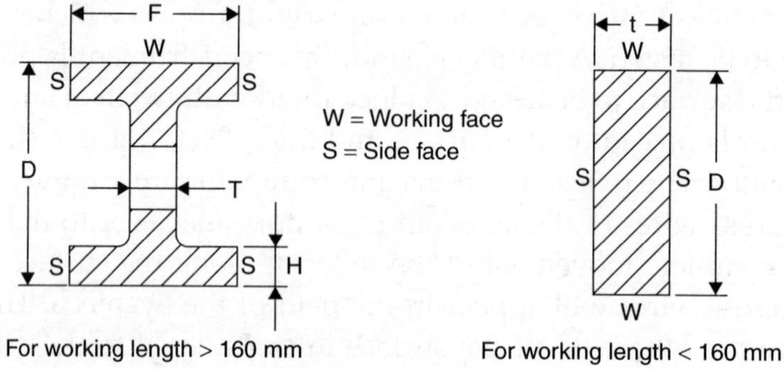

Figure 4.28 Straight edges

During measurement, when the straight edge is supported, the distance between the supports should be 5/9 of the total length of the straight edge for minimum deflection. These points are marked on one or both side faces of the straight edge. A straight edge is supported at these points on two unequal piles of slip gauges so that it is at a slight inclination to the surface to be tested. The distance between the supports is divided into a number of equal parts and marked on the straight edge. If both the straight edge and the surface are perfectly straight, the gap at each point will vary uniformly. Assume that the slip gauges used have values from 20 mm to 20.2 mm, as shown in Fig. 4.29.

Let the distance between the slip gauges be divided into five equal parts. The gap at each point will, therefore, vary by 0.04 mm (i.e., 0.2/5). Accordingly, the value of the pile of the slip gauges required for exact contact at the first position is 20 mm, at the second position is 20.04 mm, and at the third position is 20.08 mm, and so on. Insert slip gauges of appropriate value at each marked position. If there is no error, the slip gauges will make contact with both the surfaces exactly at the marked positions. If, however, there are errors in straightness, the slip gauges will not fit exactly at their marked positions, but will be displaced

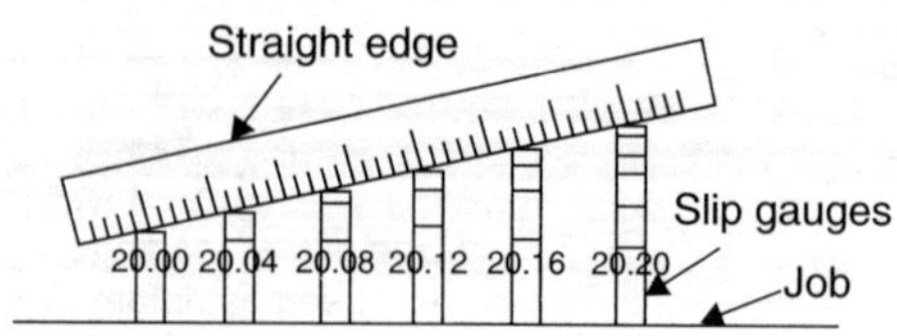

Figure 4.29 Straight edge method

one way or another, along the straight edge by amounts proportional to the errors.

4.11.4 Method of Comparison with the Surface of a Liquid

The surface of a liquid is taken as standard. A frame rests on three feet, one of which is on the axis of the micrometer. It is fitted with a micrometer head. The micrometer spindle end is tapered to a small-diameter spherical tip. The tank contains an electrical conducting liquid and rests on V blocks. Water with a small amount of soda can be used as the electrical conducting liquid. To measure the straightness, the frame is moved on the surface to be tested. The micrometer spindle is moved such that it makes contact with the liquid surface. To indicate this contact, an electric circuit can be used. The micrometer readings are noted at various positions and the errors are calculated. The set-up is shown in Fig. 4.30.

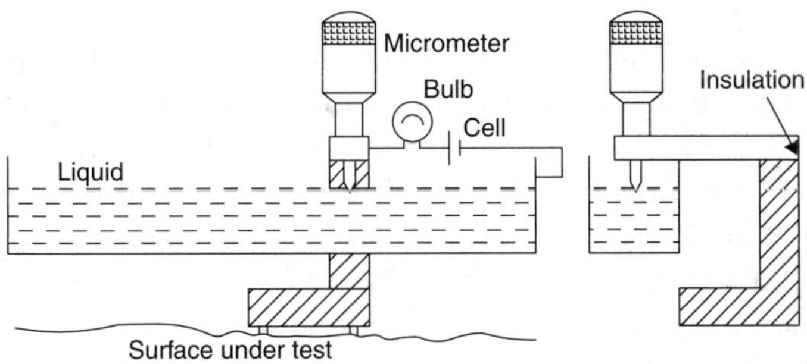

Figure 4.30 Testing straightness by comparison with the liquid surface

4.11.5 Beam Comparator

It consists of a rigid but light beam having three supporting feet with spherical ends made up of hardened steel. One supporting foot is fixed at the centre of the beam but offset from the line joining the other two. The other two supporting feet at the ends may be clamped to the beam in symmetrically disposed positions with respect to the central foot and at a distance apart depending on the size of the surface to be tested. A typical beam comparator is shown in Fig. 4.31.

As there are only three supporting feet, the beam may accidently roll over on the side not supported by the offset foot. To avoid this problem, a fourth foot in the form of a spherically ended spring plunger is provided at a position symmetrically opposite to the fixed offset foot. The fourth foot (plunger) is generally arranged to be reasonably clear of the surface plate and comes into contact with the surface plate only if the beam tends to roll. A very sensitive dial indicator supported in the central web of the beam and the axis of the contact plunger lies in the plane through the axis of the two adjustable feet.

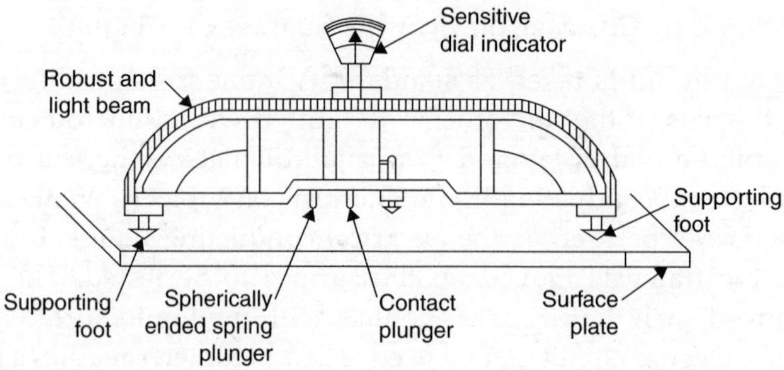

Figure 4.31 Beam comparator (front view)

Figure 4.32 shows the method used when testing a straight edge. The beam comparator is made to rest on the surface of the straight edge under test on two feet, with one foot of the dial gauge in contact with the reference surface whose straightness has been already determined. If the beam comparator is set so that the dial gauge reads zero when its foot is on the straight line through the two fixed feet, the reading *r* gives the height of point C above or below the straight line through AB. The instrument is moved along the surface under test in steps equal to the pitch of the feet, and at each position, the reading of the dial gauge is taken. Each reading is the height of the dial gauge foot above or below the line through the two points on which the fixed feet rest. Hence, the profile of the surface can be drawn as shown in Fig. 4.32.

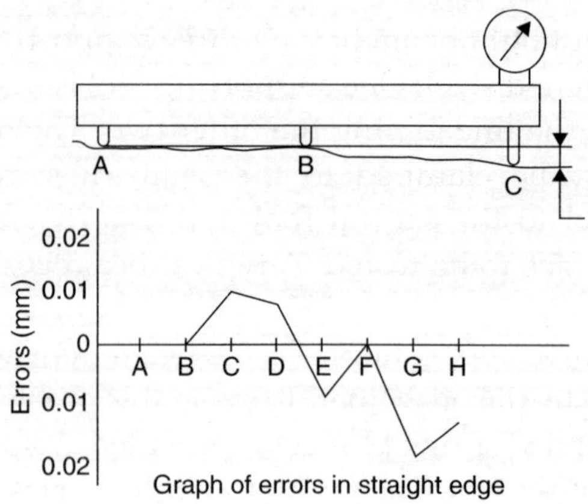

Figure 4.32 Principle of beam comparator

4.12 FLATNESS MEASUREMENT

Flatness is the condition of a surface having all elements in one plane. Flatness is defined as the minimum distance between two parallel planes that contain the surface. It is similar to straightness testing, except that measurements are made over a surface instead of a line. So, the measurements of straightness are made along a number of lines so that the whole surface is covered and then the flatness error is calculated. The methods used to measure flatness are discussed in this section.

4.12.1 Direct Comparison with the Surface Plate

This method compares the given surface whose flatness is to be checked with a surface whose accuracy is known. The cast-iron surface plate (or granite plate) is used extensively as a reference surface. First, Prussian blue or some other marking medium is applied on the surface plate. Then, the surface to be tested is placed on the surface plate. The degree of contact indicated by the marking medium when the two surfaces are rubbed will show the flatness of the surface. This method is applicable to small surfaces.

4.12.2 Spirit Level Method

This method has already been discussed for straightness measurement. In the case of straightness measurement, the readings are taken for different sections of the line to be tested. However, in flatness measurement, the readings are taken for different lines drawn at random on the surface to be tested.

4.12.3 Autocollimator Method

This method has also been discussed for straightness measurement. Here, the readings are taken for different planes (lines) of the surface to be tested.

4.12.4 Beam Comparator

This method has also been discussed for straightness measurement.

4.12.5 Comparison with the Surface of a Liquid

In this system, two cylinders connected by rubber tubing at their bases are used. They contain mercury or dilute soda solution. Both the tubes have a micrometer head with a conical point to the spindle. For testing the flatness of a surface, one of the cylinders is placed in the centre and the other cylinder is moved to different positions on the surface and the micrometer spindles are moved till contact with the liquid surface is made. An electrical circuit may be used to indicate the exact position of contact. Variations in the levels of the two surfaces will be equal to twice the reading of either micrometer head. The readings of both the cylinders are necessary to account for the effect of changes in the shape of the rubber tube. This method is suitable for testing large surfaces because of the size of the cylindrical base, particularly for levelling bed plates of heavy machinery where other methods cannot be used because of the obstructions between various points.

4.12.6 NPL Flatness Interferometer Method

This interferometer was designed by the NPL. Figure 4.33 shows the optical arrangement of the interferometer and various fringe patterns to reveal the flatness of the gauge to be tested. It consists of a mercury vapour lamp whose radiations are passed through a green filter, thus removing all other colours, and leaving a green monochromatic light whose wavelength is very close to 0.5 μm. This light is focused onto a pinhole, giving an intense point source of monochromatic light, which is in the focal plane of a collimating lens, and is thus projected as a parallel beam of light. This beam is directed onto the gauge to be tested via an optical flat so that interference fringes are formed across the face of the gauge, the fringes being viewed from directly above by means of a thick glass plate semi-reflector set at 45° to the optical axis.

It should be noted that the optical flat is mounted on an adjustable tripod, independent of the gauge base plate, so that its angle can be adjusted. Further, the gauge base plate is designed to rotate so that the fringes can be orientated to the best advantage.

4.12.7 Liquid Interference Method

A liquid surface interferometer uses the surface of a liquid as the reference surface instead of an optical flat. Light is reflected from the upper surface of the liquid corresponding to the underside of the optical flat. These methods are used for checking small lapped surfaces; for example, micrometer anvils and block gauges.

4.13 TESTING OF ROUNDNESS (CIRCULARITY)

Roundness or circularity is defined as the radius uniformity of a work surface measured from the centre line of the work-piece. The error of roundness is defined as the radial distance between the minimum inscribing circle and the maximum inscribing circle, which contains the profile of the surface at a section perpendicular to the axis of rotation.

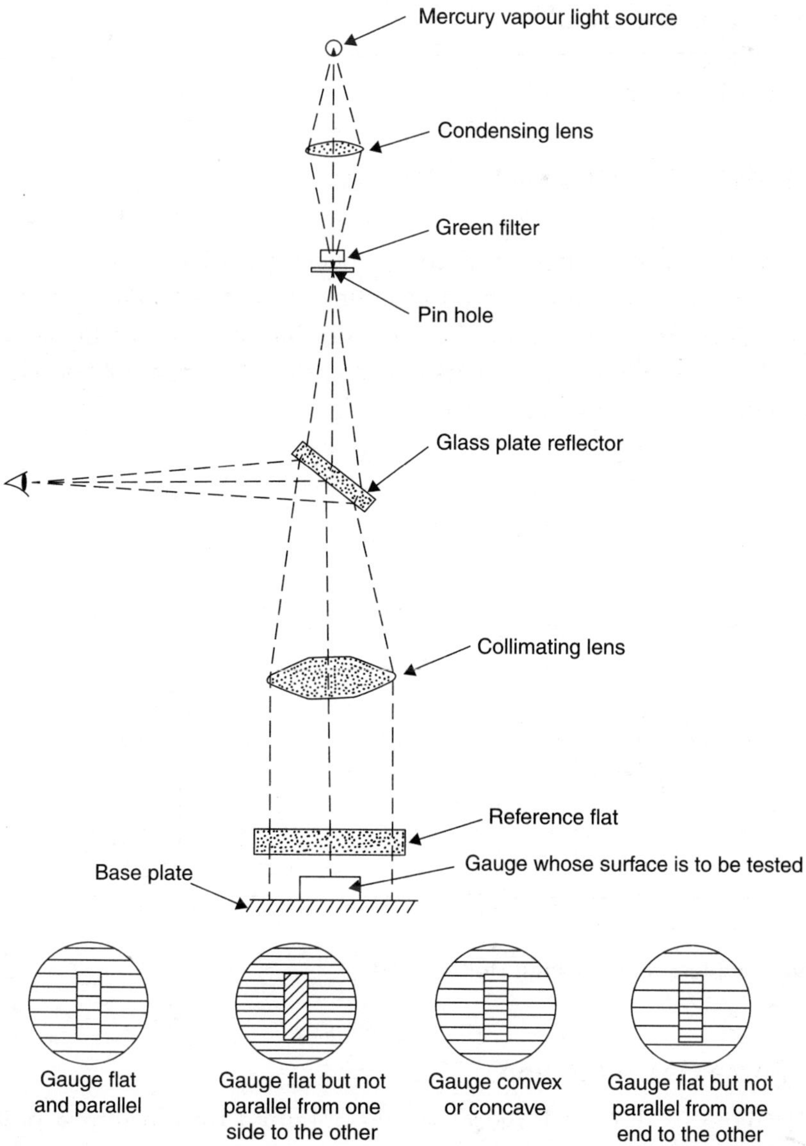

Figure 4.33 Optical arrangement of interferometer for testing flatness of surfaces

Causes of out-of-roundness (during manufacturing)

- Erratic cutting section
- Inadequate lubrication
- Poor alignment of centres
- Excessive feeds
- Presence of chips on clamping surface

- Poor bearing in the spindle
- Spindle run-out
- Clamping distortion

4.13.1 V Block and Dial Gauge Method

The set-up used for this purpose is shown in Fig. 4.34. A V block is placed on the surface plate and the work-piece to be checked is placed upon it. A sensitive dial gauge is firmly fixed in a stand and its feeler is rested against the surface of the work. The work-piece is rotated about the diameter to be checked. The dial gauge records any variation in dimensions due to out-of-roundness. This method converts the diameter measurement to a chordal height variation.

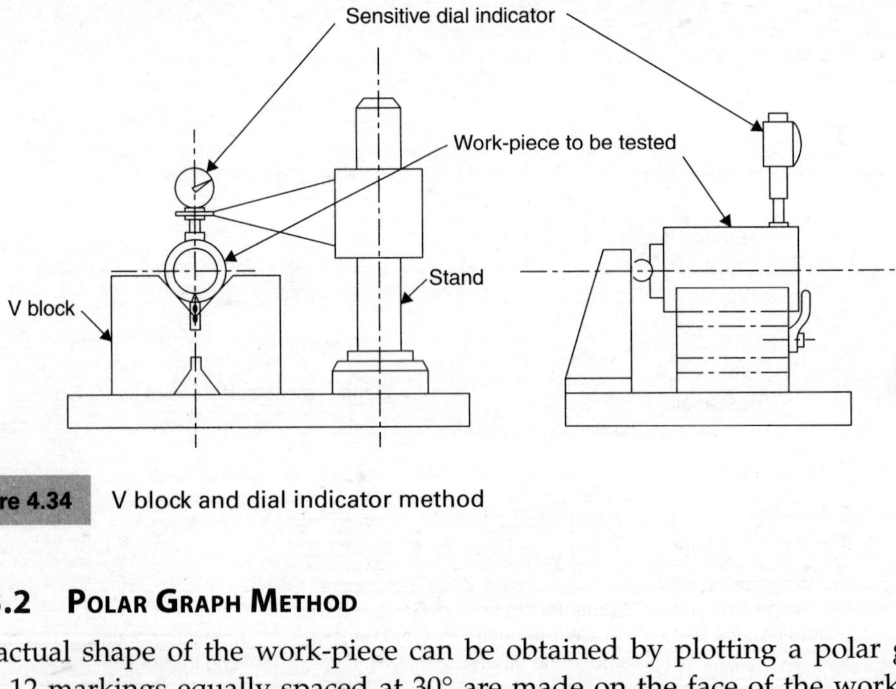

Figure 4.34 V block and dial indicator method

4.13.2 Polar Graph Method

The actual shape of the work-piece can be obtained by plotting a polar graph. First, 12 markings equally spaced at 30° are made on the face of the work-piece to be checked for roundness. The work-piece is placed on the V block after proper cleaning. The dial gauge is placed just above the work-piece so that it touches the work-piece almost at the centre of the V block. The work-piece is then rotated so that the marking on it is below the indicator plunger. A small strip-like leaf spring is placed between the indicator plunger and the surface of the work-piece. The readings of the dial indicator are noted for all the markings. For plotting the polar graph, a suitable scale is chosen, based on the maximum value of the readings. Then, a circle with a diameter nearly four times that of the maximum reading of the dial gauge is drawn and divided into 12 angular divisions, as shown in Fig. 4.35.

Inside the outer circle, another concentric circle of suitable diameter is drawn. Then, the values of the dial gauge are plotted in the radial direction, taking the smaller circle as the reference circle in order that both positive and negative readings are plotted within the prepared polar graph. Individual points are then joined by straight lines, thus providing the actual profile of the measured work-piece.

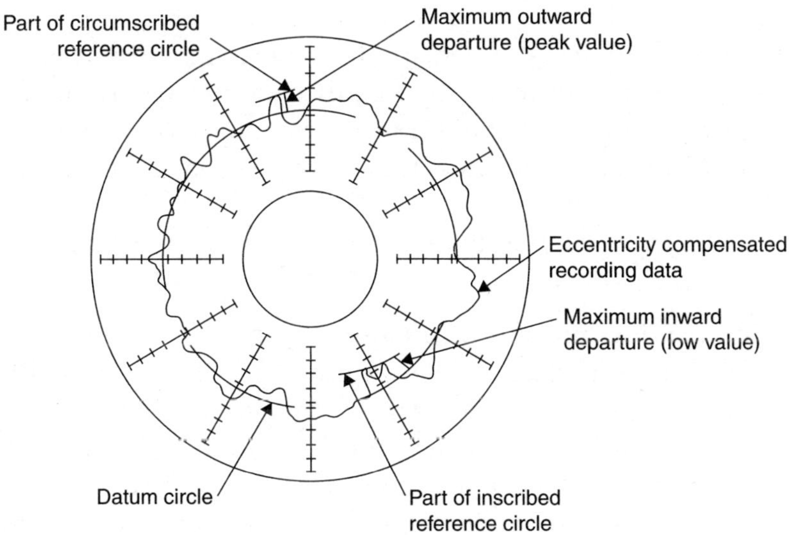

Figure 4.35 Checking roundness using a polar graph

4.13.3 Rotating on Centres Method

Shaft-like work-pieces can be tested for roundness using this method. The shaft is mounted on centres and a dial gauge is placed on the shaft, as shown in Fig. 4.36.

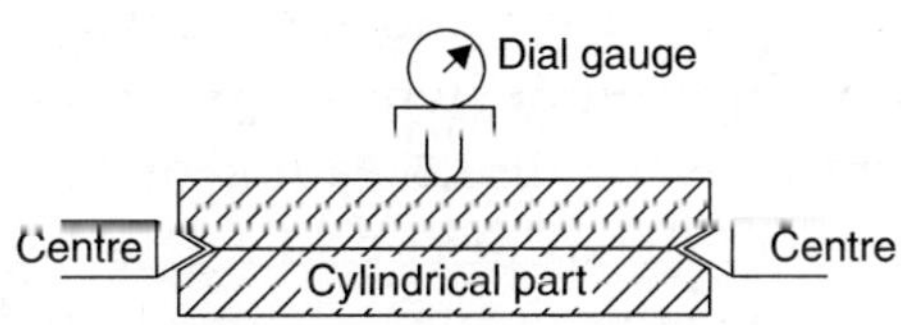

Figure 4.36 Rotating on centres method

Reliability is dependent on many factors like alignment of the centres, roundness of the centres and centre holes, and the run-out of the piece.

4.13.4 Roundness Testing Machines

If a dial gauge could be made to move in a perfectly straight line, the plunger bearing against a supposedly straight work-piece, changes in the readings of the dial gauge would consist of errors in the straightness of the work-piece plus a component caused by misalignment of the surface datum and the line of movement of the dial gauge. If the misalignment component could be eliminated, the remaining variations in readings would be errors in the straightness of the work-piece.

Similarly, if a measuring instrument can be made to move in a perfect circle, its detector or stylus bearing against a supposedly cylindrical work-piece, variations in the readings will include deviations from the roundness of the work-piece plus a component caused by misalignment of the axis of rotation of the measuring device and the axis of the work-piece.

It follows that a machine used to assess roundness must include the following features:

- **A precision bearing, the axis of rotation of which gives an accurate datum for the measurement**: Needle roller bearings are almost friction-free and, if correctly designed, are extremely stiff, i.e., the deflection per unit load is very small. It claims to revolve on a constant axis within 0.05 μm. This bearing is used by Talyrond and Optical Measuring Tools Ltd. (OMT).
- **Alignment of work and rotational axes**: This is achieved by mounting the work table on a pair of coordinate slide-ways, as shown in Fig. 4.36. The table is adjusted by means of a pair of fine thread screws, the alignment being initially set by eye, then with the machine on low magnification and finally at the desired magnification.
- **A measuring device capable of amplifying errors in roundness**: The measuring probe may move on a rotating head machine or be fixed on a rotating work machine. In either case, the measuring probe will be remote from the recording unit and will normally require flexible connections to it. The Talyrond machine uses electrical methods with their attendant advantages of easy change in amplification and an electrical output which can be used as the input to a special-purpose computer which gives a direct reading of the numerical assessment of non-roundness. The OMT machine uses an air-gauging system with two levels of pneumatic amplification which can be changed simply by adjusting a valve which changes the area of the control orifice. A second method of changing the amplification is by changing the length of the stylus arm. As the distance from the fulcrum to the measuring jet is constant, a change in stylus arm length produces a change in amplification.
- **Recording device**: The Talyrond recorder is an electrical device, as indeed is the amplifier unit. The OMT recorder is basically a large, extremely sensitive

Bourdon tube type pressure gauge whose pointer carries a tracer pen which moves over the recorder paper. Both recorders use discs of suitable paper which make one revolution as the measuring head makes one revolution, the motor drive being arranged to give just one complete revolution.

Among the four methods of assessment of roundness error, the least squares centre (LSC) is the most preferred. The least squares centre is the centre of a circle of radius R chosen so that the sum of squares of the radial distances of all points on the trace from the circle is a minimum. Thus, the circle is the equivalent of the best line on a graph drawn by the method of least squares. In fact, the least squares centre circle is not drawn. Having found its centre, the two circles are drawn from it which contain and are contained by the trace, the radial separation between these circles, divided by the magnification, representing the roundness error.

LSC is a unique position found mathematically and, therefore, all operators should obtain the same result. Its position can be found by choosing as an initial datum any point O within the trace. From this point, a number of radial lines are drawn at equi-angular spacing, as in Fig. 4.37.

From the intersection of these lines with the trace, the horizontal coordinates x_1, x_2, x_3, etc. and the vertical coordinates y_1, y_2, y_3, etc. are measured relative to point O. The coordinates X and Y of the least squares centre from this point are then found.

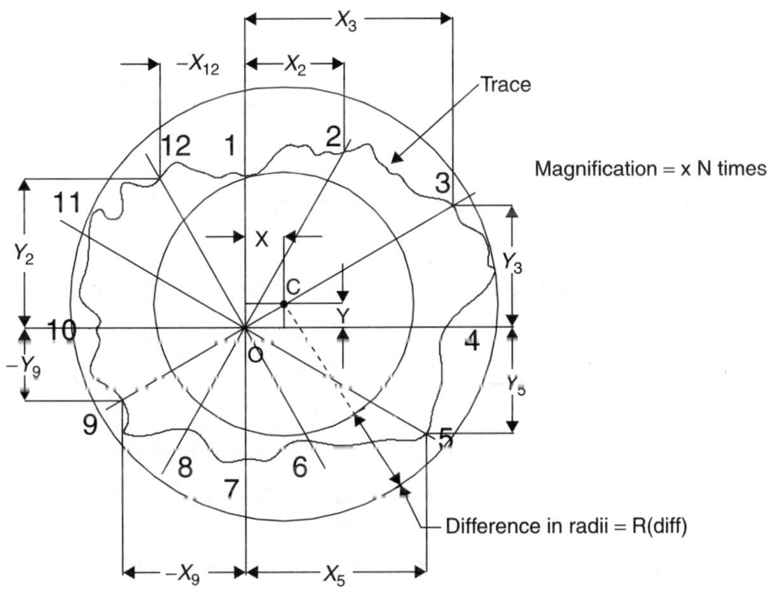

Figure 4.37 Least squares centre for calculation of roundness error

$$X = \frac{2\Sigma x}{n} \text{ and } Y = \frac{2\Sigma y}{n}$$

where n is the number of positions considered.

This calculation is best set out in tabular form, as shown in Table 4.1 referring to Fig. 4.37.

Table 4.1 Data for least squares method for testing

Position	x (mm)	y (mm)
1	0	y_1
2	X_2	y_2
3	X_3	y_3
4	X_4	0
5	X_5	$-y_5$
6	X_6	$-y_6$
7	0	$-y_7$
8	$-x_8$	$-y_8$
9	$-x_9$	$-y_9$
10	$-x_{10}$	0
11	$-x_{11}$	y_{11}
12	$-x_{12}$	y_{12}
Total	Σx	Σy

$X = \frac{2\Sigma x}{n}$ and $Y = \frac{2\Sigma y}{n}$, where n = number of positions, i.e., 12.

For example, if $\Sigma x = 60$ mm and $\Sigma y = 78$ mm for a given problem, then

$$X = \frac{2(60)}{12} = 10 \text{ mm}; \qquad Y = \frac{2(78)}{12} = 13 \text{ mm}$$

A point C is marked with the coordinates (10, 13) measured from the origin point, O. Two circles are drawn with C as the centre and radius such that one circle inscribes the trace while the other circle circumscribes the trace. The difference between the radii of the two circles is measured and represented as R(diff).

The roundness error is then calculated as:

$$\text{Roundness error} = \frac{\text{R(diff)}}{\text{Magnification}} \times 1000\ \mu\text{m}$$

$$= \frac{25}{100} \times 1000\ \mu\text{m}$$ (for the example value of R(diff) = 25 mm and magnification factor used = x 100 times)

$$= 250\ \mu\text{m}$$

4.13.5 Cylindricity, Concentricity and Run-out

Cylindricity is defined as the condition of a surface of revolution in which all the points of the surface are equidistant from a common axis. Cylindricity is similar to circularity applied on a longitudinal basis. The tolerance zone for cylindricity lies between two concentric cylinders. All the surface elements should lie within these concentric cylinders. The tolerance applies simultaneously to both the longitudinal and the circular elements of the surface.

Concentricity is the parameter used to control the relationship between the axes of two or more cylindrical features. When the axes of each cylinder fall on the same centre line, then the surface is said to be concentric. Concentricity is defined as a three-dimensional geometric tolerance that controls how much the median points of multiple diameters deviate from the specified datum axis. Alignment of centres/axes of more than one diameter of the part is called concentricity. The difference between concentricity and eccentricity is that concentricity is on the diameter whereas eccentricity is on the radius.

Concentricity can be verified with dial indicators positioned as shown in Fig. 4.38. Two dial indicators are placed diametrically opposite each other on either side of the part to be checked for concentricity. The part is rotated about the datum axis and several readings are obtained at each selected cross-section of the part along its entire length.

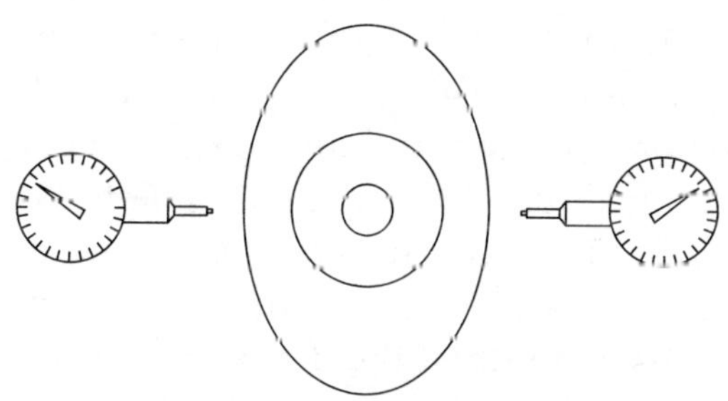

Figure 4.38 Position of dial indicators for checking concentricity

To measure the concentricity of the part, a directional compass is drawn on a piece of paper with the north, south, east and west indications. First, the dial indicators are made to touch the datum surface and positioned for zero readings. When the opposite dial indicators are located at the east and west positions against the part to be checked for concentricity, the readings on each dial indicator are noted. For example, if the east dial indicator is moving east at .005″ and the west dial indicator is moving west at 0.010″, the concentricity tolerance is the difference between these readings, i.e., 0.010″ – 0.005″ = 0.005″ (west) at that particular cross-section. When the part is rotated 90°, then the opposite dial indicators are brought to the north and south positions. The readings of the opposite dial indicators are now noted. For example, if the north dial indicator is moving south for 0.005″ and the south dial indicator is moving south for 0.010″, then the concentricity tolerance is the addition of these two readings, i.e., 0.005″ + 0.005″ = 0.010″ (south) at that particular cross-section. In short, if both the dial indicators move in the same direction, the indicator readings should be added to obtain the tolerance in the same direction, and if both the dial indicators move in opposite directions, the indicator readings should be subtracted from each other to obtain the tolerance in the direction in which one of the dial indicators has a higher value. This process is repeated at many cross-sections along the entire length of the part to be checked for concentricity. If the tolerance is within the specified range, then the concentricity of the part is said to be within acceptable levels.

Run-out tolerance is defined as a group of geometric characteristics that describes the surfaces around an axis. The tolerance zone of a run-out error is the linear distance between two circles concentric with the datum axis. The relationship between concentricity and run-out is that half the run-out value is concentricity. However, the difference between concentricity and run-out can be explained with a hexagonal part. A hexagonal part can be perfectly concentric with a cylindrical part if the axis of the hexagon coincides with the axis of the cylinder. However, for the same hexagonal part, the run-out tolerance cannot be zero as it can be specified only for cylindrical parts.

To measure the run-out, the part is mounted in a chuck or a collet. With a dial indicator in contact with the surface to be inspected, the part is rotated 360° about its simulated datum axis. This process is repeated for several positions along the length of the part. If the full indicator movement does not exceed the specified run-out tolerance, the feature is acceptable. The same preliminary checks required for circular run-out are also required for total run-out. Just as when inspecting a circular run-out, the dial indicator is in contact with the surface to be inspected, but the dial indicator is moved along the full length of the feature's profile as the part is rotated 360° about its simulated datum axis.

SUMMARY

- A screw thread is a helical ridge produced by forming a continuous helical groove of uniform section on the external or internal surface of a cylinder or cone.
- The pitch of a thread is the distance measured parallel to the axis of the thread, between corresponding points on adjacent thread forms in the same axial plane and on the same side of the axis.
- The lead of a thread is the axial distance moved by the threaded part, when it is given one complete revolution about its axis with respect to a fixed mating thread.
- The addendum of an external thread is the radial distance between the major and pitch cylinders.
- The dedendum of an external thread is the radial distance between the pitch and minor cylinders.
- The major diameter is the diameter of the major cylinder, which is an imaginary cylinder, co-axial with the screw, just touching the crest of an external thread or the root of an internal thread.
- The minor diameter is the diameter of the minor cylinder, which is an imaginary cylinder, co-axial with the screw, just touching the root of an external thread or the crest of an internal thread.
- The effective diameter is the diameter of the pitch cylinder, which is an imaginary cylinder co-axial with the axis of the screw, intersecting the flank of the threads in such a way as to make the width of the threads and the width of the spaces between the threads equal.
- Errors can occur in any of the following elements of the thread: major diameter; minor diameter; effective diameter; pitch; flank angles and profile at crest and trough.
- Progressive error of a thread is uniform, and is caused by incorrect ratio of linear velocity of the tool and angular velocity of the work-piece, pitch error of the lead of a screw of the thread-cutting machine and change in length due to hardening.
- Periodic error repeats itself at equal intervals along the thread, which is caused by non-uniform velocity of the tool; lack of squareness of the lead screw of the cutting machine; and errors in the teeth of the gears of the cutting machine
- Drunken error is similar to periodic error but repeats once per turn of the thread and is caused when the thread is not cut to the true helix.

- Erratic error varies in an irregular manner along the length of the thread, and is caused by faults in the cutting machine; disturbances in the machining set-up and non-uniformity in the material of the screw.
- The major diameter of external thread measured by bench micrometer is $D_c + (R_t - R_c)$, where D_c is the calibrated diameter of the setting cylinder; R_c is the micrometer reading on the setting cylinder and R_t is the micrometer reading on the thread.
- The minor diameter of an external thread measured by a bench micrometer is $D_c + (R_t - R_c)$, where D_c is the calibrated diameter of the setting cylinder; R_c is the micrometer reading on the setting cylinder and R_t is the micrometer reading on the thread.
- The effective diameter of the external thread measured using the two-wire method is $T + \frac{p}{2}\cot\theta - (\operatorname{cosec}\theta - 1)\, d$, where T is the measured dimension under the cylinders (wires); p is the nominal pitch; d is the cylinder diameter andis the nominal flank angle, equal to the semi-angle of the thread.
- The effective diameter of an external thread measured using the three-wire method is $M - d\,[1 + \operatorname{cosec}\theta] + \frac{p}{2}\cot\theta$, where M is the measured dimension under the cylinders (wires); p is the nominal pitch; d is the cylinder diameter andis the nominal flank angle, equal to the semi-angle of the thread.
- The pressure angle of a gear is the angle between the line of action and the common tangent to the pitch circles.
- Tooth thickness of a gear is the arc distance measured along the pitch circle from its intercept with one flank to its intercept with the other flank of the same tooth.
- Backlash of a gear is the circumferential movement of one gear of a mating pair, the other gear being fixed, measured at the pitch circle, bearing clearances being eliminated.
- Profile projector, David Brown tester, Parkinson gear tester and Gear tooth Vernier calipers are used to measure the various parameters of a gear.
- Radius of curvature of a work-piece having a portion of a circle is $\frac{(l-d)^2}{8d}$, where l is the reading over the balls and d is the diameter of the ball.
- Radius of curvature of a blunt corner of a template can be determined from the expression $\left(R - \frac{D}{2}\right)^2 + \left[\left(h - \frac{D}{2}\right) - (H - R)\right]^{2}$, where R is the radius of curvature, D is the diameter of the roller and H is the height of the template.

- Radius of curvature of a cavity when the edges are rounded up is $\frac{d^2}{8h} + \frac{h}{2}$, where d is the width of the depth micrometer and h is the depth of the cavity from the base of the depth micrometer.
- The main factors that affect the significance of control of surface finish are fatigue life; bearing properties; wear resistance; corrosion resistance; lubrication; load carrying capacity; noise reduction in case of gears; appearance and the ability to hold pressure.
- The factors affecting surface roughness are vibrations; metal cutting conditions; type, form, material and sharpness of cutting speed; type of coolant used; material of the work-piece; type of machining and rigidity of the system consisting of the machine tool, and a fixture for clamping the cutting and the work-piece.
- Roughness is surface irregularities of small wavelength, caused by the direct action of the cutting elements on the material, i.e., cutting tool shape, tool feed rate, etc.
- Waviness is surface irregularities of considerable wavelength of a periodic character, caused by inaccuracies of slides, wear of guides, misalignment of centres, non-linear feed motion, deformation of work under the action of cutting forces and vibrations of any kind.
- Tomlinson surface meter, Taylor–Hobson Talysurf, profilometer and profilograph are used to measure the surface finish.
- Straightness is the condition where one line element of an axis or a surface is in a straight line. A line is said to be straight over a given length, if the variation in perpendicular distance of the two points on the line from two planes perpendicular to each other and parallel to the line to be tested remains within the specified tolerance limits; the two points being so located on the line that they are closer to the ends of the line to be tested.
- Straightness can be measured using the sprit level method, autocollimator method, straight edge method, method of comparison with the surface of a liquid or the beam comparator method.
- Flatness is the condition of a surface having all its elements in one plane. Flatness is defined as the minimum distance between two parallel planes that contain the surface. It can be measured by direct comparison with the surface plate, spirit level method, autocollimator method, beam comparator, comparison with the surface of a liquid, liquid interference method and using the NPL flatness interferometer.

- Roundness is defined as the radius uniformity of a work surface measured from the centre line of the work-piece. The causes of out-of-roundness are erratic cutting section; inadequate lubrication; poor alignment of centres; excessive feeds; presence of chips on clamping surfaces; poor bearing in the spindle; spindle run-out and clamping distortion.
- V block and dial gauge, polar graph, rotating on centres and roundness testing machines are used to measure the roundness error.
- Cylindricity is the condition of a surface of revolution in which all points of the surface are equidistant from a common axis.
- Concentricity is the parameter used to control the relationship between the axes of two or more cylindrical features. When the axes of each cylinder fall on the same centre line, the surface is said to be concentric. Concentricity is defined as a three-dimensional geometric tolerance which controls how much the median points of multiple diameters deviate from the specified datum axis. The difference between concentricity and eccentricity is that concentricity is on the diameter whereas eccentricity is on the radius.
- Run-out tolerance is defined as a group of geometric characteristics that describes the surfaces around an axis. The tolerance zone of run-out error is the linear distance between two circles concentric with the datum axis. The relationship between concentricity and run-out is that half of the run-out value is concentricity.

EXERCISES

Multiple Choice Questions

1. The crest of a thread is defined as the prominent part of the thread, whether it be external or
   a) Right-handed b) Left-handed
   c) Internal d) None of the above

2. The straight edges that connect the crest with the root are
   a) Flank angle b) Flank of the thread
   c) Angle of the thread d) All of the above

3. The distance from the crest of the thread to the root is
   a) Depth of the thread b) Angle of the thread
   c) Helix angle d) Axial thickness

4. For an external thread, the radial distance between the major and pitch cylinder is
   a) Dedendum
   b) Adendum
   c) Major diameter
   d) Minor diameter

5. The outside diameter or crest diameter or full diameter of an external thread is referred to as
   a) Major diameter
   b) Minor diameter
   c) Effective diameter
   d) Functional diameter

6. The axial distance moved by the thread part is called
   a) Pitch
   b) Lead
   c) Lead angle
   d) Helix angle

7. The following is not a method to find the effective thread diameter:
   a) Thread micrometer
   b) Three-wire method
   c) Two-wire method
   d) V-piece method

8. The following method(s) is (are) used to measure the minor diameter of the internal thread:
   a) Taper parallel method
   b) Calibrated roller method
   c) Both a and b
   d) None of the above

9. Which method is used to test the straightness of an object?
   a) Indicator method
   b) Interference method
   c) Wedge method
   d) All of the above

10. Match Column 1 items with Column 2 items
    1. Flaw - A. Secondary texture
    2. Actual surface - B. Surface that does not exist
    3. Nominal surface - C. Primary texture
    4. Waviness - D. Surface prescribed by designer
    5. Roughness - E. Irregularities occurring at one phase
    a) 1-A, 2-B, 3-D, 4-C, 5-E
    b) 1-D, 2-E, 3-B, 4-C, 5-A
    c) 1-E, 2-D, 3-B, 4-A, 5-C
    d) 1-C, 2-A, 3-D, 4-B, 5-E

11. Why are pitch errors observed in threads?
    a) Lack of inspection
    b) Incorrect ratio of tool work velocity
    c) Interference between mating parts
    d) All of the above

12. Which of the following statements is true?
    a) Parkinson's gear tester is used to measure variation in centre distance
    b) Toolmaker's microscope is used to measure tooth thickness
    c) Teeth having wider flanks have less strength
    d) All of the above statements are true

13. Universal measuring machine is a combination of
    a) Toolmaker's microscope and Parkinson's gear tester
    b) Toolmaker's microscope and optical measuring machine
    c) Parkinson's gear tester and optical measuring machine
    d) None of the above

14. Best wire is the steel wire or cylinder whose size is chosen so that it pitches approximately
    a) Above the effective diameter
    b) Below the effective diameter
    c) At the effective diameter
    d) None of the above

15. The module of a gear tooth is
    a) Length of the pitch circle diameter per tooth
    b) Reciprocal of diametral pitch
    c) Mentioned in mm units
    d) All of the above

16. The following is used to measure any parameter of a gear:
    a) Parkinson gear tester
    b) Profile projector
    c) David Brown tester
    d) All of the above

17. Roughness is not affected by the following:
    a) Cutting tool shape
    b) Tool feed rate
    c) Vibrations
    d) Work material

18. The following method is not used to measure straightness:
    a) Autocollimator method
    b) Beam comparator method
    c) Spirit level method
    d) NPL interferometer

19. The following is not a cause of out-of-roundness of the work–piece:
    a) Poor alignment of centres
    b) Clamping distortion
    c) Spindle run-out
    d) High-speed rotation of the work-piece

20. Concentricity is
    a) The parameter to control the relationship between the axes of two or more cylindrical features
    b) A three-dimensional geometric tolerance
    c) Half of the run-out value
    d) All of the above

21. A surface gauge is used for
    a) Levelling the surface plate
    b) Checking the surface finish
    c) Finding the depth of the surface
    d) Laying out the work accurately

22. A feeler gauge is used to check
    a) Radius
    b) Screw pitch
    c) Surface roughness
    d) Thickness of clearance

23. Surface roughness on a drawing is represented by
    a) Triangles
    b) Circles
    c) Squares
    d) Rectangles

24. Straight edges are used to measure
    a) Flatness
    b) Parallelism
    c) Perpendicularity
    d) Straight length of parts

25. A collimator is simply a
    a) Source of a point of light
    b) Source of a bundle of parallel light rays
    c) Sort of alignment telescope
    d) Standard for flatness

Answers				
1. c	2. b	3. a	4. b	5. a
6. b	7. d	8. c	9. c	10. c
11. b	12. a	13. b	14. c	15. d
16. d	17. c	18. d	19. d	20. d
21. d	22. d	23. a	24. b	25. b

Review Questions

1. Differentiate primary texture from secondary texture.
2. Explain the conventional method of mentioning the surface finish.
3. Define the following with respect to a screw thread:
    a) Pitch
    b) Lead
    c) Helix angle
    d) Major diameter
    e) Minor diameter
    f) Effective diameter
    g) Addendum
    h) Dedendum
4. What is meant by 'best wire size'? Derive an expression for the same in terms of the pitch and angle of a thread.
5. What is meant by drunken thread?
6. Define straightness, flatness and concentricity.
7. The straightness of the straight edge was measured using an autocollimator and the following readings were obtained. Accuracy of the instrument is 1 sec = 0.5 μm. Compute the straightness error. 5°10′, 5°20′, 5°04′, 4°52′, 4°58′ and 5°26′.

8. Describe the elements of a thread and the types of pitch errors in a screw thread.
9. What is the helix angle of M 50 × 3, two-start thread? **Hint:** Helix angle,

$$\alpha = \tan^{-1}\left(\frac{L}{\pi d}\right) \text{ where } L = 3, d = 50$$

10. Explain how the major and minor diameters of an external thread are measured using a floating carriage micrometer.
11. Explain the two- and three-wire methods to find the effective diameter of a given thread using a floating carriage micrometer. Derive the formulae.
12. Explain how Toolmaker's microscope is used to measure all possible thread parameters.
13. The outside diameter of a gear is 110 mm and the number of teeth is 20. What is the module of gear? **Hint:** $(N + 2)m = \text{OD}$
14. What is the chordal tooth thickness at the pitch line of gear of number of teeth 50 and module (m) 4 mm. **Hint:** Tooth thickness = $\pi\, m/2$
15. Explain the constant chord and base tangent methods to check the gear profile. What is the span length over 5 teeth of gear having 45 teeth, module 4 mm and pressure angle 20°?

Hint: $\text{Nm}\cos\varphi\left[\tan\varphi - \varphi - \frac{\pi}{2N} + \frac{\pi S}{N}\right]$ where S – span length, N – number of teeth, m – module, φ – pressure angle.

16. Write any one difference between spur gear, helical gear, spiral gear and bevel gear.
17. Define the following with respect to gear:
    a) Pressure angle
    b) Module
    c) Backlash
    d) Clearance
    e) Tooth thickness
18. Explain how a profile projector can be used to measure all possible gear parameters.
19. What is meant by composite error in gear measurement?
20. Explain the working principle of the David Brown involute form tester.
21. Explain the working principle of the Parkinson gear testing machine.
22. Describe the method of inspecting the profile of a spur gear using an involute measuring machine.
23. Define the term 'cut-off length' with respect to surface roughness measurement.
24. What is meant by 'lay'? Enumerate the types and give the symbol for each type of lay.
25. Describe the working principle of the Tomlinson surface meter.
26. What is the symbol for fully defining surface roughness? Explain each term.

27. Explain how the Talysurf surface meter works, with a neat sketch.
28. Briefly explain the working principle of the profilometer.
29. Explain the working principle of the profilograph.
30. Explain the working principle of the AC laser interferometer. How is straightness measured?
31. With a neat sketch, describe the NPL flatness interferometer.
32. How does a TMM microscope differ from a profile projector?
33. How is the pitch of a screw thread and screw thread angle measured using the Toolmakers microscope? Discuss in detail.
34. What are the functions of a goniometric head?
35. What is the purpose of using a filter in the TMM projection system?
36. How does a gear-tooth Vernier calipers differ from a standard Vernier calipers?
37. Explain how the chordal width and addendum of gear could be measured using gear tooth Vernier.
38. Explain with neat diagram and suitable formula how to measure the following:
    a) Radius of curvature of a job having a portion of a circle
    b) Radius of curvature of a blunt corner of a job
    c) Radius of curvature of a concave surface of a job
39. What are the factors affecting surface finish?
40. What are the various processes used to achieve good surface finish?
41. What are the factors affecting surface roughness?
42. Differentiate roughness from waviness.
43. Explain how straightness is measured using the spirit level method.
44. Explain the principle of the autocollimator in measuring straightness.
45. With a neat sketch, briefly explain the measuring procedure for straightness using straight edges.
46. Explain how the straightness of a job can be measured by comparing with the surface of a liquid.
47. Describe the constructional details of a beam comparator and its use in measuring straightness, with neat sketches.
48. Differentiate straightness from flatness.
49. What are the factors affecting the roundness of a job.
50. Explain any one method to check the roundness of a cylindrical job.
51. Define the following:
    a) Cylindricity b) Concentricity c) Run-out

5

Advances in Metrology

5.1 INTRODUCTION

Laser was developed in the late 1950s. For measuring purposes, laser has similar properties to normal light in that it can be represented as a sine wave whose wavelength is the same for the same colours and whose amplitude is a measure of the intensity or brightness of laser light. Also, laser has been found to be more suitable than other light sources for applications involving precision length measurements on the basis of laser interferometry because of its unique property of high coherence.

In the last few years, laser has been used widely in metrology, particularly in the areas of gauging, inspection, marking, machine control and engineering measurements.

In the 1950s, two other developments took place in the field of measurement – one was the realisation of the use of diffraction gratings to measure the movement of machine tables to a high degree of accuracy, while the other was the application of almost friction-free bearings to machine slide-ways for the purpose of automatic control. This led to the design of coordinate measuring machines (CMM) with large work tables to measure work-pieces in three planes perpendicular to each other, namely, the x, y and z planes. This chapter discusses in detail the use of laser, CMM and CNC in metrology.

5.2 PRINCIPLE OF INTERFERENCE

Light is a form of energy propagated by electromagnetic waves which may be represented by a sine curve. When two rays of the same wavelength meet at some point, mutual interference occurs and the nature of this interference will depend upon the phases of the two waves at their meeting point. If the waves are in the same phase, they will reinforce each other, and the resulting intensity will be the sum of the two intensities (Fig. 5.1); if they are out of phase, the resultant intensity will be the difference of the two intensities. If both intensities are equal, then in the first case, intensity becomes twice the amplitude and the result will be a bright spot. In the second case, intensity will be zero and the result will be a dark spot (Figs 5.2 and 5.3).

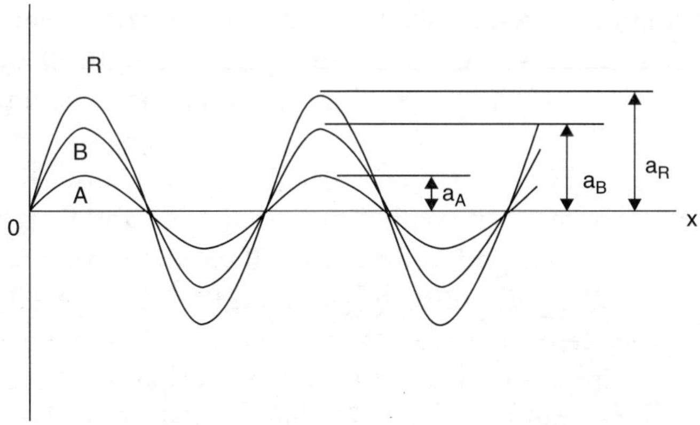

Figure 5.1 Resultant amplitude a_R of two rays A and B of different amplitudes, but in phase

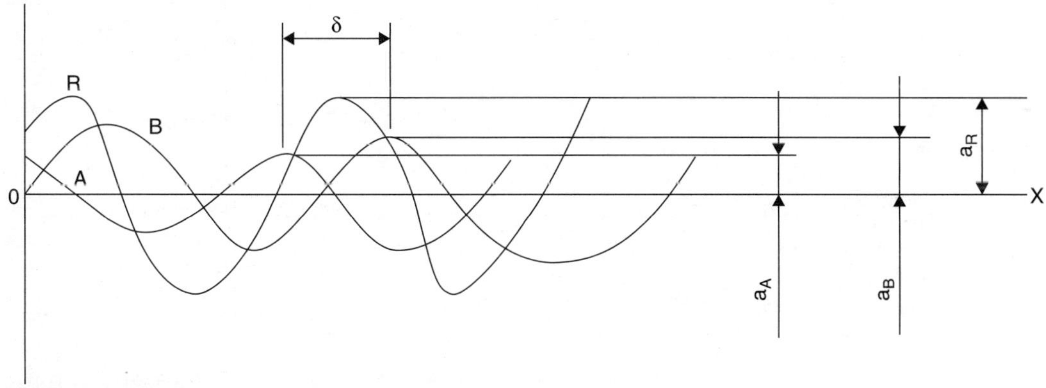

Figure 5.2 Resultant amplitude a_R of two rays A and B out of phase by δ

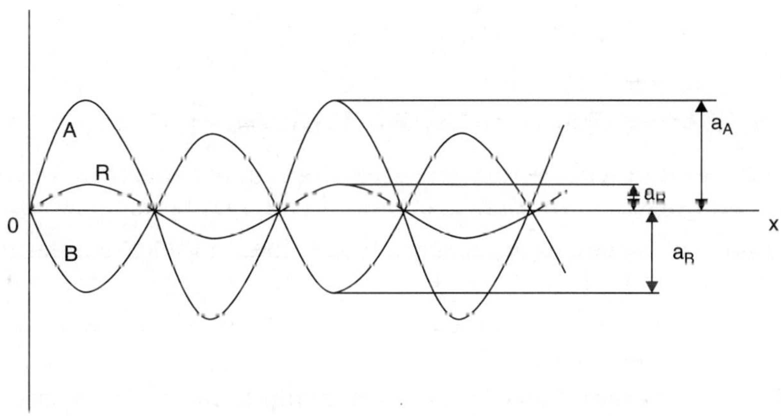

Figure 5.3 Resultant amplitude a_R of two rays A and B out of phase by $\lambda/2$

Interference can occur only when two light rays are coherent, that is, the two rays maintain their phase relationship for an appreciable length of time, and this is possible only when the two rays have originated from the same light source at the same time.

In Fig. 5.4, the paths BO and CO are exactly equal. Hence, the rays on these paths will be in phase because the path difference is zero and there will be a bright spot at O. At the point M, the path difference will be equal to a ½ wave length, i.e., CM – BM = ½ λ. When the path difference is an odd multiple of a half wavelength, the rays will be opposite in phase and a dark spot will be obtained. These dark bands are known as interference fringes. When the path difference is an even multiple of a half wavelength, the rays will be in phase, as at the points P and Q, and a bright spot will be obtained. Thus, the screen will consist of alternating dark and bright bands. There is no sharp dividing line between the bright and dark fringes as the phase relationship changes gradually.

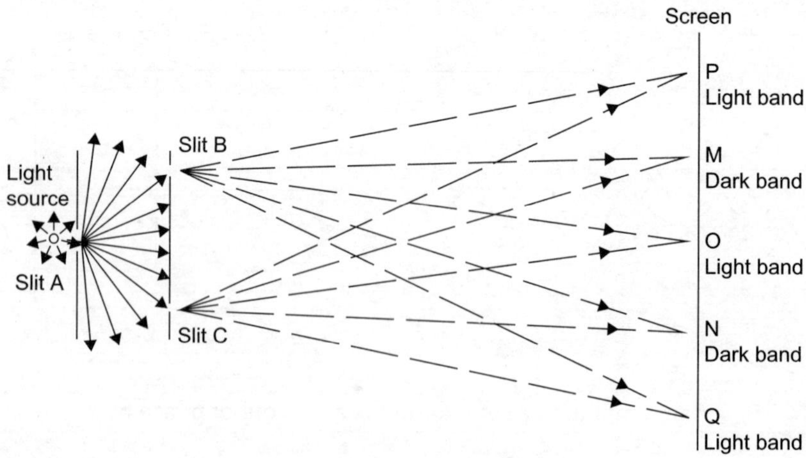

Figure 5.4 Method of producing light and dark fringes

5.2.1 IMPORTANT PROPERTIES OF LASER IN INTERFEROMETRY

- **Laser light is mono-chromatic.** The green light in the spectrum of cadmium is said to have a wavelength of 0.5086 μm. In fact, that is the average of a range of bandwidth of wavelengths close to that value. In the case of laser light, the bandwidth is much reduced, and recently, stabilised lasers of still narrower bandwidths have been developed which can be used for the interferometric measurement of slip gauges.
- **Laser light is coherent.** In a beam of normal light, the infinitely large numbers of rays making up the beam are randomly phased relative to each other. This tends to make the light relatively 'inefficient' as there is partial interference

within the beam. In the case of a laser, the rays are all in phase. If represented by a series of sine waves, the peaks and valleys of the light would all be in line.

- **Laser light is naturally collimated.** The beam of light emerging from a laser is almost perfectly parallel, with little divergence and scatter. To produce a well-collimated beam from any other source requires an optical system, and the loss due to divergence and scatter will still be greater than that from a laser.

5.3 LASER IN MEASUREMENT OF SMALL DIMENSIONS

Consider a small-diameter wire or filament placed in the path of a laser beam at a distance L from a screen. The laser beam is diffracted as it passes over the boundaries of the filament, as shown in Fig. 5.5. At points A and B, because of the coherence of the rays in the beam, all of the rays will be in phase, and therefore, the rays AE and BE will be in phase; hence, a bright spot appears at E on the screen.

At point O_1, the ray AO_1 will have travelled a distance of a half wavelength ($\lambda/2$) less than BO_1, the rays will be out of phase and interference will occur, producing a dark minimum on the screen. At point O_2, the path length difference will be two half wavelengths, giving a bright maximum, and so on up the screen.

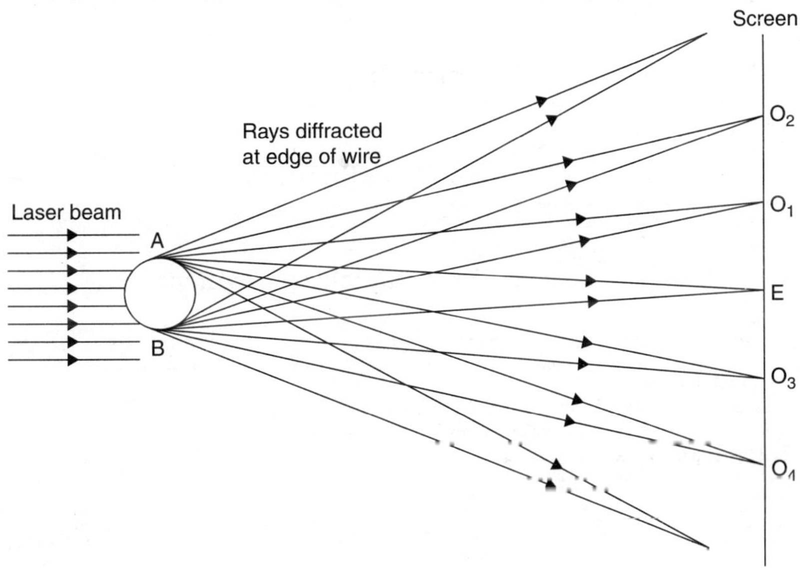

Figure 5.5 Diffraction produced by a laser beam passing over a thin wire AB

The same effect will occur on the other side of the centre line, and there will appear on the screen a series of dark and bright beads.

A narrow opaque object will create a diffraction pattern that may be used to measure the width of the object. The diffraction pattern is similar to that of a narrow slit, with the width of the object replacing the width of the slit.

In Fig. 5.6, let C be the position of the first dark minimum, at distance x_1 from B. Interference occurs at C because the path length difference between PC and RC is ½λ. From Fig. 5.6, it is seen that as long as the diffraction angle α is small, this path length difference is the arc length QR.

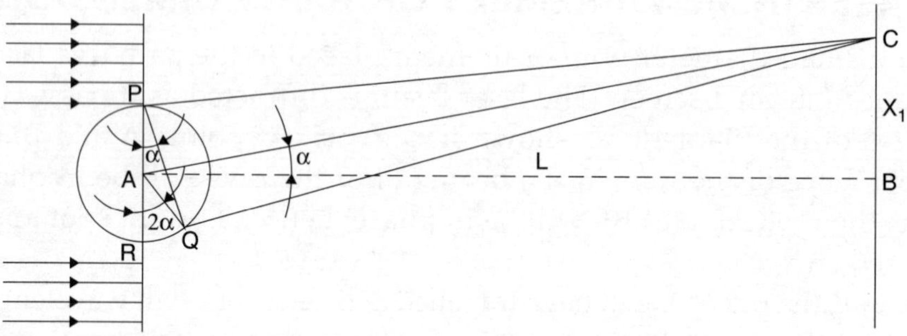

Figure 5.6 Geometry of laser diffraction over a thin wire

The angle subtended at the centre of filament A is 2α, and it follows that

$$2\alpha \text{ radians} = \frac{QR}{r} \qquad \text{in which } QR = \frac{\lambda}{2}$$

$$= \frac{\lambda}{2r} \qquad \text{where } r = \text{radius of filament}$$

$$= \frac{\lambda}{d} \qquad \text{and } d = \text{diameter of filament}$$

$$\therefore \qquad \alpha = \frac{\lambda}{2d} \text{ radians}$$

However, in triangle ABC

$$\tan\alpha = \frac{x_1}{L}$$

$$\alpha = \frac{x_1}{L} \qquad [\text{as } \tan\alpha \approx \alpha \text{ for a small angle of } \alpha]$$

$$\frac{\lambda}{2d} = \frac{x_1}{L};\ x_1 = \frac{\lambda L}{2d}$$

Similarly, at the next dark minimum, the path difference will be three half wavelengths ($3\lambda/2$) and the distance from B, at which this interference occurs, will be designated x_2. Then

$$x_2 = \frac{3\lambda L}{2d}$$

The distance between the dark minima in the diffraction pattern is given by

$$x_2 - x_1 = \frac{3\lambda L}{2d} - \frac{\lambda L}{2d}$$

$$\delta x = \frac{\lambda L}{d}$$

$$\therefore \quad d = \frac{\lambda L}{\delta x}$$

where δx is the mean distance between the dark minima on the screen.

Note that, in fact, the diffraction pattern close to the centre is obscured by the brightness of the intensity of the light from the laser. Hence, the use of the mean distance is for between the visible dark minima, rather than working outwards from the centre.

5.4 LASER INTERFEROMETER IN LINEAR MEASUREMENT (LARGE DISPLACEMENT)

Interferometers are optical instruments used for measuring flatness and determining the length of slip gauges, work-pieces, etc. They are based on the interference principle and employ wavelengths of light as their measuring units. The interferometers make use of some type of beam divider that splits an incoming ray into two parts. These two parts travel along different paths until they are recombined, usually in the same beam divider. Laser interferometers are available for the measurement of large displacements of machine slide-ways. Owing to the coherence, natural collimation and intensity of the laser beam, such measurements can be carried out over a range of many metres.

The laser interferometer consists of a fixed unit containing the laser, a pair of semi-reflectors and two photo diodes, arranged as shown in Fig. 5.7. On the unit whose displacement is to be measured is mounted a corner cube. In Fig. 5.7, the corner cube is shown as a V-block, but in practice, the corner cube is a glass disc whose back surface has three polished faces placed at right angles to each other. Regardless of the angle at which light enters such a unit, it is always turned through 180°.

Light from the laser strikes the first semi-reflector at A and is partially reflected to the second semi-reflector at B, the remainder being transmitted to the corner cube at C. This light is turned through 180° by the corner cube and is recombined with the light reflected from A to B. If the path length difference (ACDB – AB), between these two components is an odd number of half wavelengths, interference will occur at B and the diode output will be minimum. If the path length difference is an even number of half wavelengths, then the unit will give maximum brightness and the diode output will be maximum.

It follows that each time the moving unit is displaced by ½ ($\lambda/2$), the path length difference changes by $\lambda/2$, with AC and BD both being affected by the displacement. A sinusoidal voltage signal is generated by the photo diode, which is amplified and used to operate a high-speed counter whose display is modified to give the displacement in micrometres.

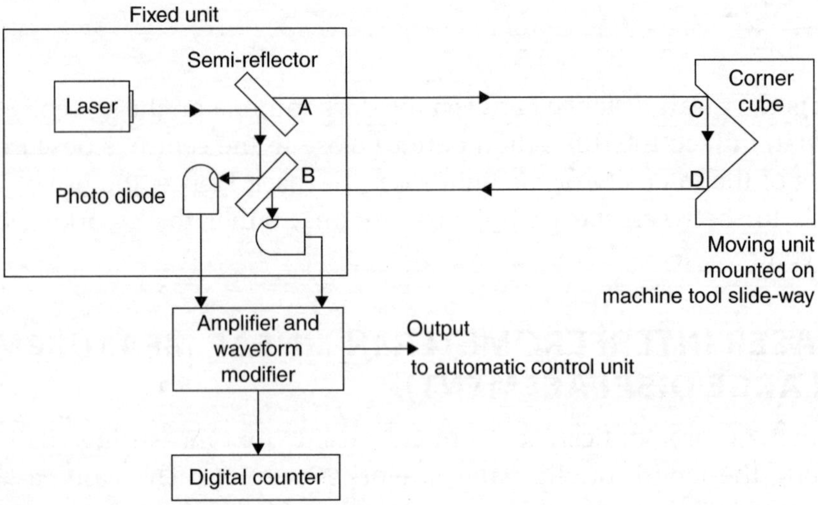

Figure 5.7 Block diagram of laser interferometer

As described, the count would only be uni-directional, but by including a second diode at position 2, as shown in Fig. 5.7, it can be arranged such that the outputs from the two diodes are out of phase by a ¼ wavelength. This enables the unit to be switched to provide bi-directional measurement, and the position of the moving member can be determined relative to a pre-set datum, regardless of the reversals of movement that have taken place.

Laser interferometers can be used to calibrate the movement of machine tables and of length bars. The bar is placed on a moving carriage and the position of its end fixed by a measuring microscope. The interferometer reading is brought to zero and the carriage is moved until the position of the other end is fixed by the microscope. The interferometer reading then gives the length of the bar.

The major advantage of laser interferometers over diffraction gratings in measurement is that the former is portable and can be brought to the machine used to calibrate its measuring system and then removed to be used in another situation, whereas diffraction gratings are built into the machine.

5.5 OPTICAL FLATS IN LASER METROLOGY

An optical flat is an optical-grade piece of glass, usually made of quartz, lapped and polished to be extremely flat on one or both sides, usually within a few millionths of an inch. It is used with a monochromatic light, usually with laser, to determine the flatness of other optical surfaces by the concept of interference. Optical flats are available in a range of sizes and shapes. The diameter of an optical flat varies from 50 to 250 mm and thickness varies from 12 to 25 mm. Optical flats are available with a coated surface. The coating is so thin that it does not affect the position of the fringe bands. The coating is composed of titanium oxide applied on the surface of the optical flat to reduce the light lost by reflection.

Types of optical flats: Optical flats are cylindrical in form and can be of two types:

- **Type A**: It has only one flat surface and is used for testing the flatness of the measuring surfaces of slip gauges, measuring tables, etc.
- **Type B**: Both its surfaces are flat and parallel to each other. This type is used for testing the measuring surfaces of micrometers, measuring anvils and similar length measuring devices for testing flatness and parallelism.

The following precautions need to be taken before using optical flats:

- It should be ensured that the surface of the work-piece to be checked and that of the optical flats are clean and free from dirt, dust and oil. Paper or chamois is used for polishing the surfaces.
- Optical flats should not be slid over the surface of the work-piece, but can be lifted from the work-piece. Sliding, creeping and wringing of optical flats over the surface of the work-piece are extremely harmful to the optical flats.
- If satisfactory interference bands are not obtained, the optical flats should be lifted up and set down again, applying pressure using the fingers at various locations on the upper surface to obtain satisfactory bands.

Interference bands by optical flats: When an optical flat is placed on another test surface and illuminated with monochromatic light (usually laser), the light rays will be reflected off both the bottom surface of the flat and the top surface of the test surface. Unless the two surfaces are exactly flat and parallel, there will be a small gap between them. These reflected rays interfere and superpose on each other and create a pattern of interference fringes (also called Newton's rings), visible as light and dark bands. The rays reflecting off the bottom surface travel on a longer path. The additional path length is equal to twice the gap between the surfaces. At all points where air gap is present, dark bands will be created in the interference fringes due to the path length difference of odd multiples of $\lambda/2$.

Two types of interference can occur: constructive and destructive. On surfaces where the path length difference between the two rays is equal to an even multiple of a half wavelength ($\lambda/2$) of the light rays, the reflected rays will be in phase, and hence, the troughs and peaks of the rays will coincide. Hence, the rays will reinforce (add) and the total light intensity will be greater, which results in a brighter area.

Similarly, at other locations where the path length difference is equal to an odd multiple of a half wavelength ($\lambda/2$), the reflected rays will be 180° out of phase. So, the trough of one ray coincides with the peak of the other ray. Hence, the rays will cancel each other out and the total light intensity will be weaker or zero, which results in a dark area.

The space between the fringes becomes smaller when the gap increases. This indicates a departure from flatness in one of the surfaces. A flat surface is indicated by a pattern of straight, parallel fringes with equal spacing, while other patterns indicate an uneven surface. Two adjacent fringes indicate a difference in elevation of half a wavelength of the light used. Hence, by counting the fringes, the differences in elevation of the surface can be measured to millionths of an inch.

5.6 PRECISION INSTRUMENTS BASED ON LASER

Laser is pre-eminent in the field of accurate measurement of the order of 0.1 μm in 100 m. The biggest advantage of a coherent laser beam is that all of the energy appears to be emanating from a very small point. Further, by a system of lenses, such a beam can be focused easily into either a parallel beam or on to a very small point.

A typical helium-neon laser source can produce a 1-mm-diameter beam of pure red light having a power of 1 MW. When such a high-power beam is focused on a point, it has very high intensity. The beam has no divergence up to a greater distance, but it then begins to expand at the rate of about 1 mm/m. The laser beam is visible on any screen and remains in position for reference and can be easily observed from a distance. The centre of the laser beam can be easily judged to an accuracy of 1 mm over 20 m.

5.7 LASER TELEMETRIC SYSTEM

The laser telemetric system is a non-contact gauge that takes measurements using a collimated laser beam. It measures at a rate of 150 scans per second. It consists of three components: a transmitter, a receiver and processor electronics. The schematic diagram of a laser telemetric system is shown in Fig. 5.8.

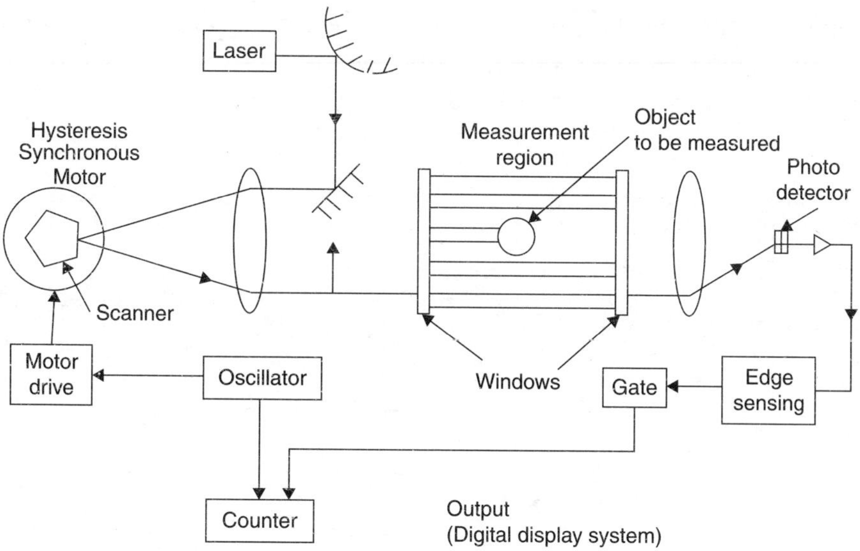

Figure 5.8 Schematic diagram of a laser telemetric system

The transmitter produces a collimated parallel laser beam that appears as a red line. The receiver collects and photo-electrically senses the laser beam transmitted past the object being measured. The processor electronics takes the received signals, converts them to a convenient form and displays the dimension being gauged. It consists of a low-power helium-neon gas laser and its power supply, a specially designed collimating lens, a hysteresis synchronous motor, a multifaceted reflector prism, a synchronous pulse photo detector and a protective replaceable window. High-speed scanning permits online gauging. Thus, it is possible to detect changes in dimensions even when the components are moving.

5.8 ANGULAR MEASUREMENT

Angular measurements, namely, pitch and yaw, are measured using a set of optics that includes an angular interferometer and an angular reflector. A schematic diagram of an angular interferometer is shown in Fig. 5.9. The angular interferometer consists of a polarised beam splitter and a turning mirror that cause the two interfering laser beams to come out parallel to each other. Two cube-corner reflectors attached to a common mount form the angular reflector. Initially, the angle between the angular interferometer and the angular reflector is set to zero and the two paths between them have the same relative length.

If either the angular interferometer or the angular reflector is turned about an axis perpendicular to the plane of the figure, the relative lengths of the two paths will change. This will cause a Doppler-shifted frequency change in the beam that is returned from the angular interferometer to the laser head. This would measure the distance l, which is given by $l = S \sin\theta$, where S is the distance between the nodal points of the cube corner and θ is the angle of rotation.

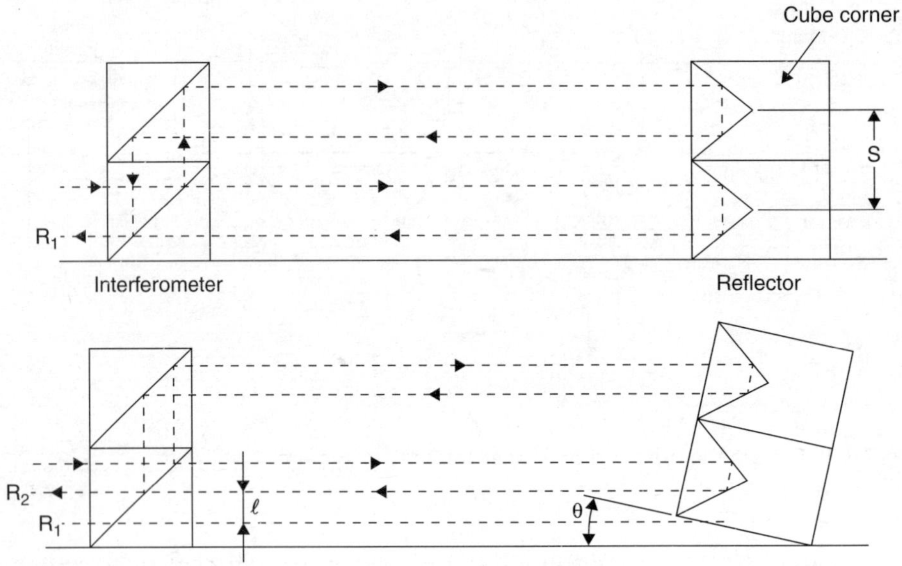

Figure 5.9 Angular measurement using a laser interferometer

5.9 COORDINATE MEASURING MACHINES (CMM)

The modern CMM has the same basic layout as that of other machines and is shown in Fig. 5.10. The table moves longitudinally (the Y coordinate) between massive portals whose cross-member carries the vertical member whose movement is in the

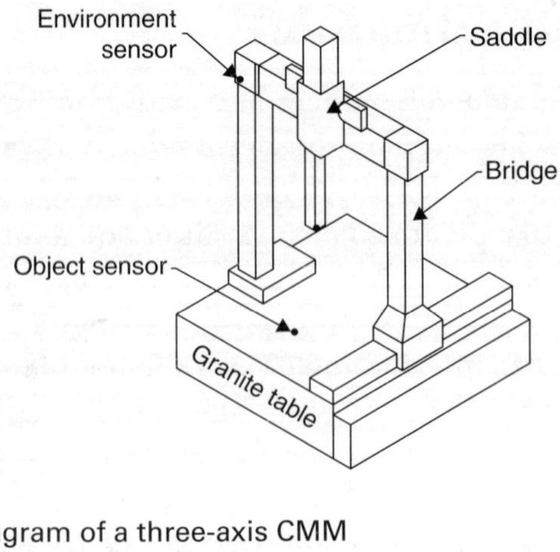

Figure 5.10 Block diagram of a three-axis CMM

Z axis. Movement of the vertical member across the beam gives the *X* coordinate. Smaller machines are of the cantilever type, and, although not as rigid as the portal type, are perfectly capable of carrying the loads required of them without machine deflections introducing unacceptable errors.

The original machines had a spindle carrying a probe with a tapered nose. The slide-ways were moved until the probe was positioned over the hole whose position needed to be determined. It was then pushed into the hole, and the force exerted by the taper was enough to move the low-friction slides to centre the hole under the probe. The position of the hole was then indicated relative to a pre-set datum on decatron counters, which have been replaced today by light emitting diodes (LEDs) that give a direct digital readout. Later, the tapered probe was replaced by the touch trigger probe. Computers are used to control the movements of the slide-ways, record data and perform any calculations previously carried out manually.

The basic configuration of a CMM comprises three mutually perpendicular axes *X*, *Y* and *Z*. Though the physical configuration of CMMs varies, all of them provide a path to move the touch trigger probe in these three axes, with respect to the position of the work table. The six basic configurations that are widely used are given in Fig. 5.11.

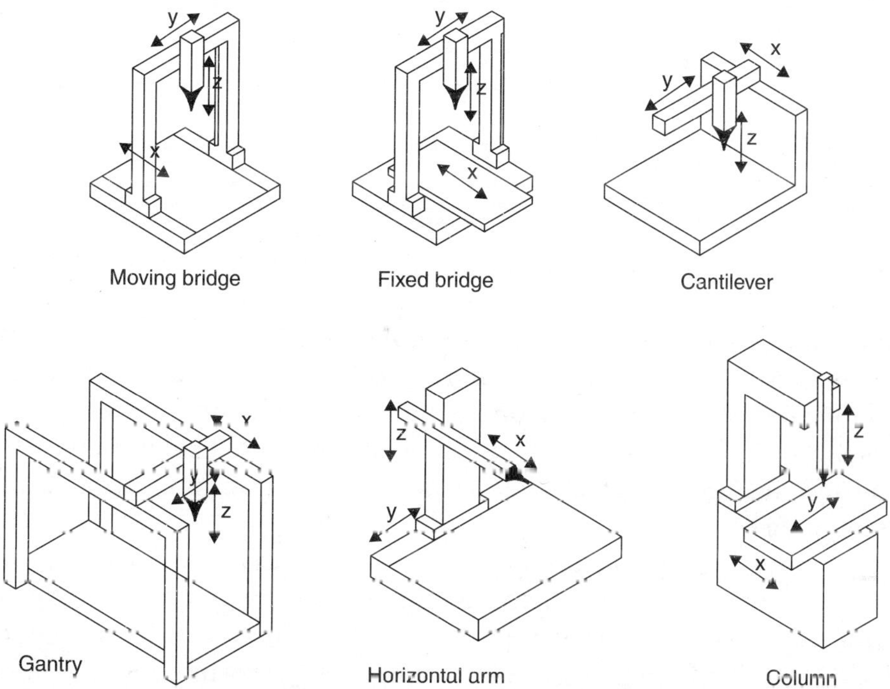

Figure 5.11 Types of CMM

5.9.1 The Touch Trigger Probe (Renishaw Probe)

The touch trigger probe is a contact probe that senses the element to be measured when it reaches the element and transmits relative data to the computer for storage and processing. As shown in Fig. 5.12, it consists essentially of a ball-ended stylus of known diameter mounted on a rod in such a way that immediate contact is made with a surface. A signal is passed to the computer to accept and store the measurement that is transmitted from the measuring system at the moment of contact. The mounting is designed so that, for the simplest probe, contact can be made in one of the three axes, *X*, *Y* and *Z*, or in any combination of these.

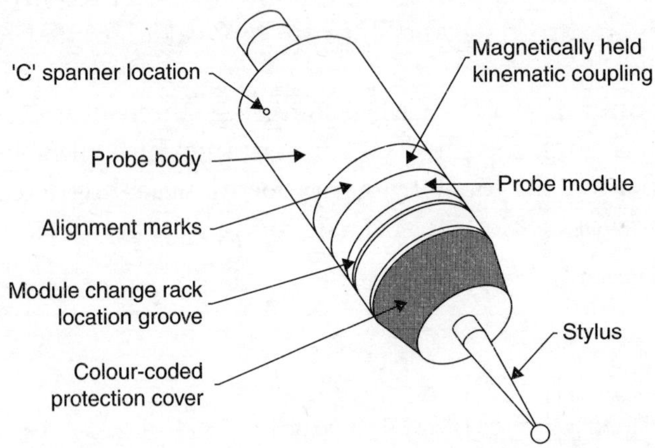

Figure 5.12 Touch trigger probe

The signal from the probe is transmitted to an interface unit, which makes the signal compatible with the computer language being used and then relays it to the computer. The probe may be directly connected to the interface unit by a flexible cable. Otherwise, it will have seven light emitting diodes around its trunk that emit an infra-red signal when the probe makes contact with the work-piece. This signal is picked up by an infra-red–sensitive unit mounted conveniently near the machine and the signal is processed by a suitable interface unit before being sent to the computer.

More complex probes have styluses in planes that are mutually at right angles, as shown in Fig. 5.13, to enable the work-piece to be approached from any direction. For complex parts, the work-piece can be mounted on a motorised rotary table so that it can be swung into the correct alignment for each measurement.

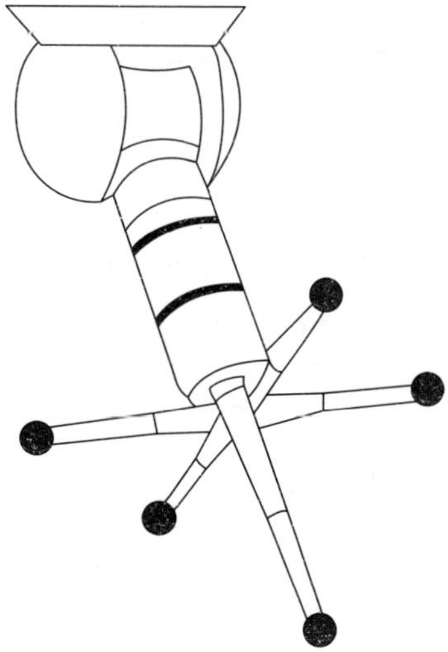

Figure 5.13 Complex probes

5.9.2 Types of CMM

CMMs can be of several types, though the bridge type is commonly used:

- Cantilever type
- Bridge (Portal) type
- Horizontal bore mill type
- Vertical bore mill type
- Floating bridge type
- Spherical coordinate (R – θ) measuring machine
- Three-axis machine
- Four-axis machine (three-axis machine fitted with a rotary table)
- Five-axis machine (three-axis machine fitted with a rotary table in which the axis of rotation can itself be inclined)
- Conventional CMM
- Computer numerical control CMM

The cantilever type is the easiest to load and unload, but is the most susceptible to mechanical error because of sag or deflection in the cantilever beam. The portal type is more difficult to load, but is less sensitive to mechanical errors. The horizontal bore mill type is best suited for large, heavy work-pieces. The vertical bore mill type is highly accurate, but slower in operation. The floating bridge type is a compromise between the cantilever and portal types and is fast in operation and simple to align.

The spherical measuring machine has motions of measuring heads in the R, Q and θ directions. It is used for inspecting parts that are basically spherical in shape. Multi-axis machines can measure various features of parts whose shapes are objects of revolutions like cones, cylinders and hemispheres.

5.9.3 Advantages and Disadvantages of CMM

The advantages of CMMs are as follows:

- Minimisation of operator error
- Improved accuracy
- Reduced operator skill requirements
- Uniform inspection quality
- Less set-up time; faster rate of inspection
- Minimum fixturing costs
- Traceability of compliance to specifications
- Reduction in off-line analysis time

The difference between conventional metrology and coordinate metrology is that the former requires manual operation and is time–consuming, and alignment of the work-piece is necessary, whereas the latter requires less time and alignment of the work-piece is not necessary.

Though CMM can be used for a variety of applications, the following errors restrict its use:

- The table of CMM may not have perfect geometric form.
- The table and probes may not be in perfect alignment.
- The probes moving up and down in the Z axis may have some perpendicularity errors.
- There may be errors in the optical readout of the digital system.

5.10 COMPUTER NUMERICAL CONTROLLED (CNC) CMM

CMM is used not only to measure slide movement, but also for automatic control of the position of the slide or measuring head, when it is interfaced with a computer. In CNC CMM, the output of diffraction gratings is fed back into the computer, programmed with the slide movements necessary for the machine to produce a part of the required size and shape. The measured position is compared with the required position and the error signal is used to direct the slide movement. This is the basis of CNC machines.

Software for the statistical evaluation of measured data, curve measurement, gear measurement and probing curved surfaces is also available. A repeat function simplifies the inspection of similar features. Using computer controls (commands), the slide-way movements can be easily adjusted. Joysticks are normally used to move the machine table. In CNC CMM, in general, the resolution of the measuring system will be 1 μm; the position accuracy will be $[1 + (5 \times L \times 10^{-3})]$ μm, where L is the measured dimension in millimetres; the measuring accuracy will be $[3 + (5 \times L \times 10^{-3})]$ μm.

The form or profile of cams, cutting dies, complex forms of gas turbine blades, etc. can be easily measured on this machine by a scanning technique that takes about fifty measurements and processes them every second. When the dimensions of any part need to be generated in a CNC machine, the part is placed in the CMM and touched by a probe on the surfaces; the coordinates are then digitised to generate CNC programming that will produce the part. This CNC programming principle finds application in reverse engineering.

5.11 DIFFRACTION GRATINGS – MEASURING SYSTEM

Diffraction gratings are used to measure the movement of machine tables to a high degree of accuracy. Many machines utilise the moire fringe concept (diffraction gratings) for measurement. Diffraction gratings are strips or discs of glass (transmission gratings) or metal (reflective gratings) ruled with fine, equally spaced lines to form the grating. They are used in pairs. One is called the fixed or scale grating or measuring scale. The other is called the moving grating or index grating.

Linear gratings: If two glass gratings with lines of equal spacing are placed close to each other so that the lines are parallel and one grating moves longitudinally relative to the other, a shuttering effect will occur. Light is transmitted when the lines are aligned and cut off when the lines on one grating align with the spaces on the other. If, however, the gratings are set at a small angle across the glass and then positioned so that the lines are angled to cross each other, as in Fig. 5.14, then a series of interference-type fringes, known as moire fringes, will be produced. These moire fringes will be more widely spaced than the lines on the gratings.

If the index grating moves along the scale grating in direction A, the fringes move across the grating in direction B. If the index grating moves by one line space, the fringes will move by one fringe space. If the index grating moves a fraction of a line space, the fringes move the same fraction of a fringe space. Thus, measurement of fringe movement can be used as measurement of grating movement. Further, if the index grating moves a considerable distance, then by counting the number of fringes that pass a given point, the distance moved by the grating can be calculated as the same number of line spaces. If light is passed through the gratings, the output can be focused on a photo-sensitive diode and

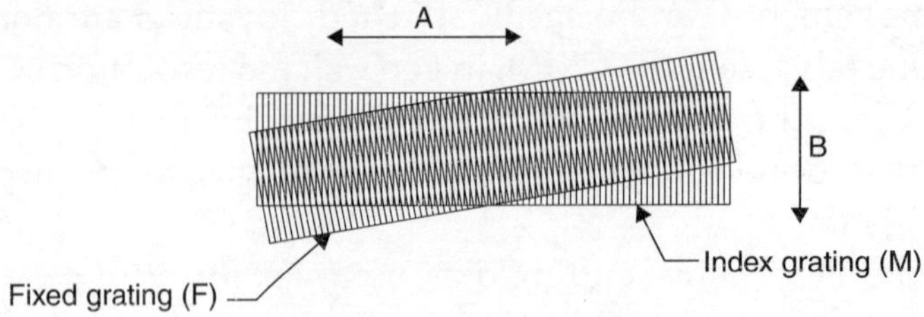

Figure 5.14 Moire fringes produced by a pair of diffraction gratings

the electrical output, which is a sinusoidal voltage signal, can be converted to pulses to provide the input to a digital output.

To detect the direction of movement, if the index grating moves in a direction to give a count of +6 fringes and then back –2 fringes, the result would be a count of 8 fringes. By using two photo diodes to scan the fringes ¼ of a fringe space out of phase, the system can be made bi-directional and can give the correct distance from a datum, regardless of the number of moves that have been made in either direction.

Radial gratings: A similar moire fringe effect can be produced with radial gratings, where glass discs are ruled with lines that are not quite radial, so that when they are placed close to each other, the rulings form a slight angle. In this case, the fringes appear as a spiral. If one grating is fixed and the other rotates, the fringes move radially, and a rotation on one line angle produces a radial movement of one fringe space.

5.11.1 Applications of Moire Fringes in Measurement

To measure the movement of a measuring machine slide, the scale grating is attached to the fixed slide-way and the index grating to the moving member. Light from a tungsten lamp is collimated and passes through the gratings as a parallel beam and through a mask with a slit set parallel to the fringes. The light from the slit is decollimated and focused on a photo-transistor, as shown in Fig. 5.15. Observing the slit as the fringes move across, it gives a visual impression of pulses of light and dark, but, in fact, the output from the transistor is sinusoidal and can be processed electronically to give a digital output in fractions of a line space and also to detect the direction of movement.

Radial gratings are used in a similar manner to measure the angular displacement of a dividing head or rotary table, one grating being attached to the rotating shaft and the other being fixed on the same axis.

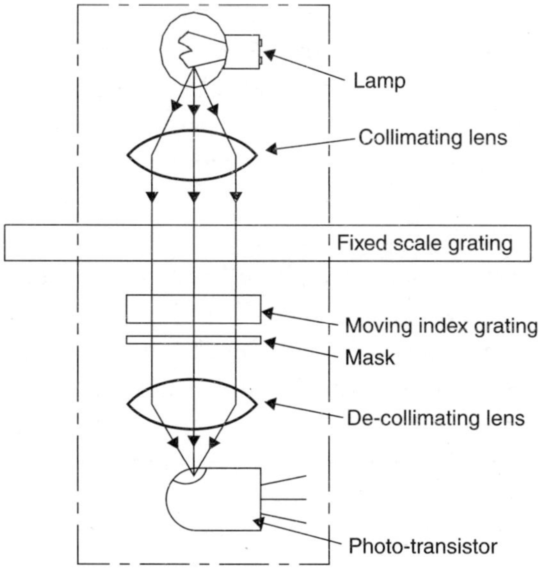

Figure 5.15 Diffraction gratings installed on a machine slide-way

5.12 LASER SCANNING MICROMETER

It is a non-contact online instrument based on laser scanning techniques used to measure the diameter, thickness and spacing associated with stationary or moving objects with high accuracy, precision and high speed of response.

When the object is placed between the scanning laser beam and the sensor, as shown in Fig. 5.16, the diameter or dimension is sensed and digitally displaced. This principle is used in the laser micrometer. A laser beam is directed at a polygonal mirror that rotates at high speed with the help of a motor in exact synchronisation with highly stable pulses from the system clock. The reflected laser beam rotates clockwise as it sweeps across the input surface of a collimating lens. The reflected laser beam changes its direction after the collimating lens' exit surface as it moves or scans downwards. This horizontal laser beam enters the measuring space

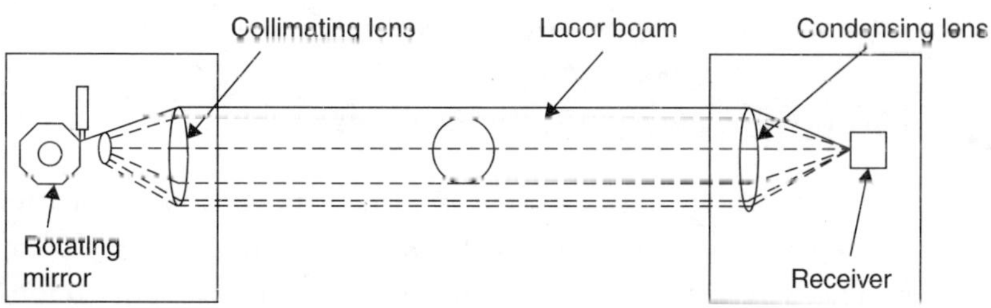

Figure 5.16 Laser micrometer

and, with no work-piece present, reaches a receiver through a condensing lens to produce an output signal. The output signal is amplified and fed into the edge detection circuit. When a work-piece is placed in the measuring space, the reflected laser beam is interrupted for a time during its sweep. The interruption time, as indicated by clock pulses while the receiver signal is absent, is proportional to the work-piece dimension in the downward direction.

Each transition between the receiver detecting the laser beam and then not detecting the laser beam, or vice versa, is called an edge. The edge marks the start and/or end of the measuring sections, called segments, so that the differences in the positions of these edges define the length of each segment. The edges and segments generated by a work-piece are numbered sequentially by the instrument, which comprises electronic RAM/ROM, counters and gates. The dimensions of the work-piece are determined using software programs based on edge and segment data. Simple and detailed block diagrams of a laser (optical) micrometer are shown in Figs 5.17 and 5.18, respectively.

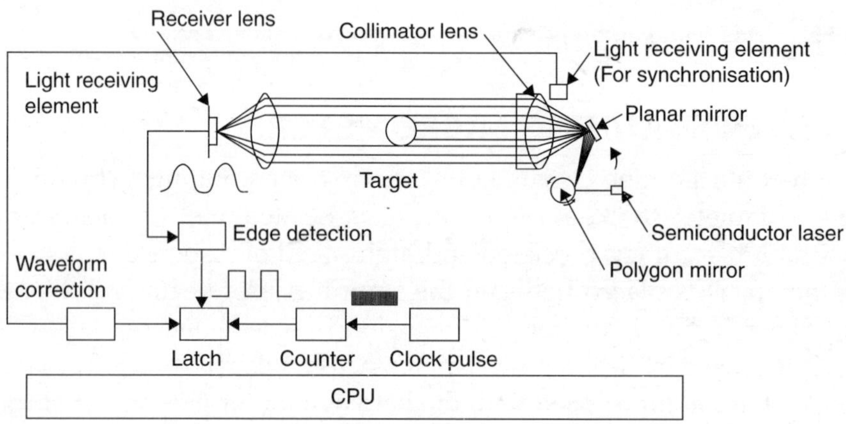

Figure 5.17 Simple block diagram of an optical micrometer

Figure 5.19 shows details of various applications of the laser micrometer. Though the laser micrometer is used in a variety of applications, it is most suitable for measuring the internal diameter of work-pieces.

5.13 COMPUTER-AIDED INSPECTION

Apart from the application of CNC CMM in computer-aided inspection, a robot arm having four degrees of freedom can be used for inspection. It incorporates various attachments, such as electronic probes, automatic tool magazines and sensors. The motion of the robot arm is controlled by a microprocessor, which performs simultaneous multi-axis control of position, speed and acceleration.

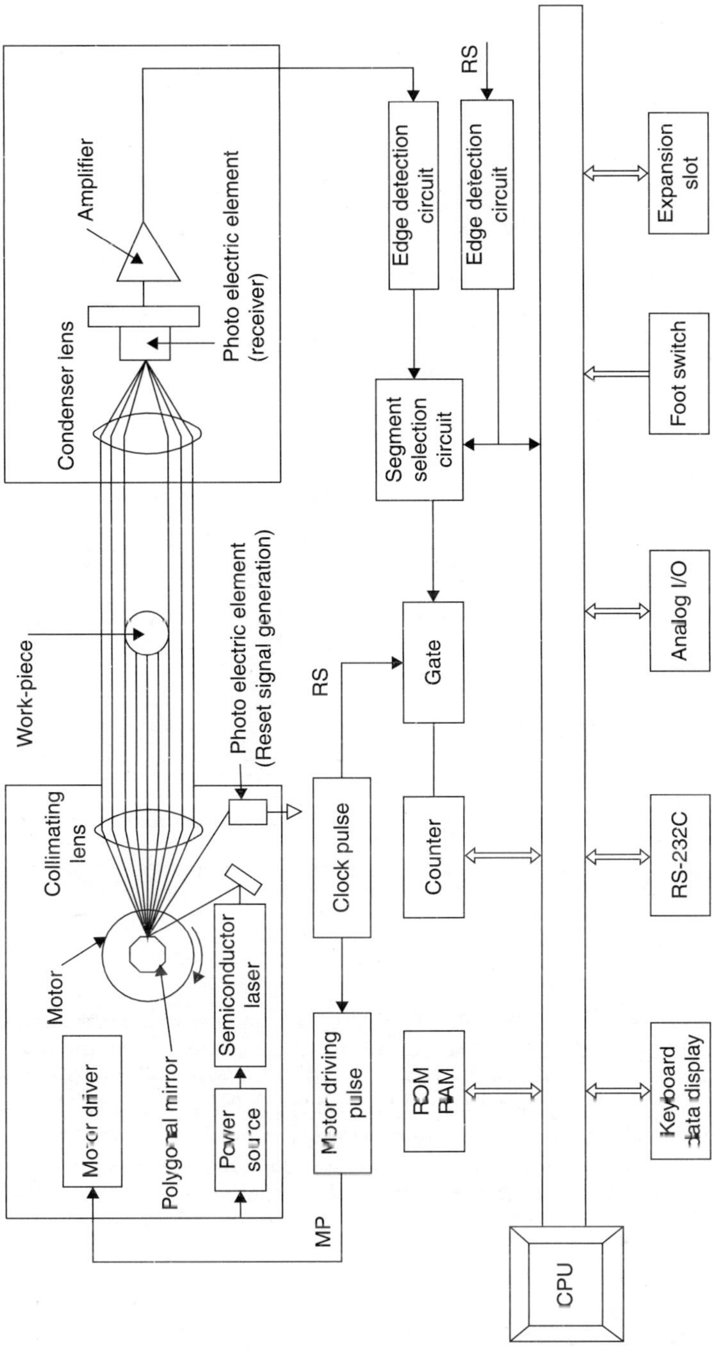

Figure 5.18 Block diagram of laser scanning micrometer

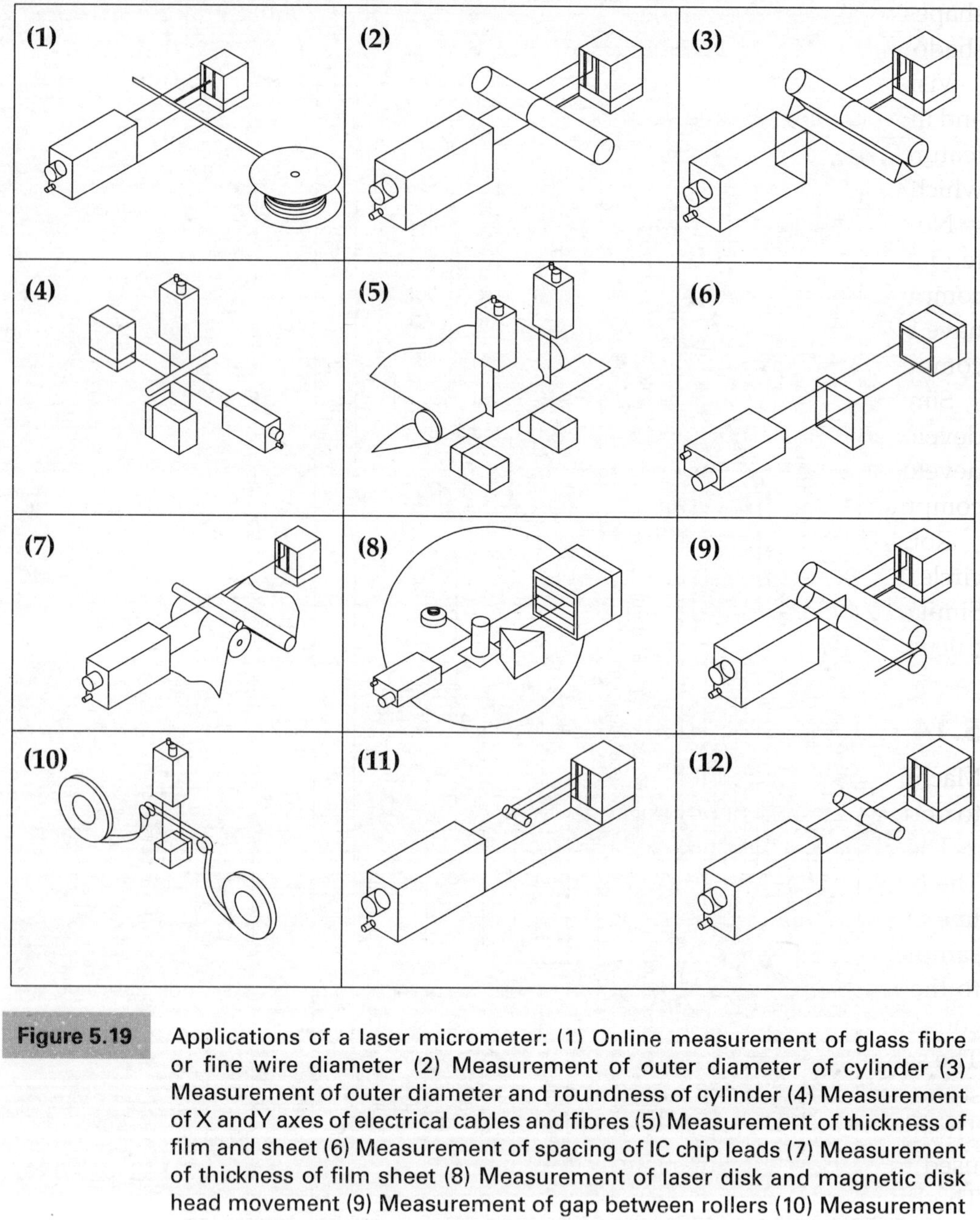

Figure 5.19 Applications of a laser micrometer: (1) Online measurement of glass fibre or fine wire diameter (2) Measurement of outer diameter of cylinder (3) Measurement of outer diameter and roundness of cylinder (4) Measurement of X andY axes of electrical cables and fibres (5) Measurement of thickness of film and sheet (6) Measurement of spacing of IC chip leads (7) Measurement of thickness of film sheet (8) Measurement of laser disk and magnetic disk head movement (9) Measurement of gap between rollers (10) Measurement of tape width (11) Measurement of outer diameter of optical connector and ferrule (12) Measurement of form

Another use of computers in inspection is non-contact dimensional measurement, in which a television camera, using charge-coupled photo diode arrays, is focused on the work-piece to be measured and interfaced to a computer, which analyses the shapes and sizes of the image. Laser illumination is commonly used for projecting shadows because the beam is well defined.

Vision systems break down images into sectors and segments and then recognize and measure their geometry. A vast array of information about various geometrical features such as critical dimensions, roughness and surface defects is stored, with which the digitised information of the processed TV image is compared.

Nowadays, most measuring instruments provide digital output that can be fed into computers for computing processes. Height gauges with built-in integral computers, and with keyboard and printer mounted on the height gauge itself, have been developed. Software programs are permanently incorporated into the computer.

Similarly, slip gauge boxes with built-in computer and printer have been developed. The computer assists in selecting various combinations of gauges to develop any dimension. The required dimension is fed into the computer and the computer displays the combination of gauges required.

Roundness measuring machines fitted with computers can easily define the circle if three points on the circumference of the round work-piece are fully defined. Similarly, gear measuring machines fitted with computers perform all the tedious calculations and produce the final results quickly and accurately.

5.14 MACHINE VISION AND IMAGE PROCESSING

Machine vision is defined as a technique that allows a sensor to view a scene and to derive a numerical or logical decision without further human intervention.

The basic principle is based that of digital image processing using a TV camera. The output of most TV cameras is a continuous video signal corresponding to a line-by-line scan of the image focused on the face of the camera tube. The signal is sampled at equally spaced intervals along each line, and a digital value is assigned to the magnitude of the signal at those points. Line by line, a sampled image is constructed within the image processing computer as a two-dimensional array. The sampling is analogous to laying a two-dimensional grid over the image of the scene and assigning numbers to the intensities within each element of the grid. The number of rows and columns that make up the grid and the range of numbers used to describe the intensities are dependent on sensor and processor memory. The number of picture elements in the grid (also called pixels) determines the spatial resolution of the stored image, and the range of numbers used to describe the brightness affects the intensity resolution.

The vision system consists of a light source, an image sensor, an image digitiser, a data compactor, a system control computer and output peripherals, as shown in Fig. 5.20. The image sensor is usually a vidicon TV camera. The image digitiser is usually a very fast, 6–8-bit analogue to digital converter that stores the

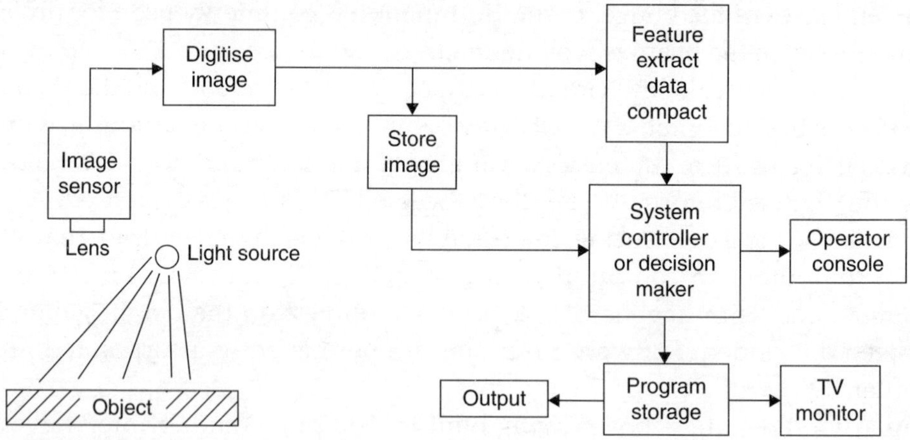

Figure 5.20 The vision system

digitised image in the main memory. Some systems use an analogue computer and a computer-controlled threshold to convert the video information to a binary image, rather than a 6–8-bit greyscale image. The data compactor provides high-speed processing of the input image data. Pattern recognition algorithms are used to generate a simple feature data set. The system control computer makes the decisions about the part being inspected.

The two basic vision systems are edge finders and correlators. Edge finders look for transitions from black to white, white to black, or from grey to non-grey level. When such a transition occurs, the edge finder system notes where it occurs. When a second transition occurs, it is also noted and the number of pixels between the transitions is noted. If a measurement is desired, then the system will have been calibrated and the physical distance between the two transitions can be computed.

Correlators are template matching systems. In this system, a complex image is stored in memory and the correlator searches the scene for a match to the image stored in memory. The template (complex image) is stored in memory as a two-dimensional matrix, called the reference. When a frame of video is loaded by the camera, the system overlays the reference on the upper-left corner of the video and calculates the number of matches.

This process, referred to as correlation, continues until the reference has been compared with the video throughout the field of view. The X and Y coordinates of the position where the best match occurs become the output. If there is no acceptable match anywhere in the field of view, another reference may be used and the process is repeated. This process continues until the object is identified and located.

For checking surface finish, the surface is treated as a kind of mirror. The vision system looks at an image of a light source reflected on the surface, and changes to that image signify a defective area. It is also possible to look for defects directly, if the defects can be relied upon to have a distinct contrast with the correct surface.

SUMMARY

- When two rays of the same wavelength meet at some point, mutual interference occurs and the nature of the interference will depend on the phases of the two waves at their meeting point. If the waves are in the same phase, they will reinforce each other and the resulting intensity will be the sum of the two intensities. If the waves are out of phase, the resultant intensity will be the difference of the two. If both intensities are equal, then in the first case, the intensity becomes twice the amplitude and the result will be a bright spot. In the second case, the intensity will be zero and the result will be a dark spot.
- A laser beam is monochromatic, coherent and naturally collimated; it is, therefore, used in interferometry.
- Laser interferometry can be used to measure even small dimensions of an object. The small diameter of an object can be computed using the expression $\lambda L/\delta x$, where λ is the wavelength of the laser beam, L is the distance between the object and the screen and δx is the mean distance between the dark minima on the screen.
- An optical flat is an optical-grade piece of glass lapped and polished to be extremely flat on one or both sides, usually within a few millionths of an inch. It is used with a monochromatic light, usually laser, to determine the flatness of other optical surfaces by the concept of interference.
- A touch trigger probe is a contact probe that senses the element to be measured when it reaches the element and transmits the relative data to the computer for storage and processing.
- The difference between conventional metrology and coordinate metrology is that the former is manual and time consuming and alignment of the work piece is necessary, whereas the latter requires less time and alignment of the work-piece is not necessary.
- The advantages of a coordinate measuring machine are minimisation of operator error; improved accuracy; reduced operator skill requirements;

uniform inspection quality; smaller set-up time; faster rate of inspection; minimum fixturing costs; traceability of compliance to specifications; and reduction in off-line analysis time.

- When the dimensions of a part need to be generated in a CNC machine, the part is placed in the CMM and touched by a probe on the surfaces; the coordinates are then digitised to generate data for CNC programming, which produces the part. This principle finds application in reverse engineering.
- A laser telemetric system is a non-contact gauge that measures using a collimated laser beam. It consists of three components: a transmitter, a receiver and processor electronics.
- Angular measurements, namely, pitch and yaw, are measured using a set of optics that includes an angular interferometer and an angular reflector.
- Machine vision is defined as a technique that allows a sensor to view a scene and to derive a numerical or logical decision without further human intervention.

EXERCISES

Multiple Choice Questions

1. Light is
   a) Propagated by electromagnetic waves b) Represented by a sine curve
   c) A form of energy d) All of the above

2. At what angle is a glass plate reflector set in the NPL interferometer?
   a) 30° b) 45°
   c) 60° d) 90°

3. Which type of fringe is formed, if a path difference of $\lambda/2$ is observed in an interference pattern?
   a) Bright fringe b) Dark fringe
   c) Both bright and dark fringes overlap d) Unpredictable

4. Which type of surface has a fringe pattern like the one shown below?

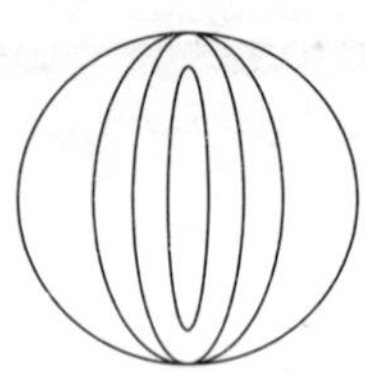

a) Cylinder
b) Conical
c) Spheroid
d) Sphere

5. Which type of light source is used in the NPL flatness interferometer?
a) Mercury vapour lamp
b) Cadmium lamp
c) Neon lamp
d) Nitrogen lamp

6. Which type of surface in a fringe pattern exhibits the movement of fringes towards the centre?
a) Convex surface
b) Concave surface
c) Flat surface
d) All of the above

7. What does the following interference band pattern indicate?

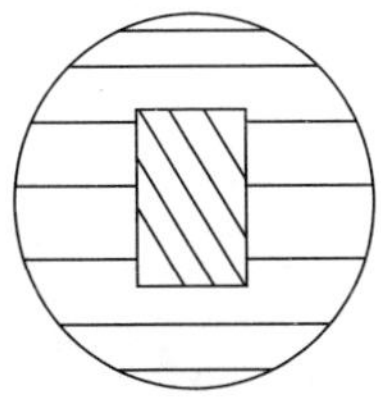

a) Flat gauge
b) Gauge has taper along its large edge
c) Gauge has taper along its short corner
d) Bands have the same direction but the pitch is different

8. Which one of the following is a property of laser?
a) Monochromatic
b) Coherent
c) Collimated
d) All of the above

9. The principle of laser is
a) Simulated emission
b) Spontaneous absorption
c) Induced emission
d) Both a and c

10. Optical flats are made of
a) Quartz
b) Glass
c) Plastic
d) Steel

11. An optical micrometer is used to
a) Measure surface profiles
b) Measure small linear displacements
c) Measure surface roughness
d) Set a very small displacement by rotating the glass block through relatively large angles

12. Optical flats are generally coated thin on the surface in order to

a) Reduce the light lost by reflection
b) Reduce the wear
c) Produce a shining surface
d) Give colour

13. An optical gauge works on the principle of
a) Reflection
b) Refraction
c) Dispersion
d) Interference of light rays

14. A stylus is
a) An input device in the CMM probe
b) An output device in the CMM probe
c) A measuring device
d) All of the above

15. Which one of the following is a contact-type automated inspection method?
a) Laser micrometer
b) Inspection probe
c) Machine vision
d) All of the above

Answers

1. d	2. b	3. b	4. c	5. a
6. b	7. c	8. d	9. b	10. a
11. d	12. a	13. d	14. a	15. b

Review Questions

1. Write down the reasons for finding laser more suitable in metrology applications.
2. Explain the principle of interference with suitable sketches.
3. What are the important properties of laser in interferometry?
4. Explain in detail how the dimensions of a small object are measured using laser interferometry.
5. Explain how larger displacement of an object can be measured using a laser interferometer.
6. Briefly explain about optical flats and their types.
7. What are the precautionary steps to be taken while using optical flats?
8. Explain how interference bands are achieved using optical flats.
9. What is the importance of using laser in precision instruments?
10. Explain how straightness is measured using an AC laser interferometer.
11. Explain in detail the various methods of testing the accuracy of a horizontal milling machine and lathe using a laser interferometer.
12. How are pitch and yaw errors in the X-axis table movements of a vertical milling machine measured? *Hint*: Pitch is rotation about the axis normal to the

saddle motion but parallel to the bed of the machine. Yaw is rotation about the normal to the saddle motion and to the face of the bed.

13. Explain how optical flats find use in metrology. Explain the principle of formation of interference fringes.
14. Write about precision instruments based on laser.
15. What is a laser micrometer?
16. Explain, with a neat block diagram, the principle of operation of an optical micrometer.
17. Explain the working principle of a laser scanning micrometer, with a block diagram.
18. Write down the various applications of a laser micrometer.
19. With the help of a schematic diagram, explain in detail the working principle of a laser telemetric system.
20. Explain how moire fringes are used in linear measurement.
21. Explain briefly how radial gratings are used in angular measurement.
22. What are the advantages of laser interferometer over diffraction gratings in linear measurement?
23. A slip gauge is checked in a flatness interferometer for its flatness. The number of fringes observed in the 0″ position and the 180″ position are 6 and 10, respectively. If the wavelength used is 0.5 um, what is the gauge taper?
24. Explain, with suitable sketches, how angular measurement can be carried out with the help of a laser interferometer.
25. Distinguish between coordinate and conventional metrology.
26. List the factors affecting the functions of CMM that are influenced by temperature.
27. What is a CNC CMM? How does it differ from conventional CMM?
28. What are the types of CMM?
29. What are the benefits of using CMM?
30. State the applications of CMM in reverse engineering.
31. Describe the steps involved in measurement using CMM.
32. How does measurement using CMM differ from conventional measurement?
33. State the possible sources of errors in CMM.
34. What are the digital devices used in advanced metrology?
35. Explain the constructional details and working principle of a three-axis CMM and a touch trigger probe.
36. Explain the need for computers in inspection.
37. Explain how image processing techniques are used in measurement.
38. What is a vision system?
39. What are the applications of vision system in metrology?
40. Describe the process of computer-aided inspection and machine vision.

Engineering Measurement

6

Measurement of Force, Strain and Pressure

6.1 MEASUREMENT OF FORCE

As defined by Newton's second law, force is the product of the mass of a body or object and its acceleration.

$$\text{Force (in newton)} = \text{Mass (in kilograms)} \times \text{Acceleration (in m/s}^2\text{)}$$

In the SI system, mass is considered as a fundamental quantity, while acceleration is a derived quantity. For the measurement of force, two basic approaches are used: direct comparison and indirect comparison.

Direct comparison methods: In the direct comparison methods, a beam-type balancing system with a null balance technique is used to measure the force. If there is no amplification (magnification) or attenuation (reduction) of the measurand force by the balancing system, in order to balance the high value of force with a small weight, the method is said to be a direct comparison.

Indirect comparison methods: In the indirect comparison methods, a smaller weight is used to balance the unknown measurand force by attenuation through a system of levers in the balancing system.

6.1.1 EQUAL ARM BALANCE

This is the simplest force or weight measurement system and it functions on the principle of comparison of the turning moments caused by the unknown force and a standard mass placed on two ends of a pivoted beam. A schematic diagram of the equal arm balance is shown in Fig. 6.1.

It consists of a beam which is pivoted on a knife edge at the centre, on which the unknown force F_1 to be determined and the known mass F_2 are placed at either end. A pointer with scale arrangement is provided at the pivot position and is perpendicular to the beam. As a result of the force applied on the arms of the balance, two moments F_1l_1 (or $m_1.g.l_1$) and F_2l_2 ($m_2.g.l_2$) are induced on the beam. As the name suggests, this system consists of arms of equal length (i.e., $l_1 = l_2$).

When null balance is read on the scale pointer, it indicates that the two forces are equal. In an equal arm balance system, F_1 (or m_1g) = F_2 (or m_2g), as $l_1 = l_2$ at true null balance. The equal arm balance system is best suited for the measurement of any unknown force (or weight) in comparison with a known force or weight. The accuracy or symmetry of the equal arm balance system can be easily verified by interchanging the two forces or weights between the two arms. If the system produces the same results after interchanging, it is understood that the arms of the beam are in symmetry about the pivot point.

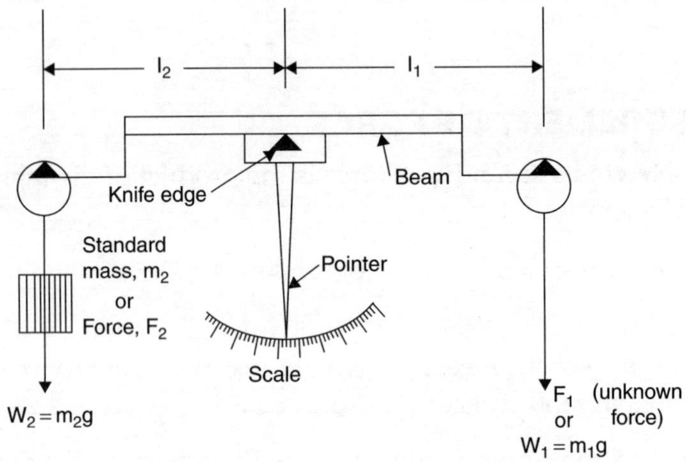

Figure 6.1 Equal arm balance

6.1.2 Analytical Balance

An analytical balance is a type of equal arm balance and is shown in Fig. 6.2. It comprises a rigid beam AB, which is pivoted by a fulcrum at point O. The arms of the balance will swing about the fulcrum, when an unbalanced condition exists between the two weights or forces applied on both ends A and B. Like in the equal arm balance, a pointer and scale arrangement is provided to read the degree of unbalance between the force or weight applied on the two arms.

Let W_G be the weight of the arm that acts on its centre of gravity *G*. W_A (known) and W_B (unknown) are the two forces or weights applied at the ends of the balance. X_A and X_B are the lengths of the two arms on the left- and right-hand sides of the pointer. If unbalance exists between the two forces, the arm will swing about its fulcrum O to an angle θ.

Considering equilibrium about O

$$W_G \cdot X_G = W_B \cdot X_B - W_A \cdot X_A \tag{6.1}$$

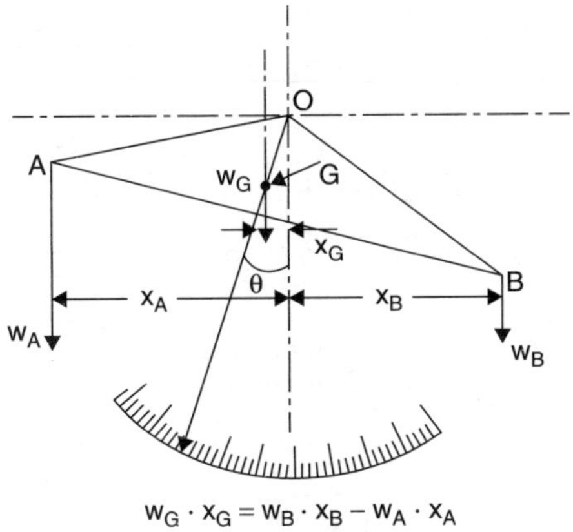

Figure 6.2 Analytical balance

For accurate measurements, the buoyant force of air acting upwards against the two weights should be corrected if it is considered significant. The sensitivity of the analytical balance is measured by,

$$S = \frac{\text{Deflection}}{\text{Unbalanced force}} = \frac{\theta}{(W_A - W_B)}$$

6.1.3 Unequal Arm Balance

The main drawback in the equal arm balance system is that it needs a set of known weights for all ranges of measurements to be performed. For example, if an unknown force or weight equal to 500 N is to be measured, a known weight of 500 N is essential for comparison. Similarly, for all individual forces to be measured, separate equivalent standard weights are required. However, in unequal arm balances, the heavier weights can be measured by using lighter weights. The schematic of the unequal arm balance is given in Fig. 6.3.

The unequal arm balance, as the name suggests, consists of two unequal arms, namely, the load arm which is shorter and the power arm which is longer. The power arm has a graduated scale and a sliding counterbalancing mass m. There are two knife edges provided at points p and q. The force or weight to be measured F_t is applied on the load arm and the counterbalancing or counterpoising force is induced with the help of an adjustable sliding mass m on the power arm to balance the beam.

Considering equilibrium about q

$$F_t \times l_1 = F_s \times l_2$$

$$F_t = F_s \times \left(\frac{l_2}{l_1}\right)$$

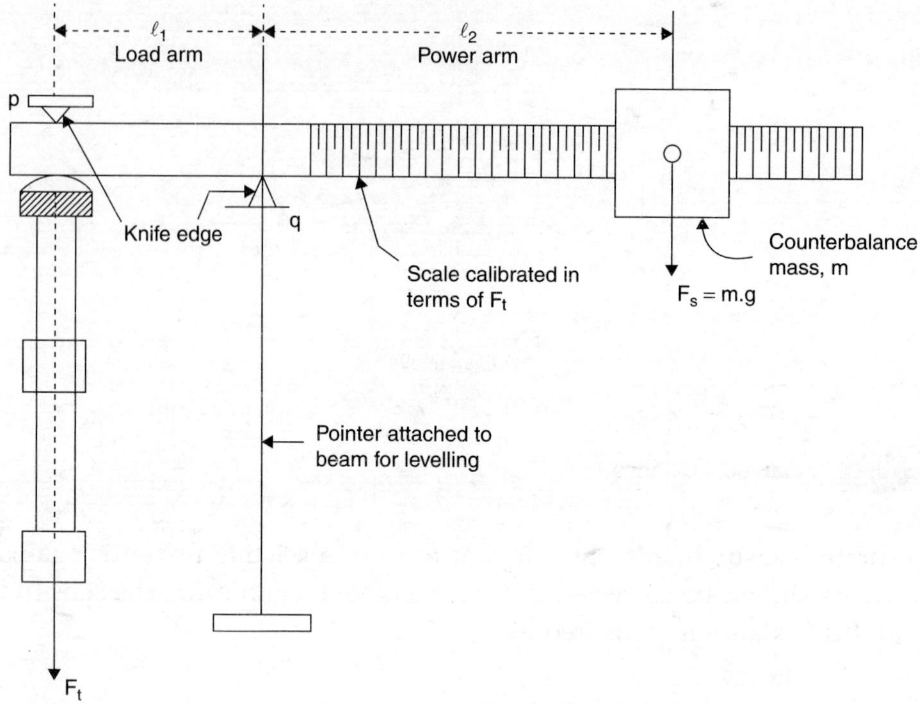

Figure 6.3 Unequal arm balance

$$F_t = m \times g \times \left(\frac{l_2}{l_1}\right) \tag{6.2}$$

If F_t is caused by an unknown mass $m_{1,}$ then

$$m_1 \times g = m \times g \times \left(\frac{l_2}{l_1}\right) \tag{6.3}$$

The distance between the pivot points p and q is always fixed. However, the distance between the pivot q and the counterbalance mass m varies according to the force or weight applied on the load arm.

Therefore, F_t = a constant value × l_2 (6.4)

This shows that the test force is proportional to the distance between the pivot q and mass m, which is variable.

6.1.4 Multiple Lever Systems

The equal arm and unequal arm balances are not capable of measuring large forces or weights. For such cases, multiple lever systems are used. A schematic diagram of a multiple lever system is given in Fig. 6.4.

In the multiple lever system, the load is applied on a horizontal platform provided at the bottom. The load applied on the platform is transferred to the measuring beam with the help of linkages. The beam is pivoted about a knife edge O, which divides it into two arms of length l_1 and l_2. The right arm has a graduated scale on which a sliding weight, known as poise weight W_p, is placed.

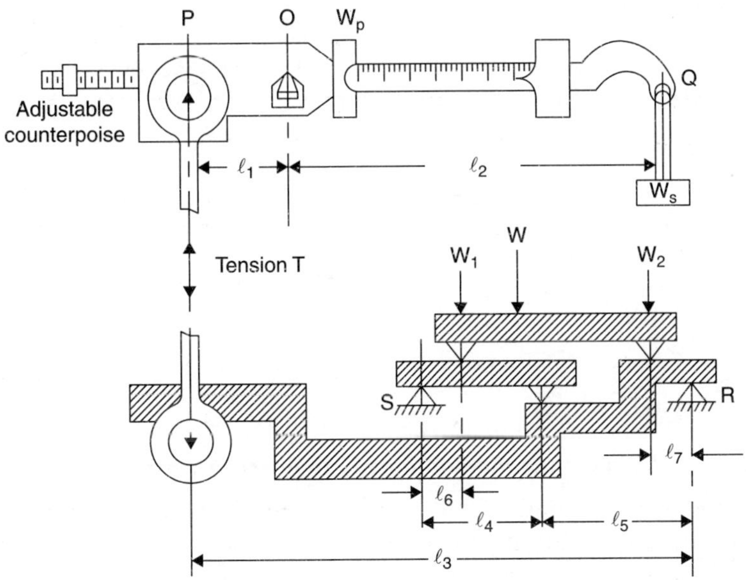

Figure 6.4 Multiple lever system

At the end of the right arm, the pan weight W_p is applied on a hanging support. As there are many levers in the multiple lever system, the mechanical advantage is high. Therefore, a large weight W is measured by a combination of the smaller poise weight W_p and pan weight W_s. An adjustable counterpoise is provided to set an initial zero balance in the system. For the purpose of analysis, let us assume that the weight W is replaced by two equivalent weights W_1 and W_2. W_p is set at the zero graduation on the beam scale and the counterpoise is adjusted for initial zero balance. The unknown load W, placed on the platform (or its equivalent W_1 and W_2), is perfectly balanced by the pan weight W_s.

Considering equilibrium about the pivot O

$$T \times l_1 = W_s \times l_2 \tag{6.5}$$

Considering equilibrium about the pivot R

$$T \times l_3 = \left(\frac{W_1 \cdot l_6}{l_4}\right) \cdot l_5 + W_2 \cdot l_7 \tag{6.6}$$

If the proportion of the linkage is arranged as

$$\frac{l_7}{l_5} = \frac{l_6}{l_4}$$

Then $T \times l_3 = l_7 (W_1 + W_2) = l_7 \cdot W$ (6.7)

From Eq. 6.7, it is understood that the weight W may be placed at any location on the loading platform without affecting the results of measurement.

Comparing Eqs 6.5 and 6.7, we get

$$\frac{W_s \cdot l_2}{l_1} = \frac{W \cdot l_7}{l_3}$$

$$W = \left(\frac{l_2 \cdot l_3}{l_1 \cdot l_7}\right) W_s$$

$$W = \text{a constant } R \times W_s \tag{6.8}$$

where $R = (l_2 . l_3 / l_1 . l_7)$ is called the **scale multiplication ratio**.

Each division on the beam scale for any poise weight W_p is graduated by W_p / l_2. Measurements in the multiple lever system are achieved by adjusting the poise and pan weights in combination.

6.1.5 Pendulum Force Measuring Scale

This is a multiple lever-type force measuring device in which the levers are replaced by ribbon- or tape-connected pivoted sectors. The schematic of the pendulum force measuring scale is shown in Fig. 6.5. The system consists of a loading platform at the bottom, which is connected to two sector arms that pivot on the two ends of an equaliser bar.

The link is made with the help of load rod and loading tapes. The sectors are pivoted to the equaliser bar in such a manner that it rotates about the fulcrum A by the application of load on the platform. A counterweight is attached to each sector. This will create effective moment on the equaliser bar according to the position of the sector at different loads.

The force to be measured is applied on the platform. Due to the applied force, the pivoted sectors rotate. This, in turn, moves the counterweight in the outward direction. Due to this movement, the effective moment of the counterweight increases until equilibrium is achieved between the applied force and the balance.

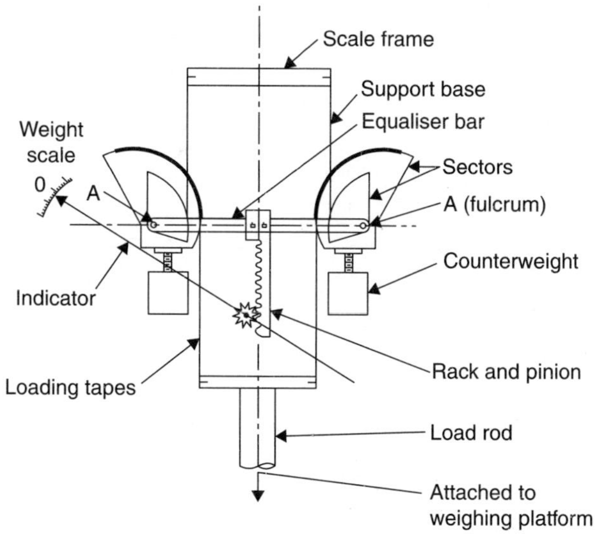

Figure 6.5 Pendulum force measuring scale

The displacement of the equaliser bar is transmitted to the indicator dial by the movement of the rack and pinion connected to it. This mechanism is best suited for force measuring systems like dynamometers.

6.1.6 Elastic Force Meters

Elastic force meters measure the deflection caused by force on the mechanical elastic member or combination of members. The deflection is always linearly proportional to the force applied, which is observed directly or through the secondary transducer that converts the input into an electrical form of output.

$$\text{Stiffness}, k = \frac{\text{Force}, F}{\text{Displacement}, y}$$

To determine the stiffness of an element, only the deflection equation is required. If the deflection is a linear function of load, then the stiffness equation is necessary. Both the deflection and stiffness equations for various types of elastic members are given in Table 6.1.

The requirements of an elastic member are:

- It should deflect considerably under load.
- It should have sensitivity in response time, to reach equilibrium at each loading.
- The elastic limit of the material should be higher than the load limit.

Table 6.1 Deflection and stiffness equations for various elastic members

Number	Figure	Elastic element	Deflection equation	Deflection constant, k
A	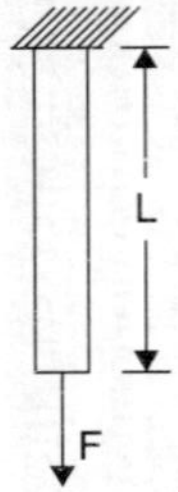	F = Load L = Length A = Cross-sectional area y = Deflection at load E = Young's modulus	$y = \frac{FL}{AE}$	$k = \frac{AE}{L}$
B	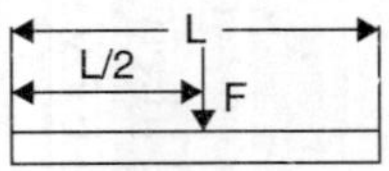	F = Load L = Length E = Young's modulus I = Moment of inertia	$y = \frac{1}{48}\frac{FL^3}{EI}$	$k = \frac{48EI}{L^3}$
C	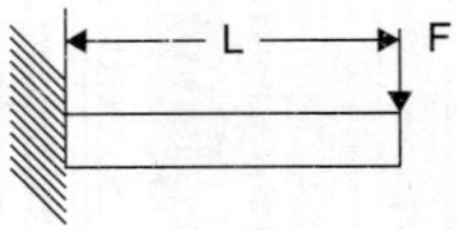	F = Load L = Length E = Young's modulus I = Moment of inertia	$y = \frac{1}{3}\frac{FL^3}{EI}$	$k = \frac{3EI}{L^3}$
D	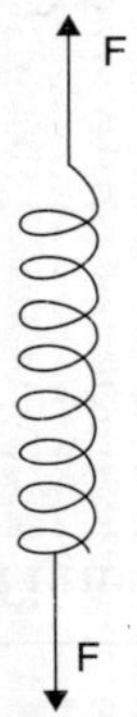	F = Load D_m = Mean coil diameter N = Number of coils E_s = Shear modulus D_w = Wire diameter	$y = \frac{8FD_m^3N}{E_SD_w^4}$	$k = \frac{E_SD_m^4}{8D_m^3N}$

(*Continued*)

Table 6.1 Continued

Number	Figure	Elastic element	Deflection equation	Deflection constant, k
E	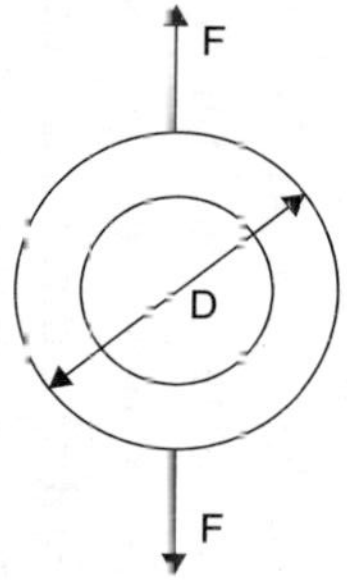	F = Load D = Diameter of the ring E = Young's modulus I = Moment of inertia of the section about the centroidal axis of the bending equation	$y = \frac{1}{16}\left(\frac{\pi}{2} - \frac{4}{\pi}\right)\frac{FD^3}{EI}$	$k = \frac{16}{\left(\frac{\pi}{2}\right) - \left(\frac{4}{\pi}\right)}\left(\frac{EI}{D^3}\right)$
F	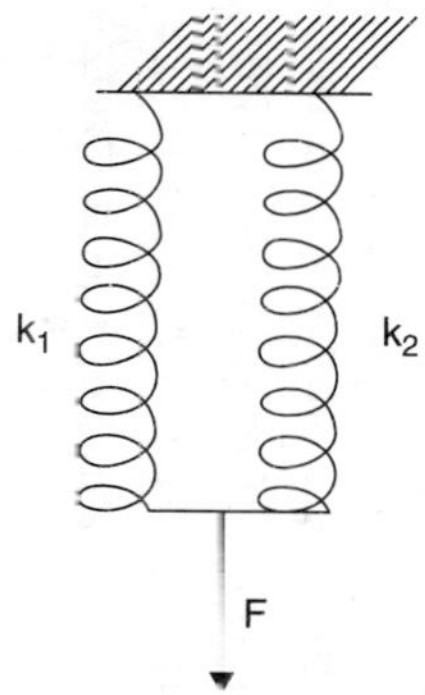	k_1 = Deflection constant of member 1 k_2 = Deflection constant of member 2 F = Load	$y = \frac{F}{k_1 + k_2}$	$k = k_1 + k_2$
G		k_1 = Deflection constant of member 1 k_2 = Deflection constant of member 2 F = Load	$y = F\left(\frac{1}{k_1} - \frac{1}{k_2}\right)$	$k = \frac{1}{\left(\frac{1}{k_1}\right) + \left(\frac{1}{k_2}\right)}$

6.1.7 Proving Ring

The proving ring consists of two main elements, the ring itself and the diameter measuring system. The proving ring is a circular ring made of steel with a thin rectangular cross-section. Deflection of the ring, when a compressive or tensile load is applied on it, is used to measure the load. The deflection is measured by precision and repeatable micrometer settings. To measure the deflection of the ring, the vibration reed is plucked and the micrometer spindle is advanced till the damping of vibration is noticed, which indicates contact of the micrometer spindle with the reed. Figure 6.6 shows a typical proving ring.

Force is applied to the ring through the external bosses. The resulting change in diameter, referred to as the deflection of the ring, is measured with a micrometer screw and the vibrating reed mounted diametrically within the ring.

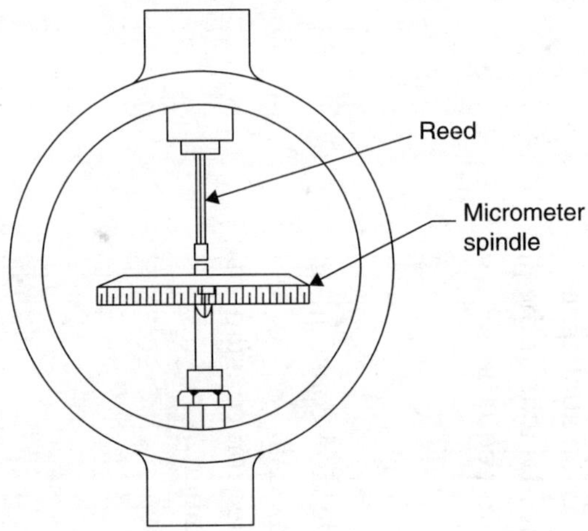

The micrometer screw and the vibrating reed are attached to the internal bosses of the ring. To read the diameter of the ring, the vibrating reed is set in motion by gently disturbing it. As the reed vibrates, the micrometer screw on the spindle is adjusted until the button on the spindle comes in contact with the vibrating reed, dampening out its vibrations. When this occurs, a characteristic buzzing sound is produced. At this point, a reading of the micrometer dial indicates the deformed diameter of the ring. The equation given for No: E in Table 6.1 can be used to calculate the load.

6.1.8 Hydraulic Load Cell

The load or force is measured by a hydraulic load cell by sensing the pressure that develops in the fluid confined inside a diaphragm to balance the load applied on other side. The pressure of the fluid is proportional to the force applied and can be determined by any pressure measuring gauge. The sectional view of a typical hydraulic cell is shown in Fig. 6.7.

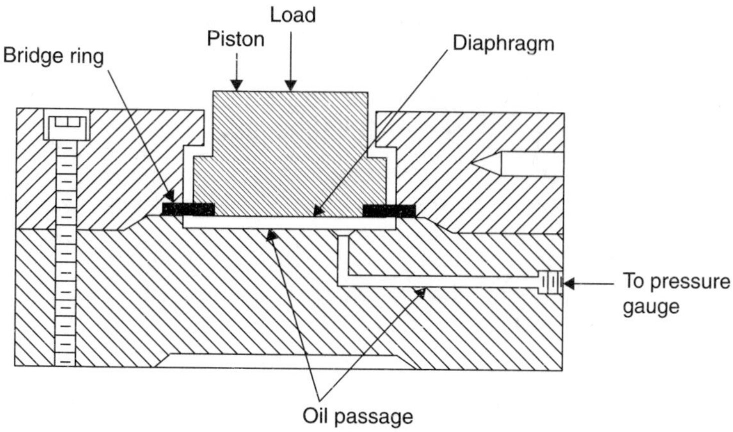

The arrangement consists of a thin elastic diaphragm. On one side of the diaphragm, a conventional piston and cylinder assembly is provided to apply load, and on other side of the diaphragm, oil is confined inside a closed space. When the load is applied on the piston, the pressure is transmitted to the pressure sensing device directly or by electric transducers which convert the input into an electric output. The piston does not have physical contact with the oil and the cylinder wall. The thin elastic diaphragm and bridge ring made of steel act as a positive seal. Mechanical stops are provided to prevent overstrain of the seal.

If the system is filled completely with oil, the piston movement may be small. This allows the system to provide very good dynamic response. The overall response of the system largely depends on the response of the pressure sensing gauges. Very high capacity and accuracy is possible with the hydraulic load cell.

6.1.9 Pneumatic Load Cell

This type of load cell is similar to hydraulic cells in the principle of operation. It consists of a closed chamber covered with a flexible diaphragm on one side. Above the diaphragm, a load plate is placed, on which the unknown force is to be applied. The chamber has a port for constant pressurised air supply from the compressor, a bleed valve, and a port for connecting the load cell with the pressure measuring instrument. Pressurised air is supplied to one side of the diaphragm and allowed to escape through the bleed valve. The position of the bleed valve is automatically governed by the movement of the flexible diaphragm. The sectional view of a typical pneumatic cell is shown in Fig. 6.8.

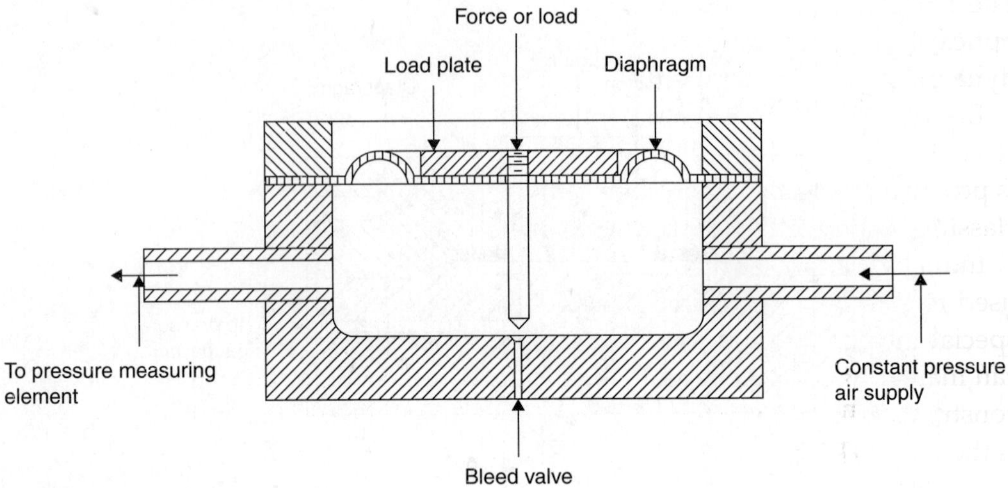

Figure 6.8 Pneumatic load cell

When the force or load is applied on the load plate, the diaphragm yields in the downward direction in proportion to the load, and the bleed valve position is automatically controlled with respect to this diaphragm position.

Therefore, the pressure in the chamber under the diaphragm is controlled both by the source pressure and by the position of the bleed valve. The pressure that develops inside the load cell will balance the force or load applied on the load plate and the pressure measured in this equilibrium condition will be proportional to the load applied. As the simple design pneumatic cell is dynamically unstable, some forms of viscous dampers are provided in commercial types to improve stability. Further, additional chambers and diaphragms are provided to counter tare adjustments.

6.2 MEASUREMENT OF STRAIN

A structural member, when subjected to any kind of external force or load, becomes deformed. The measure of such deformation in proportion to the original dimension is called strain.

For a simple loading

$$\text{Strain}, \in = \frac{\text{Change in length}, \Delta L}{\text{Original length}, L}$$

Earlier, strain was measured by measuring the displacement ΔL of the strained member with the help of an apparatus called the extensometer. This device is incorporated with a mechanical or optical lever system to sense the displacement. The Huggenberger and Tuckerman extensometers are modern mechanical and optical types, respectively. These are mostly used for static measurements. For dynamic strain measurements, electric strain gauges are the most suitable.

Electric strain gauges work on the principle that when a metal conductor is strained, a considerable change in its electric characteristics is observed and this is proportional to the magnitude of the strain. Electric strain gauges are generally classified as inductive, capacitive, piezoelectric or resistive.

Inductive strain sensitive gauges have high mass and are large, and are mostly used for permanent installations. The capacitive type of strain gauge is used for special-purpose applications. Both capacitive and inductive are more rugged and can maintain calibration for a longer duration. Piezoelectric gauges are smaller in construction but do not have long-term stability. The resistive type of strain gauge is the most widely used.

6.2.1 Metallic Resistance Strain Gauges

When the length of a metallic wire or foil is stretched, its length increases and cross-section reduces. A proportional change in electric resistance is also observed. If a metallic resistance element is attached to a strained member in such a position that the element is also strained, then the measured value of the change in resistance in the resistance element can be calibrated to the rate of strain. The relationship between the changes in electrical properties corresponding to the changes in mechanical properties is derived below.

Let us consider an electric conductor of length L having diameter D. Resistance of an electrical conductor is governed by the relation

$$R = \frac{\rho L}{A}$$

where R = resistance (in Ohms, Ω); L = length of the conductor (in meters, m); A = cross-sectional area of the conductor (in m^2); and ρ = resistivity of the conductor material (in Ω-m).

The cross-sectional area $A = \frac{\pi}{4} D^2$ for a circular cross-section.

Let us take $X = \frac{\pi}{4}$. Then, $A = X \cdot D^2$

Therefore $$R = \frac{\rho L}{A} = \frac{\rho L}{XD^2} \tag{6.9}$$

Differentiating the above equation,

$$dR = \frac{XD^2(Ld\rho + \rho dL) - 2X\rho LDdD}{(XD^2)^2}$$

$$dR = \frac{1}{XD^2}\left((Ld\rho + \rho dL) - 2\rho L\frac{dD}{D}\right) \tag{6.10}$$

Dividing Eqs 6.10 and 6.9, we get

$$\frac{dR}{R} = \frac{dL}{L} - 2\frac{dD}{D} + \frac{d\rho}{\rho} \tag{6.11}$$

Re-writing the above equation, we get

$$\frac{(dR/R)}{(dL/L)} = 1 - 2\frac{(dD/D)}{(dL/L)} + \frac{(d\rho/\rho)}{(dL/L)} \tag{6.12}$$

where $\frac{dR}{R}$ = change in resistance of conductor; $\frac{dL}{L}$ = axial stain, $\in_\ell$; and $\frac{dD}{D}$ = lateral strain, $\in_d$.

We know Poisson's ratio $\vartheta = \frac{\text{Lateral strain}}{\text{Linear strain}} = -\frac{dD/D}{dL/L}$

The ratio of change in resistance of the conductor to linear strain is defined as gauge factor F.

Therefore $$F = \frac{(dR/R)}{(dL/L)} = 1 + 2\vartheta + \frac{d\rho/\rho}{dL/L} \tag{6.13}$$

$$\in_\ell = \frac{1}{F}\cdot\frac{\Delta R}{R} \tag{6.14}$$

This is the basic equation for any resistance-type strain gauge. The value of gauge factor F and resistance R of the strain gauge is predetermined and will be provided by the instrument manufacturer. Table 6.2 shows the gauge factor and electrical parameters of various strain gauge grid materials.

Types of metallic resistance strain gauges: In general, metallic resistance strain gauges can be classified as:

- **Wire grid type**: A fine metal wire is formed in a grid pattern to minimise the length of the gauge.
- **Foil type**: The metal wire is superseded by foil.
- **Semiconductor type**: Non-metallic, made of semiconductor material.

Table 6.2 Properties of various grid materials

Grid material	Composition	Approximate gauge factor F	Approximate resistivity: Micro-ohm.cm	Approximate resistivity: Ohms per mil. foot	Approximate temperature coefficient of resistance ppm / °C	Maximum operating temperature °C (approx.)
Nichrome V*	80% Ni; 20% Cr	2	108	650	400	1100
Constantan*, Copel*, Advance*	45% Ni; 55% Cu	2	49	290	11	480
Isoelastic*	36% Ni; 8% Cr; 0.5% Mo; Fe remainder	3.5	112	680	470	–
karma*	74% Ni; 20% Cr; 3% Al; 3% Fe	2.4	130	800	18	815
Manganin*	4% Ni; 12% Mn; 84% Cu	0.47	48	260	11	–
Platinum - Iridium	95% Pt; 5% Ir	5.1	24	137	1250	1100
Monel*	67% Ni; 33% Cu	1.9	42	240	2000	–
Nickel		–12	7.8	45	6000	–
Platinum		4.8	10	60	3000	–

* Trade names

Bonded wire strain gauges: This is a wire grid-type strain gauge which consists of a fine resistance wire of about 0.025-mm diameter cemented between thin sheets in the form of a grid. The sheet material may be paper, Bakelite or Teflon. The sheet on which the grid is mounted is called the base (or carrier).

The base is bonded with adhesive to the element for which the strain has to be measured. Some of the common wire strain gauges are shown in Fig. 6.9.

The strain gauge wire should possess the following desirable properties:

- High value of gauge factor that results in high sensitivity of the gauge.
- High resistance value to overcome the effects of resistance in the measuring bridge circuit.
- Low resistance temperature coefficient.
- Good elastic limit.
- Linearity between the strain and the corresponding change in resistance.

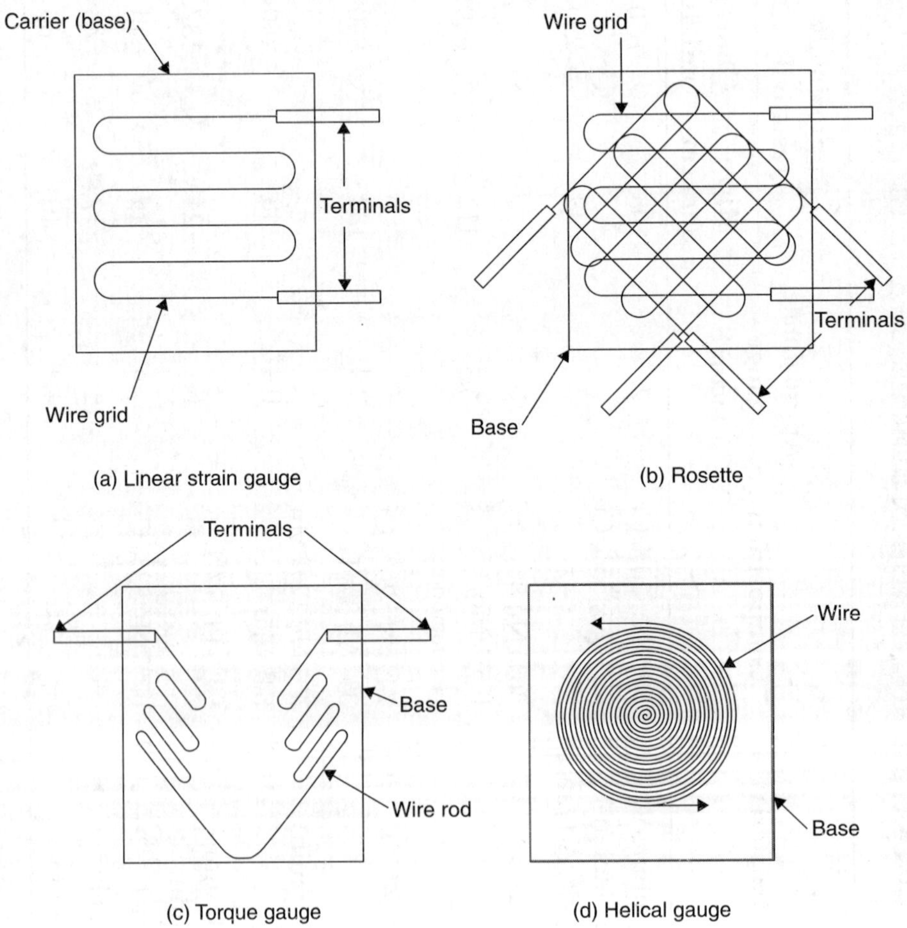

Figure 6.9 Resistance wire strain gauges

Metal foil strain gauges: This is similar to the wire grid strain gauge, except that the wire is replaced by thin foils. This type of gauge exhibits good heat dissipation capacity when compared to the wire grid type. The sensing element or foils are formed on a thin epoxy support using a photo-etching process, as is done in the manufacture of printed circuit boards.

The foils have a large surface area to cross-sectional area ratio and, therefore, possess good stability for long loading durations and extreme temperature conditions. The metal foil strain gauge is shown in Fig. 6.10.

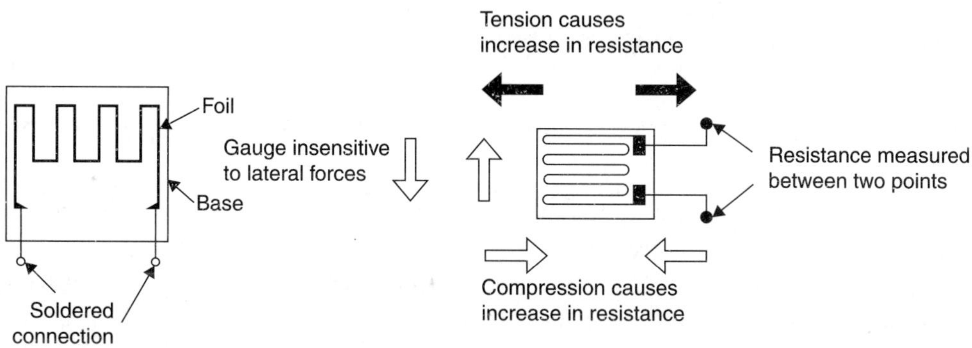

Figure 6.10 Metal foil strain gauge

Semiconductor strain gauges: Semiconductor materials exhibit accurate and considerable change in electrical resistance when subjected to mechanical strain. Semiconductor materials such as silicon or germanium, which are sliced into filaments, are used as sensing elements. The gauge factor of this type may be as high as 150 when compared to metallic gauges that have a range of 4 or 5. Micro-strains can be measured because of this high gauge factor.

The resistance of strain gauges made of P-type semiconductor material increases with increase in tensile strain, whereas that of N-type material decreases with increasing tensile strain. A schematic of a typical semiconductor strain gauge is shown in Fig. 6.11.

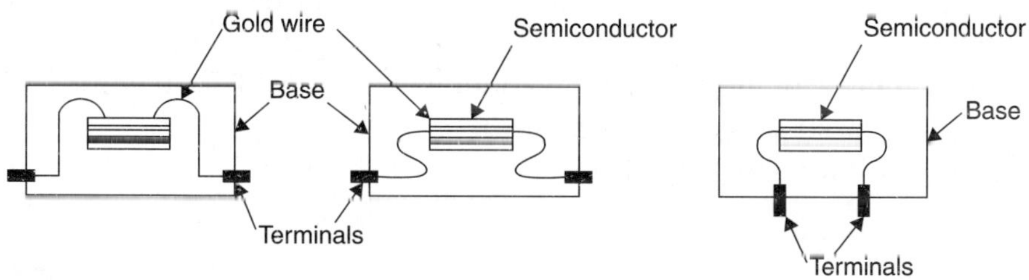

Figure 6.11 Semiconductor strain gauge

Though the semiconductor strain gauge has the advantage of high change in resistance, it is accompanied by the disadvantage of temperature sensitivity and linearity with strain. For example, if the gauge factor is 130 at 0.2 percent strain, it is 112 at 0.4 percent strain. This linearity problem can be improved by doping the semiconductor material used. Semiconductor strain gauges are more expensive and difficult to affix to the strained member being studied. The fatigue life of a semiconductor strain gauge is very high.

6.2.2 Circuitry for Strain Gauges

The sensitivity of a strain gauge is defined by

$$\varepsilon = \frac{1}{F} \cdot \frac{\Delta R_g}{R_g} \tag{6.15}$$

where ε = strain; F = gauge factor and R_g and ΔR_g = strain gauge resistance and change in resistance, respectively.

Let us consider a typical commercial gauge: the gauge factor is 2 and the resistance of the gauge in normal conditions is 120 Ω. If the strain to be measured is 1 ppm (0.000001), the corresponding change in resistance ΔR_g that must be measured in the gauge is calculated as

$$\Delta R_g = F \cdot R_g \cdot \varepsilon = 2 \times 120 \times 0.000001 = 0.00024\ \Omega$$

The percentage change in resistance is $\dfrac{0.00024}{120} \times 100 = 0.0002\%$

In order to measure such a small change in resistance, a circuit with high sensitivity is required.

Strain gauge ballast circuit: The schematic of a simple strain gauge ballast is shown in Fig. 6.12. If i is the current flowing through the circuit, then according to Ohm's law

$$i = \frac{e_i}{R_b + R_g} \tag{6.16}$$

where i = current; e_i = input exciting voltage; e_o = output voltage; R_b = ballast resistance; and R_g = strain gauge resistance.

Let e_o be the voltage across R_g. Then

$$e_0 = i \cdot R_g = \frac{e_i R_g}{R_b + R_g} \tag{6.17}$$

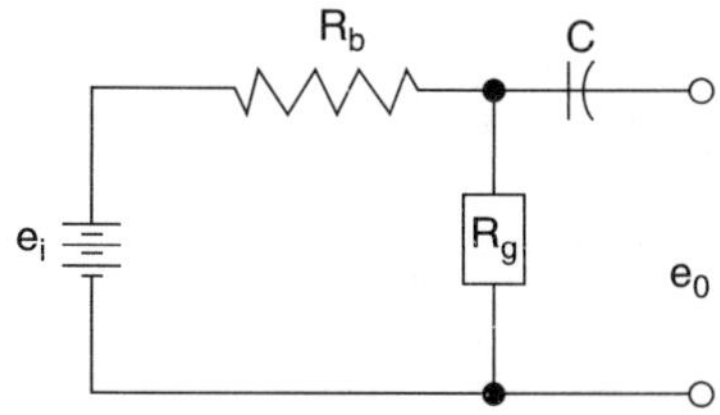

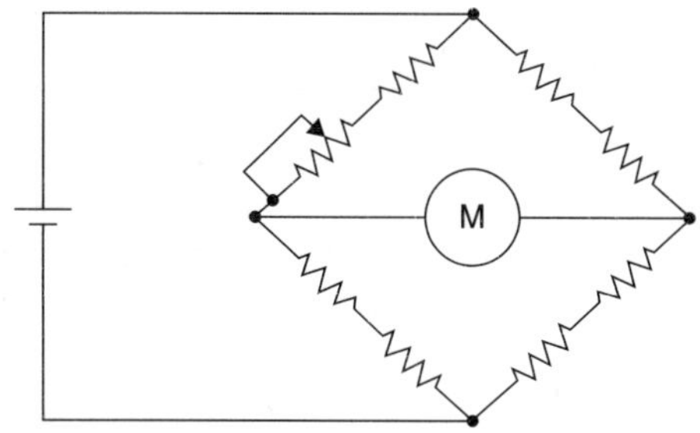

Figure 6.12 A simple strain gauge ballast circuit

Differentiating

$$de_0 = \frac{e_i R_b dR_g}{(R_b + R_g)^2} = \frac{e_i R_b R_g}{(R_b + R_g)^2} \frac{dR_g}{R_g} = \frac{e_i R_b R_g}{(R_b + R_g)^2} F\varepsilon \tag{6.18}$$

The limitation of the above ballast circuit is analysed with a numerical example. Let us consider a gauge resistance of 120 Ω (= R_g = R_b, a condition for maximum sensitivity), input exciting voltage of 8 V and gauge factor of 2.

Then

$$e_0 = \frac{(8)(120)}{120 + 120} = 4V$$

$$de_0 = \frac{(8)(120)(120)(2)(\varepsilon)}{(120 + 120)^2} = 4\varepsilon V$$

Therefore, for a strain of 1 μ, the gauge must sense a variation of 4 μV, which is 0.00010% when compared with an output voltage of 4 V.

For this critical reason, the ballast circuit is not suitable for the measurement of static strain. However, the circuit could be used to measure dynamic strain if a capacitor is placed in the output lead. This will arrest the DC input exciting voltage and allow only the variable component of the voltage to pass through.

Strain gauge bridge circuit: The most common type of circuit used in measuring systems is the bridge circuit. A Wheatstone bridge is a typical example of a bridge circuit and is shown in Fig. 6.13.

A Wheatstone bridge consists of four resistor arms (R_1, R_2, R_3 and R_4), a voltage source (e_i) and a reading meter (a current-sensitive galvanometer). One or more resistors of the arms are replaced with resistance strain gauges whose resistance is to be determined when used in the application.

The measurements may be achieved by *balancing the bridge;* i.e., by balancing the voltage across the meter to zero using known resistance adjustments in one or

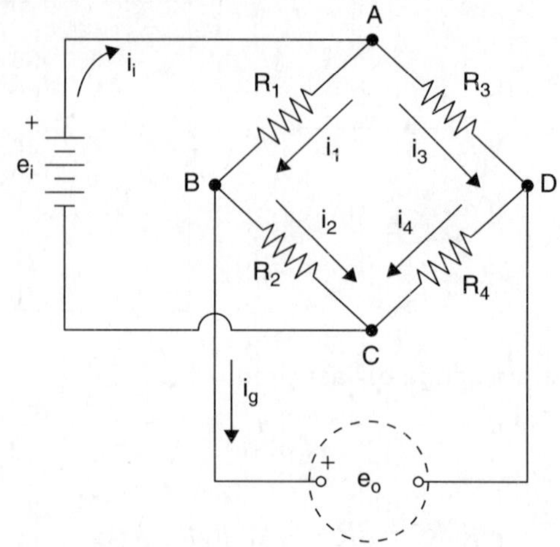

Figure 6.13 A Wheatstone bridge circuit

more arms, or *by determining the output voltage of the circuit due to unbalance in the resistance of the circuit*. These two methods are generally called null-type bridge and deflection-type bridge, respectively.

Null-type bridge: In Fig. 6.13, let us assume that R_1 is the resistance transducer of the strain gauge whose value has to be measured, and it varies according to variation in the strain. When the circuit is balanced, the voltage across the meter becomes zero and no current will flow through the circuit. Therefore, i_g is zero and the voltage at the junctions B and D will be equal; i.e., $e_b = e_d$ (or $i_1R_1 = i_3R_3$ and $i_2R_2 = i_4R_4$). Thus, we get

$$\frac{i_1R_1}{i_2R_2} = \frac{i_3R_3}{i_4R_4} \tag{6.19}$$

Also, it is understood in this condition that $i_1 = i_2$ and $i_3 = i_4$.

$$\text{Therefore} \quad \frac{R_1}{R_2} = \frac{R_3}{R_4} \tag{6.20}$$

Usually, the ratio of resistance R_3 to R_4 will always be fixed, and the bridge circuit is balanced by varying resistance R_2. When the three resistances are known, the resistance of R_1 may be calculated with the help of the above equation. To balance the arms, the resistance must be adjusted with the help of some arrangement so that the voltage in the circuit is balanced. If a large change in resistance in the strain gauge must be balanced, the sliding contact variable resistance transducers in a series arrangement, as shown in Fig. 6.14, may be used.

To accommodate small resistance changes in the strain gauges, the shunt balance may be used, as shown in Fig. 6.15.

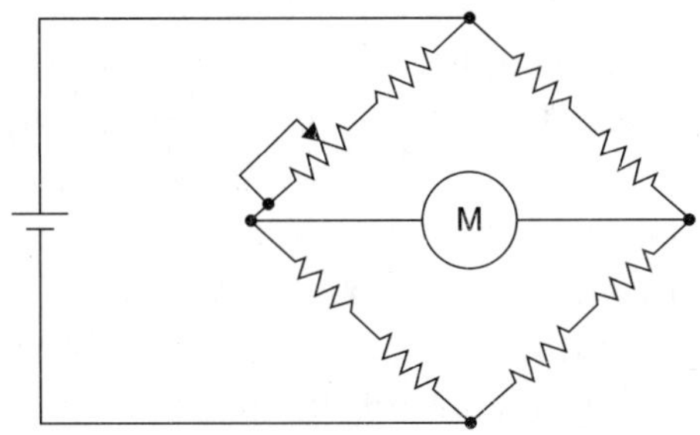

Figure 6.14 A series balance resistance bridge

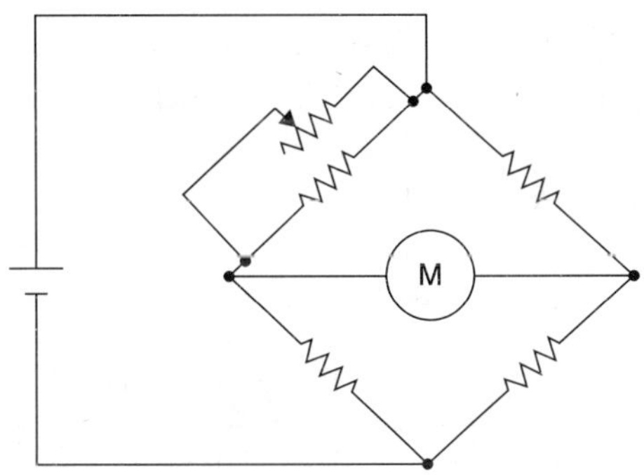

Figure 6.15 A shunt balance resistance bridge

The selection of the type of balance is based on the bridge sensitivity required for the application. To accommodate a range of varying resistances, a bridge with both series arrangement and shunt balance may be used.

Deflection-type bridge: The null-type bridge circuit is most suitable for static strain measurement. However, it would be difficult to measure frequently varying strain using this circuit. Therefore, for dynamic strain measurement, the deflection-type bridge is the most suitable.

In a deflection bridge, when the input resistance R_1 varies, it creates an unbalance in the circuit, which is measured by the deflection of the output meter. The scale of the meter is calibrated to indicate the measured value of resistance. Though balancing of the bridge is not required to measure the change in resistance at R_1, the deflection bridge circuit is provided with initial balancing arrangements, such as series or shunt balance (discussed earlier), to overcome the unbalance arising from manufacturing variations or for other reasons.

If the deflection bridge is used to measure the static strain, an ordinary voltmeter may be used as an output display meter, whereas for measuring dynamic strain, an oscilloscope or analog-to-digital convertor may be used. The output of the deflection bridge may be connected to a low-impedance or high-impedance device.

If the output is connected to a galvanometer (low-impedance output circuit), a significant amount of current i_g will flow through the output. Such a type of bridge is called a current-sensitive bridge. In certain cases, the output signal needs to be amplified by an amplifier circuit (high-impedance output circuit), and this will make the output current $i_g = 0$. As discussed earlier, this is a case when an oscilloscope or analog-to-digital convertor is used as an output meter.

Voltage-sensitive Wheatstone bridge: Consider Fig. 6.13, where the voltage difference between the two junctions B and D is given by

$$e_0 = e_B - e_D \tag{6.21}$$

Let us assume that due to high impedance of the meter connected to the circuit, the bridge output $i_g = 0$. Therefore, $i_1 = i_2$ and $i_3 = i_4$ and

$$e_0 = i_1 R_1 - i_3 R_3$$

$$i_1 = \frac{e_i}{R_1 + R_2} \quad \text{and} \quad i_3 = \frac{e_i}{R_3 + R_4}$$

$$\text{Therefore } e_0 = e_i \left[\frac{R_1}{R_1 + R_2} - \frac{R_3}{R_3 + R_4} \right] = e_i \left[\frac{R_1 R_4 - R_2 R_3}{(R_1 + R_2)(R_3 + R_4)} \right] \tag{6.22}$$

Let us assume that the resistance R_1 is used as a strain gauge transducer and its value varies by a small quantity ΔR_1.

$$e_0 + \Delta e_0 = e_i \left[\frac{(R_1 + \Delta R_1)R_4 - R_2 R_3}{(R_1 + \Delta R_1 + R_2)(R_3 + R_4)} \right]$$

$$e_0 + \Delta e_0 = e_i \left[\frac{(1 + \Delta R_1/R_1) - R_2 R_3/R_1 R_4}{(1 + \Delta R_1/R_1 + R_2/R_1)(1 + R_3/R_4)} \right] \tag{6.23}$$

If the resistances are assumed to be equal, i.e., $R_1 = R_2 = R_3 = R_4 = R$, then e_0 becomes zero and

$$\frac{\Delta e_0}{e_i} = \frac{\Delta R_1/R}{4 + 2(\Delta R_1/R)} \tag{6.24}$$

It is understood from the equation that the output voltage e_0 for any change in resistance ΔR_1 will be non-linear. If the actual change in resistance is too small $[2(\Delta R_1/R) << 1]$, the bridge will provide a linearised output.

$$\frac{\Delta e_0}{e_i} = \frac{\Delta R_1}{4R} \tag{6.25}$$

Current-sensitive Wheatstone bridge: As discussed earlier, when the deflection bridge output is connected to a low-impedance device, such as a current-sensitive galvanometer, there will be sufficient flow of current in the output of the bridge. To calculate the output current of the bridge i_g, which flows to the galvanometer, the resistance of the galvanometer R_g should be considered.

$$i_g = \frac{i_i(R_2R_3 - R_1R_4)}{R_g(R_1 + R_2 + R_3 + R_4) + (R_2 + R_4)(R_1 + R_3)} \tag{6.26}$$

If we assume that the transducer resistance of the strain gauge is R_1 and its value varies by a resistance $\Delta R_{1,}$ the above equation may be re-written as

$$\frac{\Delta i_g}{i_i} = \frac{-\Delta R_1/R}{4[1 + (R_g/R)] + [2 + (R_g/R)](\Delta R_i/R)} \tag{6.27}$$

6.2.3 Temperature Compensation for Strain Gauges

In general, all resistive strain gauges are sensitive to the prevailing temperature in the testing condition. Errors in measurement of strain may occur due to varying resistance of the wire grid of the strain gauge with respect to temperature and due to differential expansion between the strain gauge and the specimen material.

For accurate measurement of static strain, it is very important that proper temperature compensation be provided in the bridge circuit. The compensation for temperature effects can be accomplished in two ways: by providing adjacent-arm balancing or compensation gauges and by self-temperature compensation.

Adjacent-arm balancing gauge: Figure 6.16 shows a simple resistance bridge arrangement that has two strain gauges, of which one is sensitive to strain and the other is a dummy gauge. We know that the equation for bridge balance is

$$\frac{R_1}{R_2} = \frac{R_3}{R_4} \tag{6.28}$$

Let us assume that both the gauges 1 and 2 are similar and placed on similar materials so that there exists the same resistance change ΔR_t due to temperature variation. Then, the equation of bridge balance will become

$$\frac{R_1 + \Delta R_t}{R_2 + \Delta R_t} = \frac{R_3}{R_4} \tag{6.29}$$

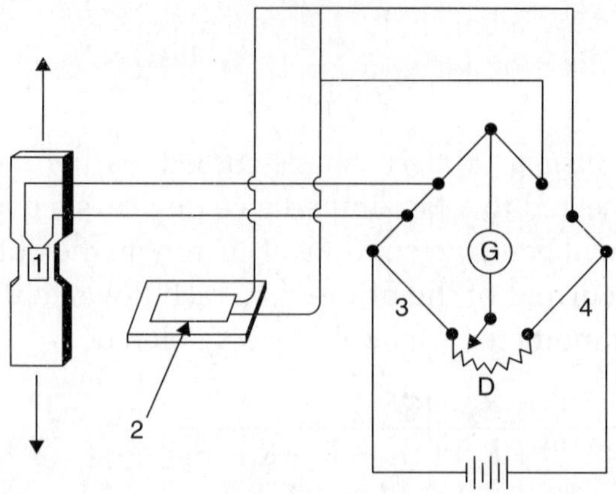

Figure 6.16 A simple resistance bridge arrangement

Thus, the ratio is unaltered; i.e., the resistance change due to temperature in the active gauge is compensated for by the resistance change in the dummy gauge.

Self-temperature compensation: In certain cases, when the temperature gradient or the temperature variation on the test surface is very high, it will become undesirable to place two gauges at a similar temperature. In some other cases, it may be difficult to use the bridge circuit itself for strain measurement, and a ballast circuit may instead be utilised for this purpose. In such situations, the possibility of providing adjacent-arm compensation is ruled out, and self-temperature compensation is the only possible way. Basically, there are two types of self-temperature compensated gauges: selected melt gauges and dual element gauges.

In the selected melt gauge, the temperature sensitivity of the grid material is controlled by selecting proper alloy materials, and its processing is carried out by a special cold-working process.

When the proper preparation method is used, the grid material exhibits very low apparent strain with respect to temperature change over certain temperature ranges. Therefore, for various temperature conditions, different manipulations of alloys with different strain properties must be selected.

In the dual element gauge, two different elements with different temperature characteristics are selected so that the overall strain effect due to temperature is minimised when the gauge is mounted for testing.

6.2.4 Strain Gauge Rosettes

A wire strain gauge can effectively measure strain in only one direction. To determine the three independent components of plane strain, three linearly independent strain measures are required, i.e., three strain gauges positioned in a rosette-like layout. A strain gauge rosette is an arrangement of two or more closely positioned gauge grids, separately oriented to measure normal strains along different directions in the underlying surface of the test part.

- **Two-element rosette gauge**: This type of rosette is formed by placing two strain gauges at 90° to each other. It is used to measure biaxial stress when the direction of principal stress is already known.
- **Three-element rosette gauge**: Where the directions of the principal strains are not known, a three-element rectangular or delta rosette is required. Three-element strain gauge rosettes are used when it is desired to measure ε_x, ε_y and ε_{xy}, induced at a point (or equivalently, when the principal strains and principal directions are unknown). Referring to the X–Y coordinate system in Fig. 6.17, the normal strain induced in an arbitrary direction from the X axis, defined by angle θ (strain ε_θ), is related to the strains in the X–Y coordinate system according to:

$$\varepsilon_\theta = \varepsilon_x \cos 2\theta + \varepsilon_y \sin 2\theta + \varepsilon_{xy} \cos\theta \sin\theta \tag{6.30}$$

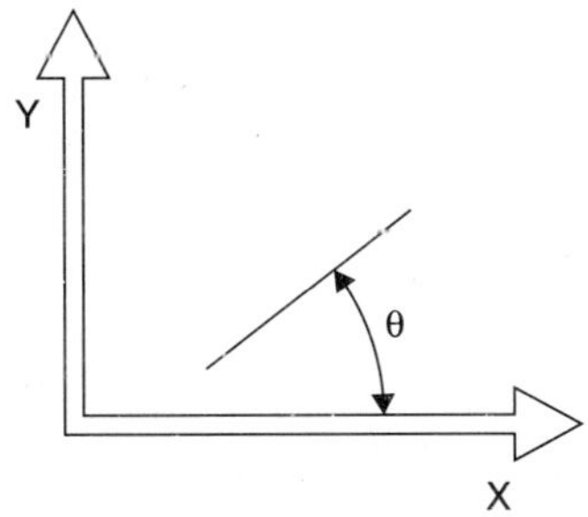

Figure 6.17 Single strain arbitrarily oriented to the X–Y coordinate system

The strain ε_θ can be measured by simply mounting a strain gauge in the direction defined by angle θ. Solving the above equation, ε_θ and θ are known, but there are still three unknowns, $\varepsilon_{x,}$ ε_y and Υ_{xy}. Thus, to solve for the unknowns, two more equations are required (i.e., three equations for three unknowns). If three strain gauges (Fig. 6.18) are mounted in three distinct and arbitrary directions (θ_1, θ_2 and θ_3), the same equation can be applied such that

$$\varepsilon_{\theta 1} = \varepsilon_x \cos^2\theta_1 + \varepsilon_y \sin^2\theta_1 + \Upsilon_{xy} \cos\theta_1 \sin\theta_1 \tag{6.31a}$$

$$\varepsilon_{\theta 2} = \varepsilon_x \cos^2\theta_2 + \varepsilon_y \sin^2\theta_2 + \Upsilon_{xy} \cos\theta_2 \sin\theta_2 \tag{6.31b}$$

$$\varepsilon_{\theta 3} = \varepsilon_x \cos^2\theta_3 + \varepsilon_y \sin^2\theta_3 + \Upsilon_{xy} \cos\theta_3 \sin\theta_3 \tag{6.31c}$$

These three equations are called general rosette equations. If θ_1 is set to 0, θ_2 is set to 45° and θ_3 is set to 90°, the resulting strain gauge configuration is called a rectangular rosette (Fig. 6.19). The resulting equations are

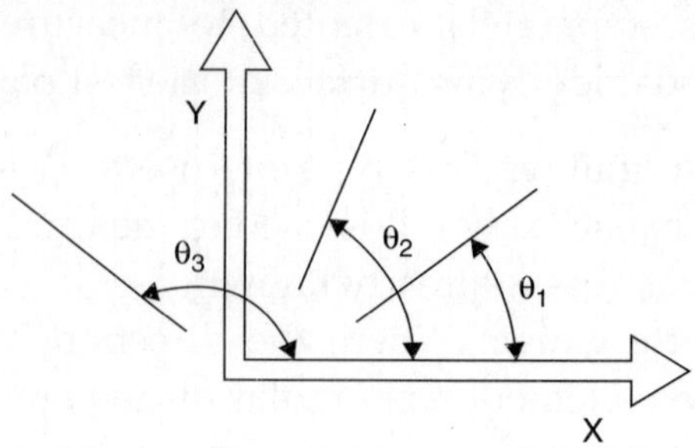

Figure 6.18 Three strains arbitrarily oriented to the X–Y coordinate system

$$\varepsilon_{0^\circ} = \varepsilon_x \cos^2(0^\circ) + \varepsilon_y \sin^2(0^\circ) + \Upsilon_{xy} \cos(0^\circ)\sin(0^\circ)$$

$$\varepsilon_{45^\circ} = \varepsilon_x \cos^2(45^\circ) + \varepsilon_y \sin^2(45^\circ) + \Upsilon_{xy} \cos(45^\circ)\sin(45^\circ)$$

$$\varepsilon_{90^\circ} = \varepsilon_x \cos^2(90^\circ) + \varepsilon_y \sin^2(90^\circ) + \Upsilon_{xy} \cos(90^\circ)\sin(90^\circ)$$

Finally

$$\varepsilon_x = \varepsilon_{0^\circ}$$

$$\varepsilon_y = \varepsilon_{90^\circ}$$

$$\Upsilon_{xy} = 2\varepsilon_{45^\circ} - (\varepsilon_{0^\circ} + \varepsilon_{90^\circ})$$

Similarly, if θ_1 is set to 0, θ_2 is set to 60° and θ_3 is set to 120°, the resulting strain gauge configuration is called a delta rosette. The resulting equations are

$$\varepsilon_{0^\circ} = \varepsilon_x \cos^2(0^\circ) + \varepsilon_y \sin^2(0^\circ) + \Upsilon_{xy} \cos(0^\circ)\sin(0^\circ)$$

$$\varepsilon_{60^\circ} = \varepsilon_x \cos^2(60^\circ) + \varepsilon_y \sin^2(60^\circ) + \Upsilon_{xy} \cos(60^\circ)\sin(60^\circ)$$

$$\varepsilon_{120^\circ} = \varepsilon_x \cos^2(120^\circ) + \varepsilon_y \sin^2(120^\circ) + \Upsilon_{xy} \cos(120^\circ)\sin(120^\circ)$$

Finally

$$\varepsilon_x = \varepsilon_{0^\circ} \tag{6.32a}$$

$$\varepsilon_y = \frac{1}{3}[2(\varepsilon_{60^\circ} + \varepsilon_{120^\circ}) - \varepsilon_{0^\circ}] \tag{6.32b}$$

$$\Upsilon_{xy} = \frac{2\sqrt{3}}{3}[\varepsilon_{60^\circ} - \varepsilon_{120^\circ}] \tag{6.32c}$$

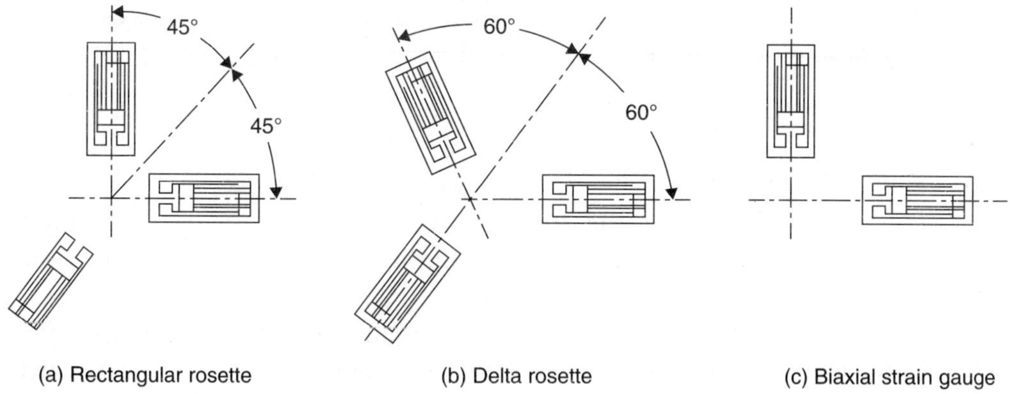

Figure 6.19 Strain gauge rosettes

6.3 MEASUREMENT OF PRESSURE

Pressure is defined as the normal force exerted by a fluid on a unit area of surface with which it is in contact. In SI units, it is expressed as N/mm^2. Pressure is measured with reference to absolute zero pressure (a complete vacuum), which is called **absolute pressure,** or to atmospheric pressure, which is called **gauge pressure**. Figure 6.20 shows the relationship between absolute pressure, gauge pressure and atmospheric pressure.

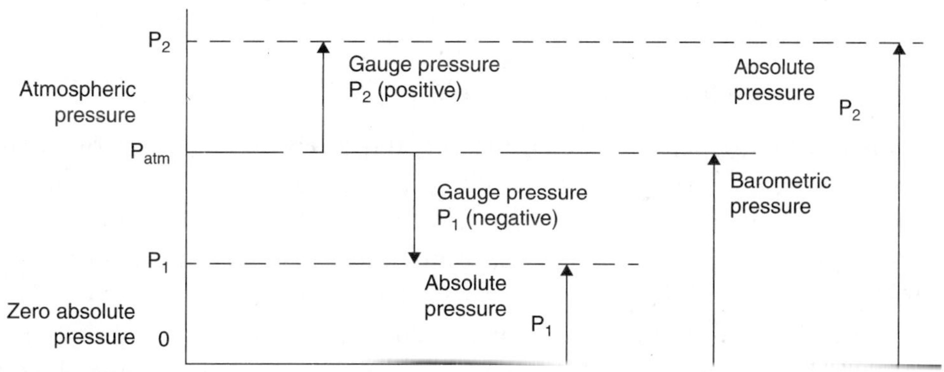

Figure 6.20 Relationship between absolute pressure, gauge pressure and atmospheric pressure

$$\text{Absolute pressure, } P_2 \text{ (Positive)} = \text{Atmospheric pressure} + \text{Gauge pressure}$$

$$\text{Absolute pressure, } P_1 \text{ (Vacuum)} = \text{Atmospheric pressure} - \text{Gauge pressure}$$

Atmospheric pressure may also be called barometric pressure, which is the pressure that prevails in the atmosphere. The standard atmospheric pressure is given as

1.01325 bar or 760 mm of mercury column. As gauge pressure is measured with reference to atmospheric pressure, zero of gauge pressure is equal to atmospheric pressure.

6.3.1 Static and Dynamic Pressure

Pascal's law of static fluid states that the pressure at any point on a particular plane inside a static fluid (fluid at rest) is equal in all directions and it is the function of depth from the free surface.

Imagine that a small object, say a cube of very small dimensions, is submerged in a fluid. All the faces of the object, whether it faces downwards, upward, leftwards/rightwards, will experience the same pressure according to the hydrostatic law of pressure. This is known as static pressure. When an object is placed in a fluid that is set in motion, the face perpendicular to the direction of flow and the face along the direction of flow will experience different pressures. The flowing fluid will isentropically decelerate to zero velocity on the face perpendicular to the direction of flow and the resulting loss in kinetic energy is converted into an additional pressure, called dynamic pressure, whereas the face that is along the direction of flow will experience only static pressure.

The pressure exerted on the face perpendicular to the direction of flow is called total or stagnation pressure.

Total or stagnation pressure = Static pressure + Dynamic pressure
(loss in kinetic energy)

6.3.2 Methods of Pressure Measurement

As pressure is defined as force per unit area, the unknown pressure or pressure difference to be measured may be converted into a force and, in turn, the force may be measured in two ways: by comparing with the known weight of an object or by producing a deflection on a diaphragm or an elastic member and measuring the deflection using a mechanical gauge or an electrical transducer.

Manometry is a method for measuring pressure, in which the unknown pressure force is balanced against the hydrostatic pressure of a liquid column of known height. In the case of dynamic pressure measurement, which varies frequently with respect to time, very highly responsive transducers such as electromechanical transducers must be used.

6.3.3 Manometers

Manometers are simple, inexpensive devices used for pressure measurement. In manometers, the unknown pressure or pressure difference is measured by balancing it against the hydrostatic pressure produced by a liquid column of known height. The main disadvantages of this method are its poor sensitivity to frequent variations in pressure with respect to time and that it cannot be used to measure very high pressure.

Simple U-tube manometer: Figure 6.21 shows the schematic arrangement of a simple U-tube manometer. The manometer consists of a U-shaped tube made of transparent glass. The tube is partially filled with manometric fluid whose specific gravity is higher than that of the fluid whose pressure is to be measured. Usually, mercury or water is used as the manometric fluid. The left limb of the manometer is connected to the fluid whose pressure is to be measured. The right limb is either kept open to the atmosphere or connected to any other reference fluid pressure. A vertical scale is mounted along the U-tube to measure the height of the fluid column.

Consider the datum XX in Fig. 6.21. The pressure at points A and B will be the same as per the hydrostatic law of pressure. Applying the pressure equilibrium, we get

$$p_A + \Upsilon_A(h_1 + h_2 + \Delta h) = p_B + \Upsilon_B \cdot h_2 + \Upsilon_m \cdot \Delta h \tag{6.33}$$

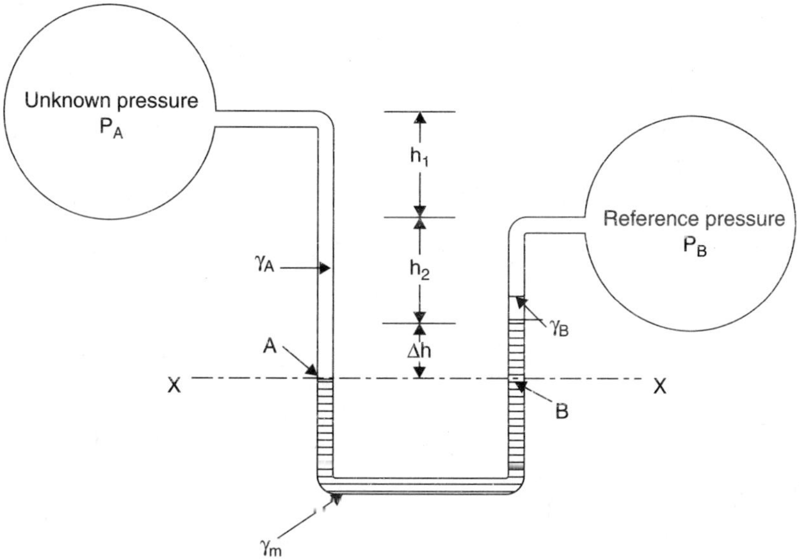

Figure 6.21 A simple U-tube manometer

where P_A and P_B = unknown pressure connected to the left limb and reference pressure connected to the right limb, respectively; and Υ_A, Υ_B and Υ_M = specific weight of fluid in the left limb, right limb and manometric fluid, respectively.

From Eq. 6.33, the differential pressure $P_A - P_B$ can be calculated. If the reference pressure P_B is zero, the absolute pressure P_A can be calculated; or if the right limb is open to atmospheric pressure (i.e., the reference pressure is atmospheric pressure), then $P_A - P_{atm}$ will provide the gauge pressure. The U-tube manometer should never be inclined and the reading should be carefully recorded, to avoid errors.

Well-type manometer: Figure 6.22 shows the schematic arrangement of a simple well-type manometer. In contrast to the U-tube manometer, well-type manometers have only one limb, and instead of the right limb, a well or small reservoir is present. The cross-sectional area of the well is very large when compared to that of the tube. Therefore, a small variation in the level of fluid in the well due to the pressure source connected to B will create greater displacement of the liquid level in the tube.

The pressure balance can be calculated by neglecting variations in the fluid level in the well as

$$p_A - p_B = \Upsilon_m (h + h') \tag{6.34}$$

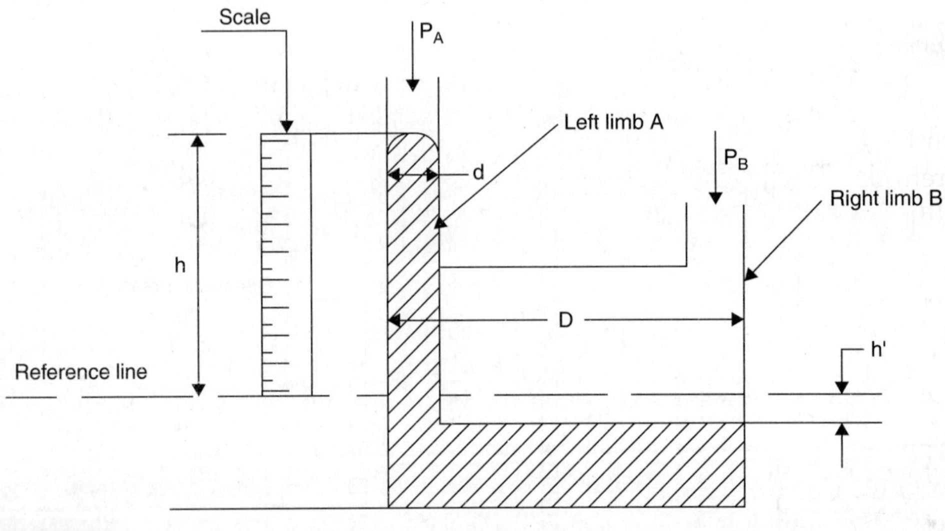

Figure 6.22 Well-type manometer

Also, the volume of liquid displaced in the well is equal to the volume that rises up in the tube.

$$h' \cdot A = h \cdot a \tag{6.35}$$

where A and a refer to the cross-sectional area of the well and tube, respectively.

Therefore $$p_A - p_B = \Upsilon_m \cdot h\left(1 + \frac{a}{A}\right) \tag{6.36}$$

Inclined tube manometer: To improve the accuracy of measurement, the vertical tube of the well-type manometer is inclined, with the angle of inclination ranging between 10° and 20° to the horizontal. Even a small rise in the level of fluid in the vertical direction will provide more displacement of fluid along the tube. Therefore, an inclined tube manometer shows higher sensitivity when compared to any of the vertical tube manometers. Figure 6.23 shows the schematic arrangement of an inclined tube manometer.

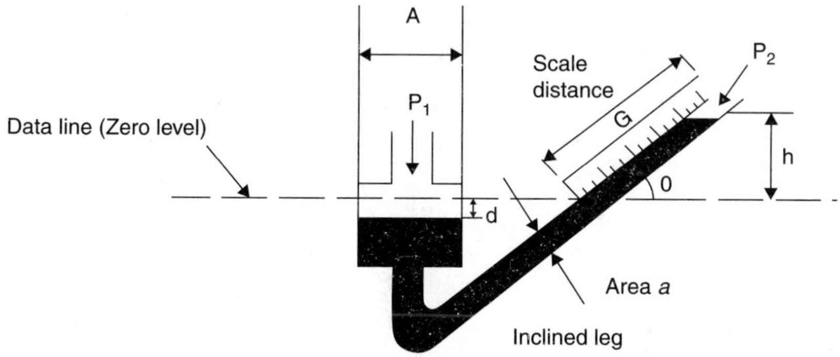

Figure 6.23 Inclined tube manometer

From Fig. 6.23, it can be seen that the vertical displacement of fluid $h = G \cdot \sin\theta$. Therefore, the pressure balance equation used in the well-type manometer is modified as follows:

$$p_A - p_B = \Upsilon_m \cdot G \cdot \sin\theta\left(1 + \frac{a}{A}\right) \tag{6.37}$$

6.3.4 Bourdon Tube Pressure Gauge

This pressure gauge was first developed by E Bourdon in the year 1849. Figure 6.24 shows the schematic of a Bourdon tube pressure gauge. It consists of a C-shaped hollow tube with elliptical or oval cross-section. One end of the tube is connected to the pressure source that needs to be measured and the other end, generally called the tip, is sealed and left free. When pressure is applied, the elliptical cross-section tends to regain its circular shape. This will build stress on the tube and it will tend to uncoil itself. Therefore, the free end of the tube will move away from its position in proportion to the amount of pressure applied. The free end of the tube is connected to a sector arm by a small linkage, and the sector arm is coupled with a pinion on which a needle pointer is fixed. The movement of the free end will cause a rotary motion on the pinion, and the pressure reading is indicated by the pointer.

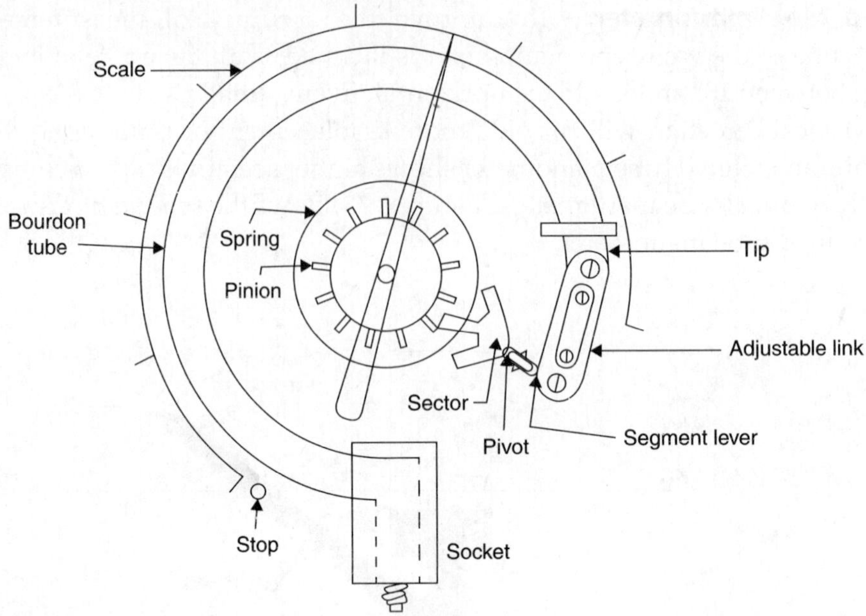

Figure 6.24 Bourdon tube pressure gauge

Commercial Bourdon tube gauges can measure up to 150 MPa. Such gauges are mostly used for static pressure readings and hold good for moderate variations in pressure. The materials most commonly used for the manufacture of Bourdon tube pressure gauges are phosphor bronze, brass and beryllium copper.

6.3.5 Bellows Pressure Gauge

The bellows-type pressure gauge is made of an elastic bellows element that is axially flexible so that the element can expand and collapse based on the application of pressure. The element is composed of a ductile thin metal or alloy with many convolutions or folds, as shown in Fig. 6.25. Pressure is applied on one side of the bellows and it is balanced by a spring provided inside the bellows. The movement of the bellows is transferred to a calibrated pointer scale arrangement to obtain the pressure readings.

The following important parameters influence the manufacture of a bellows pressure gauge:

- Range of pressure
- Hysteresis
- Fatigue due to repeated operation
- Sensitivity to fluctuating pressure

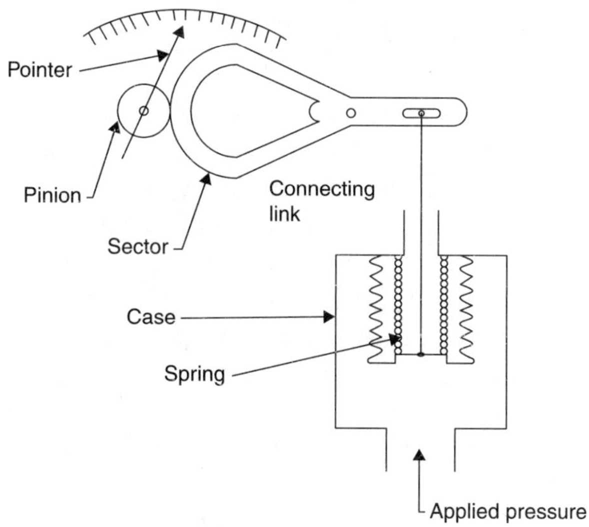

Figure 6.25 Bellows pressure gauge

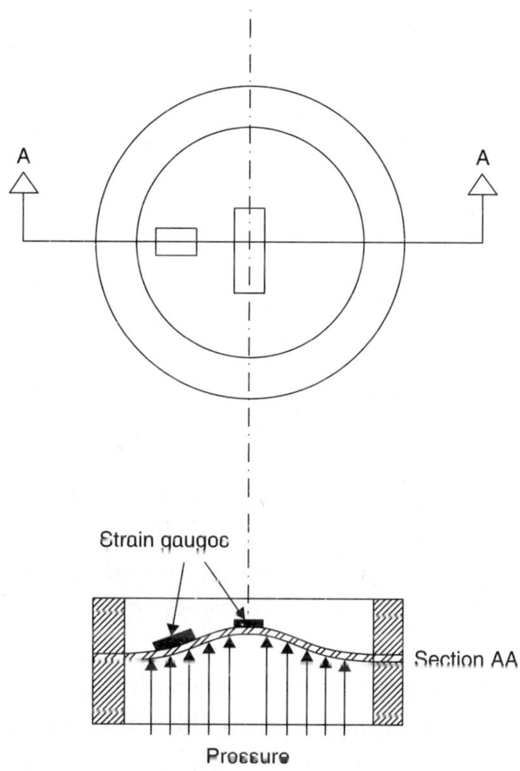

Figure 6.26 Diaphragm pressure gauge

6.3.6 Diaphragm Pressure Gauge

The diaphragm pressure gauge uses a thin diaphragm or membrane as the primary pressure transducer. Elastic deformation of the diaphragm due to the pressure applied is measured by a secondary transducer, such as mechanical linkages which in turn are connected to a calibrated dial indicator, strain gauge or linear variable differential transformer (LVDT). The schematic of a diaphragm pressure gauge is shown in Fig. 6.26.

Diaphragm gauges have very good sensitivity to frequently varying pressure and can be used as a dynamic pressure measuring device. The requirement of any elastic pressure gauge is that its output be linear. In order to meet the requirement, two types of metal diaphragms are used: flat and corrugated. When the diameter of the diaphragm needs to be large, the corrugated metal diaphragm design is desirable.

6.3.7 Low-Pressure Measurement

Pressure values below atmospheric pressure are termed low pressure or vacuum pressure. Vacuum pressure is measured in torr, and 1 Torr is equivalent to 1 mm of Hg. The range of vacuum pressure is shown in Fig. 6.27.

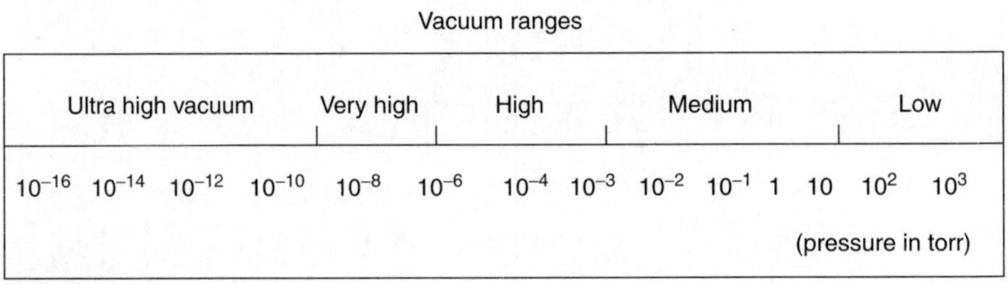

Figure 6.27 Range of vacuum pressure

Low pressure can be measured by any of the direct pressure measurement methods discussed earlier or by indirect or inferential methods in which the pressure of the fluid is obtained by measuring the varying properties of the fluid, such as volume and thermal conductivity, in proportion to the pressure of the fluid.

McLeod gauge: This was invented by Herbert McLeod in 1874. The schematic of the McLeod vacuum gauge is shown in Fig. 6.28. The gauge works on the principle of Boyle's law. A known volume of gas (V_1) at an unknown pressure (P_1) that needs to be found is isothermally compressed to a smaller volume (V_2) by applying a known pressure (P_2) on it. By Boyle's law

$$P_1 = \frac{P_2 \cdot V_2}{V_1} \tag{6.38}$$

The unknown pressure to be measured (P_1) is connected to the reference column of the McLeod gauge. The piston is forced on the mercury reservoir and causes the mercury level to rise up to the cut-off point (shaded grey). The unknown pressure applied to the gauge completely fills the bulb, measuring capillary, reference capillary and reference column.

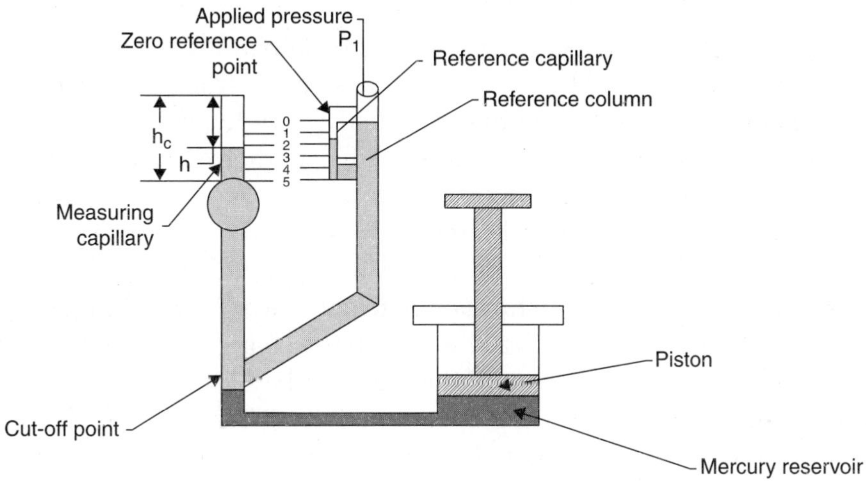

Figure 6.28 Schematic of McLeod gauge

In this condition, the volume of fluid (V_1) whose pressure is to be measured fills the bulb. The mercury level is increased further to the zero reference point by forcing the piston, so that the fluid trapped in the bulb and measuring capillary tube is compressed. Now, the volume of the gas in the measuring capillary tube is read directly from the scale. The difference in height h of the measuring capillary and the reference capillary is the measure of the volume V_2 and pressure P_2 of the trapped fluid.

Therefore, V_1, V_2 and P_2 are determined, and the applied pressure P_1 can be calculated using Boyle's law.

Let the volume of the bulb between the cut off point and the measuring capillary be V, the cross-sectional area of the measuring capillary tube be a, and the height of the measuring capillary tube be h_c. Let V_1 be the initial volume of the gas entrapped in the bulb plus the measuring capillary tube.

Therefore $\quad V_1 = V + a \cdot h_c$

After the mercury is forced to reach the zero reference point in the reference capillary, let the height of the compressed gas in the measuring capillary tube be h. Therefore, the final volume of the gas $V_2 = a \cdot h$.

Let P_1 be the unknown pressure of the fluid to be found and P_2 be the pressure of gas after compression. Then $P_2 = P_1 + h$.

By Boyle's law, $P_1V_1 = P_2V_2$. Therefore, $P_1V_1 = (P_1 + h) \cdot a \cdot h$, from which the unknown applied force P_1 is calculated.

The main limitation of McLeod gauge is that the fluid whose pressure is to be measured should obey Boyle's law, and suitable moisture traps must be provided to prevent any condensable vapour from entering the gauge.

6.3.8 Thermal Conductivity Gauges

An electric conductor that conducts electricity generates heat because of the resistance in the conductor. The rate of heat dissipation from the conductor to the surroundings depends on the thermal conductivity of the surrounding medium. If the density of the surrounding medium is low, its thermal conductivity will be low and will, therefore, affect (reduce) the heat dissipation rate from the conductor. The conductor will thus become hotter and will further increase its resistivity. Ultimately, the resistance of the conductor will depend on the density of the medium surrounding it, which in turn, depends on the pressure of the surrounding medium. Therefore, by measuring the resistance change in a conducting wire surrounded by a fluid, the pressure of the fluid can be measured.

A Pirani gauge is a type of thermal conductivity gauge, shown in Fig. 6.29. It consists of a single platinum element housed in a chamber and a compensating cell to compensate for variations caused by ambient temperature changes. These are mounted on two arms of a Wheatstone bridge.

A constant current is passed through the platinum filament housed in the Pirani gauge chamber. The resistance of the filament heats up the filament and this resistance value is measured using the bridge. Now, the unknown pressure

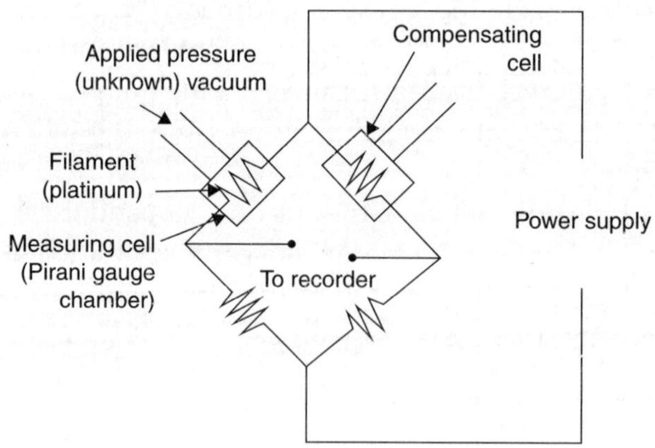

Figure 6.29 Schematic of a Pirani thermal conductivity gauge

to be measured (applied pressure) is connected to the Pirani gauge chamber. This changes the density of the surrounding medium in the chamber.

This affects the thermal conductivity of the surroundings and, subsequently, the heat dissipation rate of the filament. Therefore, the filament becomes hotter and its resistance changes. This change in resistance of the filament is measured using the bridge, which is calibrated to the applied pressure.

The main limitation of the thermal conductivity gauge is that below the level of 1 Pa, the effect of thermal conductivity of the surrounding medium becomes very minimal.

6.3.9 Ionisation Gauges

An ionisation gauge can be used to measure very low pressure (i.e., very high vacuum) in the threshold of 10^{-6} Pa to 0.1 Pa. It is also called a hot filament or hot cathode ionisation gauge. It is typically a vacuum triode that possesses a heated filament as a cathode, a positive biased grid or ion collector and a negatively biased plate to collect the free electrons. The schematic of the ionisation gauge is shown in Fig. 6.30.

The cathode is heated by a supply of very high voltage of around 400 V, and it emits electrons. The electrons emitted from the cathode are collected by the plate, and a current i_p flows through the external circuit. Now, the gas whose pressure is to be measured is admitted inside the evacuated triode. Therefore, the electrons collide with the gas molecules and ionise them. The gas molecules become positive due to loss of free electrons, and these positive ions are collected by the positively biased grid or ion collector. This current is measured as i_g. The rate of production of positive ions for a particular gas is directly proportional to the number of gas molecules per unit volume, which, in turn, is proportional to the gas pressure.

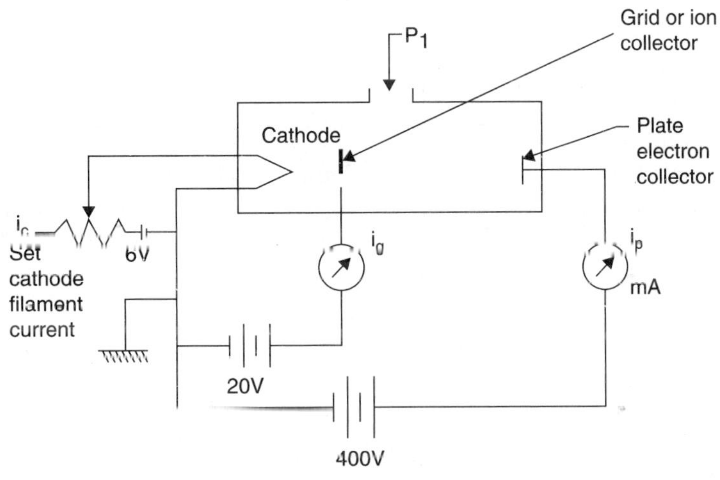

Figure 6.30 Schematic of ionisation gauge

$$p = \frac{i_g}{S \cdot i_c} \tag{6.39}$$

S is called the sensitivity of the gauge. If the pressure of the gas inside the triode falls, the ion current will also decrease in proportion.

The main limitation of this type of gauge is that the filament will be adversely affected when it is heated at high gas pressure. Also, the presence of air in the gas will lead to burning of the filament.

6.3.10 Knudsen Gauge

The schematic of the Knudsen gauge is shown in Fig. 6.31. It consists of a movable vane placed between two fixed plates. The movable vane hangs in suspension with the help of springs. The distance between the plates should be maintained at a distance less than the mean free path of the gas. A small mirror is positioned on the movable plane. A light source is provided, which will project a beam of light on the mirror, and the angle of the deflected light is measured by a pointer scale.

The fixed plates are heated to a temperature higher than that of the gas and the movable vane is maintained at the temperature of the surrounding gas. When gas molecules strike the fixed heated plates, they bounce back with greater momentum when compared with the colder movable plates. This creates a net force effect on the movable plate and it will tend to deflect in proportion to this net force acting on it. The deflection can be determined by measuring the light beam deflection from the mirror attached to the movable vane. The absolute temperature of the fixed plate and movable vane are measured as T_f and T_v. The net force produced

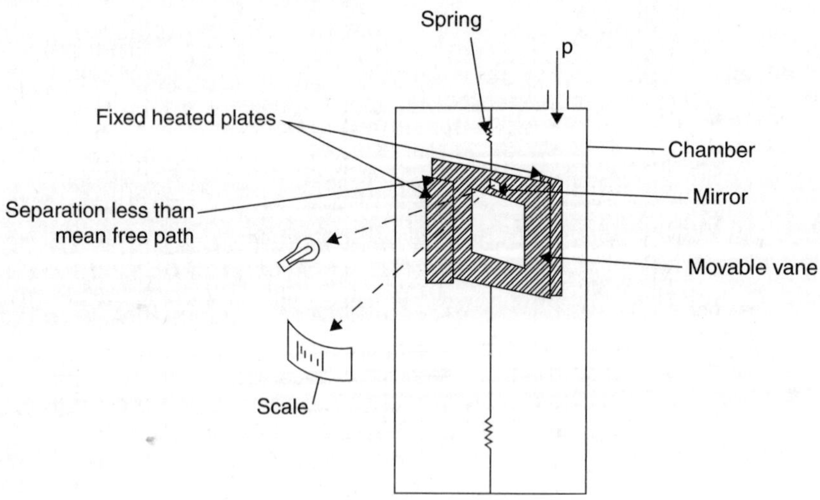

Figure 6.31 Schematic of the Knudsen gauge

on the movable vane is directly proportional to the pressure at a particular value of T_f and T_v.

$$p = \frac{K \cdot F}{\sqrt{\frac{T_f}{T_v} - 1}} \tag{6.40}$$

where K is a constant and F refers to force.

This type of gauge is used to measure pressure in the range of 10^{-8} to 10^{-2} Torr. It is insensitive to the composition of the gas used for measurement.

6.3.11 Pressure Cells

When a closed container is pressurised, it is placed under strain and this will result in change in the dimensions of the container. These changes can be measured using a secondary transducer, such as a strain gauge, and the results can be calibrated as the measurement of the applied pressure. Such a system is called a pressure cell. Two types of pressure cells are available: flattened tube and cylindrical tube.

Flattened tube pressure cell: The schematic of the flattened tube pressure cell is shown in Fig. 6.32. It consists of a flat elastic tube that is pinched at the ends. Two strain gauges are placed on the two flat surfaces of the tube, i.e., one at the top and the other at the bottom. One end of the tube is open to the pressure source, whereas the other end is sealed. When the tube is pressurised, the resulting strain on the gauges will create a change in the resistance, and this change can be measured by properly connecting the gauges to a bridge circuit. The measure of the change in resistance is proportional to the applied pressure. In order to avoid errors in measurement due to temperature, two temperature compensating gauges should be fixed at a suitable unstressed location on the tube. This type of pressure cell is used for low-pressure measurement.

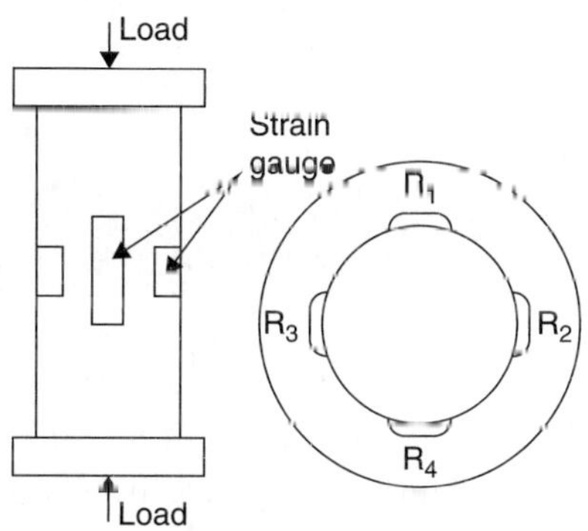

Figure 6.32 Schematic of a flattened tube pressure cell

Cylindrical tube pressure cell: The schematic of a cylindrical tube pressure cell is shown in Fig. 6.33. In this, two strain gauges are fixed on the circumferential surface of a cylindrical tube at diametrically opposite locations, and two dummy gauges are provided at the ends of the tube which act as temperature compensating gauges. The working of this type of pressure cell is very similar to that of the flattened tube pressure cell; the only difference is the construction. This type of pressure cell is used to measure medium pressure. The change in resistance, which is a measure of the applied pressure, is measured by the following relationship:

$$\frac{\Delta R}{P} = \frac{2FRD^2}{E}\left[\frac{2-\vartheta}{D^2-d^2}\right] \tag{6.41}$$

where ΔR = change in resistance; P = applied pressure; F = sensitivity of the gauge; R = original resistance of the gauge; E = Young's modulus; υ = Poisson's ratio; and D and d = outside and inside diameters, respectively, of the cylindrical tube.

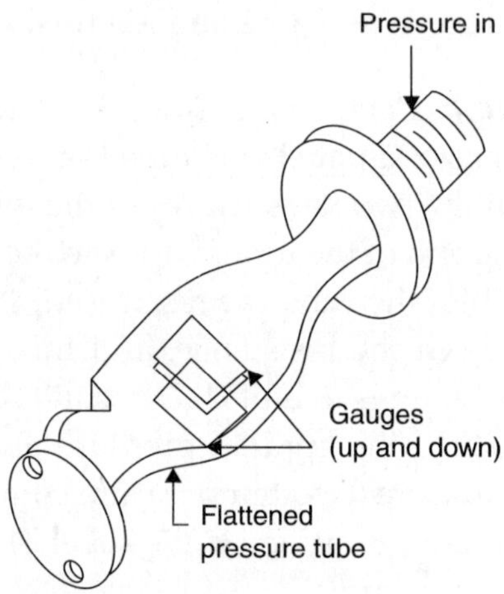

Figure 6.33 Schematic of a cylindrical tube pressure cell

SUMMARY

❖ By Newton's second law, force is defined as the product of the mass of a body or object and its acceleration. Two approaches are followed for the measurement of force: direct comparison method and indirect comparison method. In the direct comparison method, the measurand force will not be amplified or attenuated to be compared with the known force or weight. In the indirect comparison method, the measurand force is attenuated with the help of levers and linkages in the measuring system so that a very small known weight is used to measure a huge force.

- Equal arm balance and analytical balance are types of beam balance systems in which a set of known standard weights are required to measure the unknown force or weight. Unequal arm balance, multiple lever systems and pendulum force measuring scale are instruments in which a smaller force is used to balance a huge force, with the help of levers or linkages. Elastic force meters and proving rings are used to determine force based on the deflection caused on the loading member of the measuring system. In hydraulic and pneumatic load cells, the force or load is measured by sensing the pressure induced in the fluid confined inside a diaphragm because of the load applied to it.
- When force is applied on a structural member, it will undergo deformation. The ratio of the change in dimension to the original dimension of the loaded element is called the strain. Electric strain gauges are very popular devices used for strain measurement. They work on the principle that if the metallic wire or foil is stretched, the dimension of its length and cross-section changes with a proportionate change in the electric resistance of the wire. The measure of such variation in electric resistance is used to measure the strain on a body. Bonded wire and metal foil strain gauges are different types of metallic resistance strain gauges. Semiconductor materials such as germanium and silicon also exhibit variation in resistance when filaments or wafers made of such materials are strained. Such semiconductor strain gauges have a high gauge factor when compared to metallic gauges. The gauge factor is a measure of the sensitivity of a gauge.
- In order to measure the variation in the resistance of strain gauges, an electric circuit with greater sensitivity is required. Two basic types of circuits are utilised for this purpose: ballast circuit and bridge circuit. The Wheatstone bridge circuit is frequently used. In general, strain gauges are sensitive to temperature, which needs to be compensated for to avoid error in measurement of the strain. This is done by providing adjacent arm balancing or by self temperature compensation. To determine the strain that occurs in more than one direction, strain gauge rosettes are used. They are made of two or more closely positioned strain gauge grids, oriented to measure the strains independently along different directions.
- Pressure is defined as force per unit area. Therefore, the unknown pressure or pressure difference to be measured may be converted to a force and, in turn, the force may be measured by any of the force measuring methods: by comparison with the known weight of an object or by producing deflection on a diaphragm

or an elastic member and measuring the deflection directly using a mechanical gauge or indirectly using an electrical transducer.

- ❖ Manometry is a method used for measuring pressure, in which the unknown pressure force is balanced against the hydrostatic pressure of a liquid column of known height. Simple U-tube manometer, well-type manometer and inclined tube manometer are some of the types of manometers used for pressure measurement.
- ❖ Bourdon tube pressure gauges have a C-shaped hollow tube of elliptic cross-section sealed at one end while pressure is applied on the other end. The pressure will tend to move the free end of the tube, and measurement of this displacement with suitable arrangements will provide the pressure measurement. Bellows and diaphragm gauges have an elastic element, such as bellows and thin diaphragm membrane, respectively, which elongate when pressure is applied. The elastic deformation is proportional to the pressure.
- ❖ For measurement of very low pressure (vacuum), the McLeod gauge is suitable; it works on the principle of Boyle's law. In thermal conductivity gauges, pressure measurement is achieved by measuring the heat dissipation from an electric conductor, which varies with the thermal conductivity of the gas that surrounds it. This, again, is proportional to the pressure of the gas. In ionisation gauges, the gas for which pressure needs to be measured is ionised in a triod and the rate of production of positive ions for a particular gas is proportional to the gas pressure. In the Knudsen gauge, the measure of the deflection of a heated movable plate placed near a fixed plate at a distance equal to the mean free path of the gas is used to calculate the pressure of the gas. The gas that is heated by the temperature of the hot plate will impose a net force on the movable plate, as it cannot move the fixed plate and will therefore create a deflection.
- ❖ Pressure cells use strain gauge transducers to measure the deformation on a pressurised closed container. The flexible container closed at one end is connected to the pressure source and the quantity of deformation is proportional to the pressure applied. Flattened tube and cylindrical tube are some of the common types of pressure cells.

EXERCISES

Multiple Choice Questions

1. The SI unit for force is
   a) Newton　　b) Kilogram
   c) Newton metre　　d) Watt

2. The accuracy or symmetry of the equal arm balance system can be verified by
a) Placing the same weight on both the arms
b) Measuring with different weights
c) Interchanging the weights on both the arms
d) None of the above

3. The effect of buoyancy force on force measurement
a) Will be less for the measurement of unknown masses with high density
b) Will be high for the measurement of unknown masses with low density
c) Will not be affected by the density of the material
d) None of the above

4. Buoyancy force will be zero if force measurement is carried out in
a) Vacuum
b) Ambient air with low humidity
c) Ambient air with high humidity
d) None of the above

5. The sensitivity of an analytical balance is
a) $S = \frac{\text{Deflection}}{\text{Unbalanced force}}$
b) $S = \frac{\text{Deflection}}{\text{Known force}}$
c) $S = \frac{\text{Deflection}}{\text{Unknown force}}$
d) $S = \frac{\text{Unknown force}}{\text{Deflection}}$

6. In an unequal arm balance, the unknown test force is proportional to
a) $\frac{\text{Length of load arm}}{\text{Length of power arm}}$
b) $\frac{\text{Length of power arm}}{\text{Length of load arm}}$
c) $\frac{\text{Length of load arm}}{\text{Overall arm length}}$
d) $\frac{\text{Length of power arm}}{\text{Overall arm length}}$

7. Which balance is most suitable for measuring large loads?
a) Equal arm balance
b) Unequal arm balance
c) Multiple lever balance
d) All the above

8. If the multiplication factor of a multiple lever system is 100, this means that ____________ of 100 kg is required to measure 1 kg of the unknown mass.
a) Poise weight
b) Pan weight
c) Counterpoise weight
d) None of the above

9. Stiffness is given by
a) $\frac{\text{Force}}{\text{Thickness}}$
b) $\frac{\text{Force}}{\text{Displacement}}$
c) $\frac{\text{Force}}{\text{Elastic limit}}$
d) None of the above

10. Stiffness of the elastic member used in an elastic force meter should be
    a) Less than the load limit
    b) Equal to the load limit
    c) Greater than the load limit
    d) None of the above

11. If the stiffness of the elastic element is very high, then the sensitivity of the elastic force meter will be
    a) Very high
    b) Moderate
    c) Low
    d) None of the above

12. The vibrating reed is an integral part of
    a) Equal arm balance
    b) Unequal arm balance
    c) Multi lever balance
    d) Proving ring

13. Strain is defined as
    a) $\frac{\text{Change in length}}{\text{Original length}}$
    b) $\frac{\text{Change in length}}{\text{Load applied}}$
    c) $\frac{\text{Original length}}{\text{Change in length}}$
    d) $\frac{\text{Load applied}}{\text{Deflection}}$

14. Electric strain gauges are generally classified as
    a) Inductive
    b) Capacitive
    c) Piezoelectric
    d) All the above

15. Which type of the following strain gauges is most suited for dynamic measurement of strain?
    a) Optical
    b) Electric
    c) Mechanical
    d) All the above

16. The gauge factor of a strain gauge is defined as
    a) Ratio of linear strain to change in resistance of the conductor
    b) Ratio of change in resistance of the conductor to the linear strain
    c) Ratio of original resistance to change in resistance of the conductor
    d) None of the above

17. The strain gauge will have good sensitivity if the gauge factor is
    a) High
    b) Low
    c) Less than 1
    d) Equal to 2

18. A good strain gauge should posses
    a) Low resistance temperature coefficient
    b) Good elastic limit
    c) Linearity between the strain and the corresponding change in resistance
    d) All the above

19. The gauge factor of metallic strain gauges will be in the range of
 a) 4 to 5 b) 0.001 to 1
 c) 150 to 160 d) None of the above

20. The gauge factor of semiconductor strain gauges will be in the range of
 a) 4 to 5 b) 0.001 to 1
 c) 150 to 160 d) None of the above

21. The resistance of strain gauges made of P-type semiconductor material increases
 a) With increase in tensile strain b) With decrease in tensile strain
 c) With decrease in shear strain d) None of the above

22. The ballast circuit is not suitable for the measurement of
 a) Static strain b) Dynamic strain
 c) Both the above d) None of the above

23. Compensation strain gauges are provided to
 a) Increase the gauge factor b) Nullify the effect of temperature
 c) Measure the principal strain d) All the above

24. Which of the following gauges is used to measure the strain in three different directions?
 a) Wire strain gauge b) Metallic strain gauge
 c) Rosette strain gauge d) Semiconductor strain gauge

25. The zero reading in a pressure gauge is equal to
 a) Absolute pressure b) Atmospheric pressure
 c) Static pressure d) Dynamic pressure

26. The sum of atmospheric pressure and gauge pressure is called
 a) Stagnation pressure b) Static pressure
 c) Absolute pressure d) Vacuum pressure

27. The sum of static pressure and stagnation pressure is called
 a) Stagnation pressure b) Atmospheric pressure
 c) Absolute pressure d) Vacuum pressure

28. The specific gravity of the manometric fluid should be ____________ in comparison to the measuring fluid.
 a) Low b) High
 c) Equal d) None of the above

29. The advantage of using an inclined manometer over a vertical tube manometer is that the former is
    a) More sensitive
    b) More accurate
    c) Both (a) and (b)
    d) None of the above

30. Which of the following pressure measuring devices requires a strain gauge to measure?
    a) Manometer
    b) Bourdon gauge
    c) Bellows
    d) Diaphragm gauge

31. The McLeod gauge works on the principle of
    a) Charles' law
    b) Boyle's law
    c) McLeod's law
    d) Hydrostatic law

32. The limitation of the McLeod gauge is
    a) Measuring fluid should obey Boyle's law
    b) Condensable vapours should not enter the gauge
    c) Both the above
    d) None of the above

33. An example of a thermal conductivity gauge is
    a) McLeod's gauge
    b) Pirani gauge
    c) Bourdon gauge
    d) Ionisation gauge

34. The fixed plates of the Knudsen gauge should be heated to a temperature
    a) Higher than the gas temperature
    b) Equal to the gas temperature
    c) Lower than the gas temperature
    d) None of the above

35. The temperature of the moving vane of a Knudsen gauge should be maintained
    a) Higher than the gas temperature
    b) Equal to the gas temperature
    c) Lower than the gas temperature
    d) None of the above

36. The distance between the fixed and moving plates of a Knudsen gauge should be maintained at a distance
    a) Less than the mean free path of the gas
    b) More than the mean free path of the gas
    c) Equal to the mean free path of the gas
    d) None of the above

37. The measuring range of the ionisation gauge is
    a) 10^{-6} Pa to 0.1 Pa
    b) 0.1 bar to 1 bar
    c) 0.1 Pa to 1 Pa
    d) None of the above

38. Which of the following pressure gauges is used for low-pressure measurement?
   a) McLeod's gauge
   b) Knudsen gauge
   c) Ionisation gauge
   d) All the above

39. In McLeod's gauge
   a) High pressure fluid is expanded to low pressure and measured by a manometer
   b) Low pressure fluid is compressed to high pressure and measured by a Bourdon tube
   c) High pressure fluid is expanded to low pressure and measured by a Bourdon tube
   d) Low pressure fluid is compressed to high pressure and measured by a manometer

40. The SI unit for pressure is
   a) N/m^2
   b) $N\text{-}m^2$
   c) N
   d) None of the above

Answers

1. a	2. c	3. a	4. a	5. a
6. a	7. c	8. b	9. b	10. c
11. c	12. d	13. a	14. d	15. b
16. b	17. a	18. d	19. a	20. c
21. a	22. a	23. b	24. c	25. b
26. c	27. a	28. b	29. a	30. d
31. b	32. c	33. b	34. a	35. a
36. a	37. a	38. d	39. d	40. a

Review Questions

1. List the various methods followed for measuring force.
2. Discuss any three direct comparison methods for force measurement.
3. Explain in detail any two indirect comparison methods for force measurement.
4. What is a proving ring?
5. Define sensitivity of an analytical balance.
6. What is the main drawback of the equal arm balance system?
7. With a neat sketch, explain force measurement using a multiple lever system.
8. What are the requirements for the elastic member of an elastic force meter?

9. What is a load cell? Explain the working of hydraulic and pneumatic load cells.
10. What are the types of strain gauges?
11. Write the principle of electric resistance strain gauge.
12. Define the sensitivity of an electric resistance strain gauge.
13. Define gauge factor.
14. Explain the working of a strain gauge ballast circuit.
15. Explain with a circuit diagram, the principle and working of a strain gauge bridge circuit.
16. What are the advantages of metal foil strain gauge over wire grid strain gauge?
17. Why is temperature compensation required in strain gauges?
18. Explain the methods followed for temperature compensation in strain gauges.
19. What is a strain gauge rosette? Explain how two-element and three-element strain gauge rosettes are formed.
20. What is the advantage of semiconductor strain gauge?
21. Define pressure and name different methods of pressure measurement.
22. Define the following: i) Absolute pressure ii) Gauge pressure iii) Vacuum.
23. Define: i) Stagnation pressure ii) Static pressure iii) Dynamic pressure.
24. What is a manometer? Mention the different types of manometers.
25. Explain the construction of a simple U-tube manometer and how pressure is measured.
26. Explain the construction and working of a Bourdon tube pressure gauge.
27. With a neat sketch, explain the working of the bellows pressure gauge.
28. Explain the working principle of McLeod's gauge for low-pressure measurement.
29. Describe: i) Thermal conductivity gauge ii) Ionisation gauge.
30. What is a pressure cell? Explain its types.

7

Measurement of Torque and Power

7.1 INTRODUCTION

Torque is an important parameter in many equipment. Torque is a rotational force or a force through a distance. It can be classified as static or dynamic. It is considered static if it has no angular acceleration. The torque exerted by a clock spring is a static torque as there is no rotation, and hence, no angular acceleration. In the case of dynamic torque, a dynamic force involving acceleration plays a role. For example, the force required to stop a car with its substantial mass would be a dynamic force, as the car must be decelerated. However, the force exerted by the brake calipers to stop the car would be a static force because there is no acceleration of the brake pads involved.

The torque produced by a car engine would be both static and dynamic, depending on where it is measured. If the torque is measured in the crankshaft, there will be large dynamic torque fluctuations as each cylinder fires and its piston rotates the crankshaft. However, if the torque is measured in the drive shaft, then the torque will be of static nature because the rotational inertia of the flywheel and transmission will dampen the dynamic torque produced by the engine.

Torque measurement is often associated with determination of mechanical power, either the power required to operate a machine or the power developed by the machine. In this connection, torque measuring devices are commonly referred to as dynamometers. As the power available at the crankshaft is the product of the torque transmitted by the crankshaft and the speed of the engine, the power is calculated from the measurement of torque and the corresponding speed of the engine. The following topics discuss how torque is measured – and hence, the power – using various techniques.

7.2 MECHANICAL TORSION METER

The mechanical torsion meter is based on the principle that when a shaft is connected between a driving element – for example, an engine – and a driven load, a twist (angular displacement) occurs on the shaft between the ends of the shaft. This angle of twist is measured and calibrated in terms of torque.

Figure 7.1 shows the constructional details of a mechanical torsion meter. It comprises a shaft that has two drums and two flanges mounted on the ends of the shaft. One of the drums carries a pointer and the other carries a torque-calibrated scale. A stroboscope is used to take the readings on the rotating shaft.

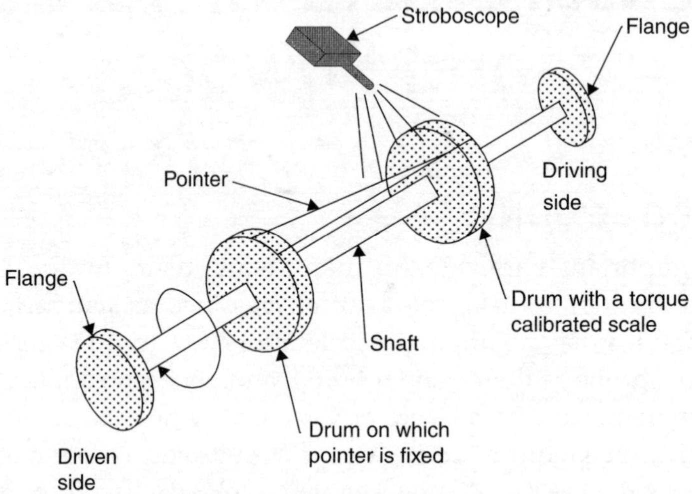

Figure 7.1 Mechanical torsion meter

Working principle of the mechanical torsion meter: One end of the shaft of the torsion meter is connected to a driving element and the other end to the driven load. In this configuration, an angle of twist is experienced by the shaft along its length between the two flanges, which is proportional to the torque applied to the shaft. A measure of the angle of twist is a measure of torque, when calibrated. The angular twist caused is observed on the torque-calibrated scale that corresponds to the position of the pointer. As the scale on the drum also rotates along with the shaft, the reading cannot be measured directly. Therefore, a stroboscope is used to flash a light on the calibrated scale; the flashing frequency is adjusted till a stationary image of the scale is obtained. The scale reading is then taken and is considered a measure of the angle of twist that occurred on the shaft.

Torque is then calculated using the formula

$$\text{Torque } T = \frac{GJ\theta}{L}$$

where G is the shear modulus of the material of the shaft; J is the second moment of inertia of the shaft; θ is the twist of the shaft measured; and L is the length of the shaft.

Excepting the twist of the shaft, all other parameters are constant for the given material and size of the shaft, and hence, the angle of twist is a direct measure of the torque on the calibrated scale.

Advantages

- Simple and inexpensive.
- The power of the shaft can also be calculated using the formula, Speed × Torque, as the flashing frequency of the stroboscope gives information about the speed of the shaft.

Disadvantages

- Poor accuracy due to small displacement of the pointer on the calibrated scale.
- Sensitivity of measurement is reduced even due to small variations in the speed of the shaft.
- Applicable only when the shaft is rotating at a constant speed.

7.3 OPTICAL TORSION METER

The basic principle of the optical torsion meter is that torque is measured using optical means, wherein the angular deflection of light rays is proportional to the twist. It consists of a shaft on which two castings M and N are connected at a known distance. The shaft of the torsion meter is coupled to the rotating shaft of the engine. A tension strip connects the two castings fixed on the shaft. Two mirrors are fixed at an alignment to the castings. The arrangement is shown in Fig. 7.2. A light beam is made to fall on the mirrors – an optical system – which, in turn, reflects the light beam to fall on a torque-calibrated scale.

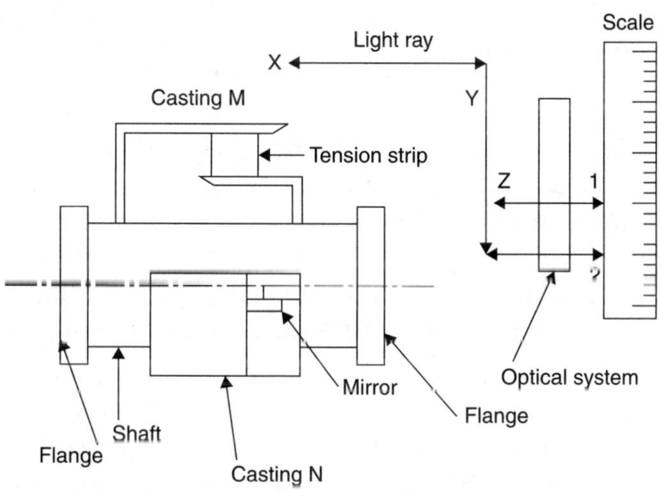

Figure 7.2 Optical torsion meter

When the shaft of the engine transmits the torque, it does so to the shaft of the torsion meter. A relative movement occurs between the castings *M* and *N* due to the twist on the shaft, and thus, there is a change in the position of the two mirrors as they are attached to the respective castings. As the mirrors are constantly made to reflect the light beam on the torque-calibrated scale, the change in the position of the *M* and *N* castings – in other words, the twist of the shaft – is directly measured from the torque-calibrated scale.

7.4 ELECTRICAL TORSION METER

It consists of two slotted discs attached on either side of the shaft connected to a driving engine and a driven load. A magnetic or photo-electric transducer is fixed near the shaft to count the pulses from the slotted disc. The construction details are shown in Fig. 7.3.

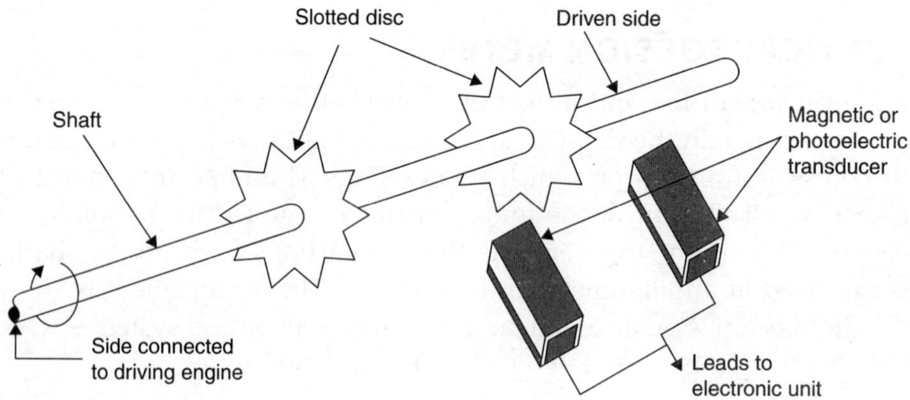

Figure 7.3 Electrical torsion meter

When there is no rotation on the shaft, no torque is generated, and hence, the teeth of the slotted discs align perfectly with each other; therefore, the voltage pulses produced in the transducers will have zero time difference. When the engine rotates, the shaft is subjected to torque, and hence, there is a twist in the shaft. This angular shift creates a relative displacement between the two slotted discs. Due to this relative displacement, the teeth of both the discs will not align with each other, and hence, a phase shift occurs between the pulses generated by the transducers. When these pulses are connected to an electronic unit, the time lapse between the two pulses can be determined. This time lapse is proportional to the torque, and hence, the twist of the shaft. When the time lapse is calibrated for the torque, the torque produced by the shaft can be directly determined. As there is rubbing action against the rotating motion of the shaft, no noise is created in this torsion meter.

7.5 STRAIN GAUGE TORSION METER

Four strain gauges are mounted onto the shaft whose torque needs to be measured, as shown in Fig. 7.4. Brushes and slip rings are used to carry the electrical connection to the four strain gauges and complete the electrical circuit. The schematic diagram of the strain gauge torsion meter is shown in Fig. 7.4.

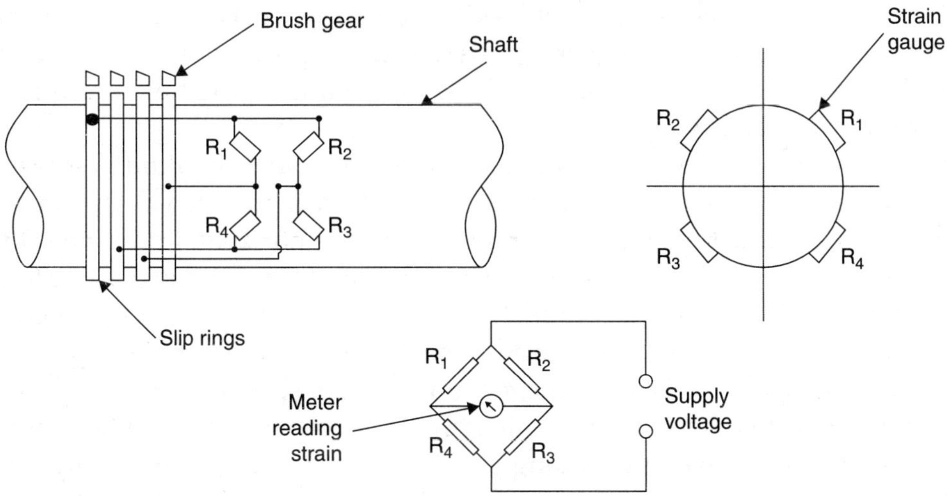

Figure 7.4 Strain gauge torsion meter

When the engine transmits the rotation through the shaft, it is subjected to a twisting action, i.e., torque. This results in a change in the resistance of the strain gauge system or bridge. A frequency converter attached to the shaft converts this change in resistance to a corresponding frequency change. The frequency signal is then transmitted without contact to a digital frequency receiver. When the digital frequency receiver is calibrated for the torque, the torque applied to the shaft can be directly measured from the readings of the frequency change, and hence, the change in resistance due to change in strain. As there are no rubbing surfaces in the system, wear and tear of the torsion meter is not a factor, and hence, this meter requires less maintenance.

7.6 DYNAMOMETER – HYDRAULIC

A hydraulic dynamometer works on the principle of dissipation of power in fluid friction rather than in dry friction. The brake power of an engine is the useful power available at the crankshaft. It is measured by running the engine against some form of absorption brake.

In high-speed motor vehicle engines, the type of brake widely used is the Heenan and Froude hydraulic dynamometer. The construction of the hydraulic dynamometer is similar to that of a fluid flywheel. As shown in Fig. 7.5, it consists of an impeller or rotor coupled to the output shaft of an engine. The impeller rotates in a casing filled with fluid, generally a coolant flowing steadily through the inlet and outlet pipes. The casing is freely mounted on bearings fitted to trunnion brackets.

At the periphery of the impeller is a series of semi-elliptical pockets. When the impeller is driven by the engine, fluid is flung out of its pockets by the centrifugal force that develops and this is then transferred to the pockets of the outer casing. Thus, the casing with the impeller tends to turn but is resisted by a torque arm supporting the balance weight.

The frictional force generated between the impeller and the fluid is measured by a spring balance fitted on the casing. A hand wheel is provided at the top of the spring balance frame to adjust the torque arm to the horizontal position. This is facilitated by a small pointer, as shown. The amount of load, and hence, the torque absorbed by the hydraulic dynamometer can be varied by controlling the flow of fluid. This is usually done by operating another hand wheel, which slides thin metal plates (sluice gates) between the impeller and the casing pockets, thus blanking-off a number of effective pockets. As the length of the torque arm, together with the static load and spring balance reading, is known, the torque on the casing can be easily determined. The heat released due to dissipation of power in the hydraulic dynamometer is carried away by the continuous supply of fluid.

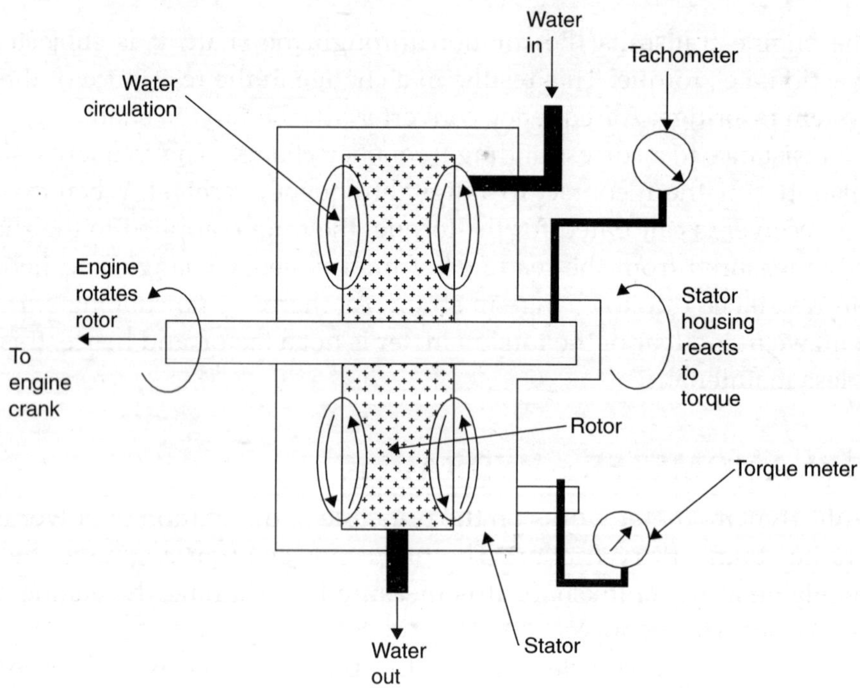

Figure 7.5 Hydraulic dynamometer

The torque transmitted by the engine T = Effective load (W), in kg × Length of the torque arm (L), in m

$$\text{Effective load, } W = \text{Spring balance reading} + \text{Static load}$$

Brake power $P_b = 2\pi NT/60$ in watts, where N is the speed of the engine measured with a tachometer (in rpm).

If the constant terms, $2\pi R/60$ is written as $1/K$, and the K value is stamped on the name plate attached to the hydraulic dynamometer, the brake power formula can be reduced to the simple form of $P_b = WN/K$.

Power can be expressed in horsepower, using the expression $HP = W/745.7$.

7.7 DYNAMOMETER – MECHANICAL

Pony brake dynamometer: This is one of the simplest mechanical dynamometers for measuring the power available at the crankshaft of an engine. It converts power into heat by dry friction. It attempts to stop the engine by means of a brake on the flywheel and then measures the weight supported at the end of the arm attached to the brake.

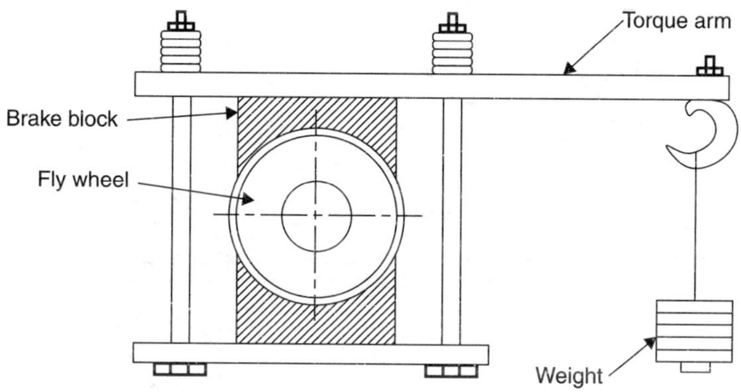

Figure 7.6 Pony brake dynamometer

The pony brake dynamometer consists of a wooden block, frame, rope, brake shoes and flywheel, as shown in Fig. 7.6. Spring-loaded bolts are provided to increase the friction by tightening the wooden block. When the engine rotates, the flywheel also rotates. When the wooden blocks are tightened against the rotation of the flywheel, by pressing them against the flywheel, the whole of the power absorbed is converted into heat due to friction. The torque produced by the engine is absorbed at the flywheel in terms of heat due to friction. This torque is balanced by the load applied at the end of the arm attached to the brake system through the wooden blocks. A lot of frictional heat is produced between the flywheel drum and the wooden blocks; therefore, the rotating drum is required to be cooled frequently.

Brake power can be determined using the formula $P_b = 2\pi NT/60$, where N is the speed of the engine measured with a tachometer (in rpm); T is the torque produced at the crankshaft ($W \times L$); W is the weight applied at the end of the arm (in kg); and L is the distance of the weight from the axis centre of the flywheel (in m).

Rope brake dynamometer: This is another type of mechanical dynamometer used to measure the brake power of an engine. The construction details are given in Fig. 7.7. It consists of a number of turns of rope wound around a rotating drum attached to the output shaft of an engine. One side of the rope is connected to a spring balance and the other to a loading device. The working principle of this dynamometer is similar to that of the pony brake dynamometer, except for the loading mechanism. In the rope brake dynamometer, the power is absorbed between the rope and the rotating drum, in terms of heat due to friction. Due to heat dissipation, the drum in the rope brake dynamometer also needs to be cooled frequently.

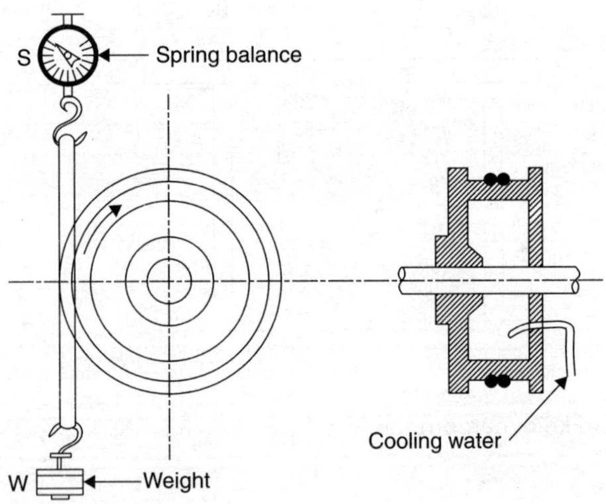

Figure 7.7 Rope brake dynamometer

The formula to find the brake power is given by $P_b = \pi DN (W - S)$ in watts, where D is the brake drum diameter (in m); W is the weight of the load (in kg); and S is the spring balance reading (in kg).

The changing friction coefficient of the rope with change in temperature affects the accuracy of power measurement in the rope brake dynamometer.

7.8 DYNAMOMETER – ELECTRICAL

Almost any form of rotating electric machine can be used as a driving dynamometer, as an absorption dynamometer, or both. The eddy current dynamometer is a type of electrical dynamometer. However, it is strictly of the absorption type. It is incapable of driving a test machine such as a pump or a compressor, and hence, it is useful for measuring the power from a source, such as an internal combustion engine or electric motor.

The principle of operation of this dynamometer is as follows. When a conducting material moves through a magnetic flux field, voltage is induced, causing current to flow. If the conductor is a wire forming a part of a complete circuit, current will pass through the circuit and, with some form of commutating device, a type of AC or DC generator can be set up. If the conductor is simply an isolated piece of material, such as a short metal bar, and not a part of a complete circuit, voltage will not be induced. However, only local currents may flow in practically short circuit paths within the metal bar itself. These currents, called eddy currents, will be dissipated in the form of heat. The heat produced, and thus, the eddy currents produced is directly proportional to the rotating speed of the shaft whose output power needs to be measured.

The schematic diagram of the dynamometer is shown in Fig. 7.8. It consists of a metal disk or wheel that is rotated in the flux of a magnetic field. The magnetic field is produced by coils excited by an external source and attached to the housing of the dynamometer. As the metal disk rotates, eddy currents are induced and the reaction with the magnetic field tends to make the complete housing in the trunnion bearings rotate.

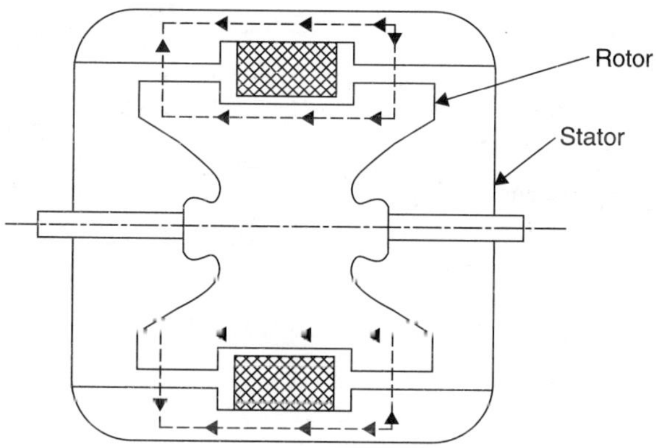

Figure 7.8 Eddy current dynamometer

The eddy currents produced are dissipated in the form of heat, and hence, the dynamometer requires a cooling arrangement. Torque is measured in exactly the same manner as in other types of absorption dynamometers, i.e., with the help of the moment arm. In this dynamometer, the current in the electromagnets is regulated to control the load.

Advantages

- As the development of eddy current is smooth, the torque produced in the rotor shaft is also smooth, and hence, continuous in all conditions.
- It produces relatively higher torque under low-speed conditions.
- The level of field excitation is below 1% of the total power being handled by the dynamometer, and hence, it is easy to control the operation.
- It has no intricate rotating parts, except the shaft bearing, and hence, less maintenance is required.

7.9 TRANSMISSION DYNAMOMETER

Transmission dynamometers may be thought of as passive devices, neither appreciably adding to nor subtracting from the energy involved in the test system. For example, any gear box system used for gear change is subjected to reaction torque, equal to the difference between the output and input torques. When the reaction torque of a gear box is measured, a function of input or output torque can be obtained. More common forms of transmission dynamometers are based on the calibrated measurement of unit or total strains in elastic load-carrying members. The beam and strain gauge transmission dynamometer is of this type. This type of dynamometer uses bonded strain gauges applied to a section of torque-transmitting shaft, as shown in Fig. 7.9. This type of dynamometer is also called a torque meter.

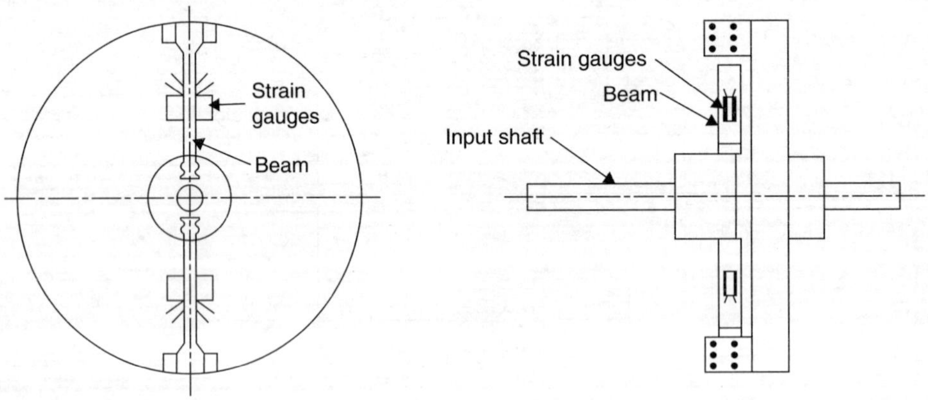

Figure 7.9 Transmission dynamometer

This dynamometer is used as a coupling between the driving and driven machines or between any two portions of a machine. A four-arm bridge circuit is used, incorporating modulus gauges to minimise temperature sensitivity. Electrical connections are made through slip rings, with provision to lift the brushes when they are not in use, thereby minimising the wear. A common strain gauge indicator is used to interpret the output. When the input shaft transmits the torque, part of the torque is transmitted to the dynamometer, which in turn is subjected to a twisting action. This angular twist causes a strain in the form of bending of the beam of the dynamometer. When the strain gauge indicator is calibrated for the torque, the torque of the rotating shaft is obtained directly.

SUMMARY

- Torque is a rotational force applied through a distance. Static torque does not have angular acceleration, whereas a dynamic force involving acceleration plays a role in dynamic torque.
- Both torsion meters and dynamometers are used to measure the torque, and thus, the power transmitted by the driving shaft.
- A mechanical torsion meter works on the principle that when a shaft is connected to a driving engine and a driven load, a twist occurs between the ends of the shaft. This twist is calibrated in terms of torque.
- An optical torsion meter works on the principle that the angular deflection of light rays from the two castings fixed on the rotating shaft is proportional to the twist on the shaft.
- An electrical torsion meter works on the principle that the teeth of the two slotted discs align with each other when there is no rotation, and the teeth of the two slotted discs misalign with each other at the time of rotation, and thus, the torque is transmitted. The change in the phase of the two pulses is calibrated for the torque to be measured.
- In a strain gauge torsion meter, there is a strain on the shaft when the torque is transmitted by the shaft. This strain results in a change in the resistance of the strain gauge system. This change in resistance is converted to a change in frequency, using a frequency converter, and is calibrated for torque.
- A hydraulic dynamometer works on the principle of dissipating the power in fluid friction rather than in dry friction.
- In a pony brake dynamometer, the power available at the crankshaft of the engine is converted into heat by means of dry friction, by applying a brake on the flywheel of the engine, using wooden blocks.

- The rope brake dynamometer is similar to the pony brake dynamometer, except that the wooden blocks are replaced by ropes rotating around the rotating drum attached to the flywheel of the engine.
- The electrical dynamometer works on the principle of producing eddy currents in the meter fitted with electromagnets, directly proportional to the torque produced in the shaft of the rotor.
- The beam and strain gauge transmission dynamometer uses strain gauges fitted on a beam attached to the flywheel of the shaft. The strain produced on the beam due to angular displacement of the shaft is proportional to the torque transmitted by the engine shaft.

EXERCISES

Multiple Choice Questions

1. Which of the following instruments is used for measuring torque?
   a) Manometer　b) Flowmeter
   c) Dynamometer　d) All of the above
2. Torque can be
   a) Static　b) Dynamic
   c) Both a and b　d) None of the above
3. Torque is directly proportional to the following, for a fixed shaft:
   a) Shear modulus of shaft
   b) Second moment of inertia of shaft
   c) Length of shaft　d) Twist of shaft
4. Which of the following is correct?
   a) $\Phi = TL/JG$　b) $\Phi = JG/TL$
   c) $\Phi = TG/JL$　d) $\Phi = JL/TG$
5. The Torque/Weight ratio of an instrument indicates
   a) Sensitivity　b) Selectivity
   c) Accuracy　d) Fidelity
6. Which parameter(s) can be measured using a dynamometer?
   a) Force　b) Torque
   c) Power　d) All of the above

7. Hydraulic brake functions on the principle of
   a) Law of conservation of energy
   b) Law of conservation of momentum
   c) Pascal's law
   d) None of the above

8. Hydraulic dynamometer works on the principle of dissipating the power in
   a) Fluid friction
   b) Dry friction
   c) Hydraulic machine
   d) Lifting a load

9. Double block brake is a type of
   a) Band brake
   b) Internal expanding shoe brake
   c) Shoe brake
   d) None of the above

10. Which types of brake have wooden blocks placed inside flexible steel bands?
   a) Pivoted block brake
   b) Block brake
   c) Band brake
   d) Band and block brake

11. Pony brake dynamometer works on the principle of converting power into heat by means of
   a) Wet friction
   b) Dry friction
   c) Mechanical power
   d) All of the above

12. Which of the following is a type of absorption dynamometer?
   a) Torsion dynamometer
   b) Tatham dynamometer
   c) Pony brake dynamometer
   d) Epicyclic dynamometer

13. Pony brakes are used for testing
   a) Small engines
   b) Slow speed engines
   c) High speed engines
   d) Large capacity engines

14. Which of the following is not a drum dynamometer?
   a) Disc brake
   b) Internal expanding brake
   c) External contracting brake
   d) All of the above

15. The following affects the accuracy of power measurement considerably in rope brake dynamometer:
   a) Rope length
   b) Changing friction coefficient of the rope
   c) Cooling water passing through the brake drum
   d) All of the above

16. Why does the rotating drum need to be supplied with flowing water frequently in mechanical dynamometer?
   a) To apply braking effect
   b) To clean the drum

c) To remove heat produced between drum and brake
d) To produce hot water

17. Eddy current dynamometer is of
a) Absorption type
b) Driving type
c) Generator type
d) All of the above

18. Strain gauge is insensitive to
a) Tensile force
b) Compressive force
c) Lateral force
d) None of the above

19. With all the four strain gauges active, the most sensitive bridge will have
a) One strain gauge in compression and three strain gauges in tension
b) One strain gauge in tension and three strain gauges in compression
c) Two strain gauges in tension and two strain gauges in compression
d) All four strain gauges in either tension or compression

20. Strain gauges can be used as primary or secondary transducers for measuring
a) Pressure
b) Force
c) Torque
d) All of the above

Answers

1. c	2. c	3. d	4. a	5. a
6. d	7. c	8. a	9. c	10. d
11. b	12. c	13. a	14. a	15. b
16. c	17. a	18. c	19. c	20. d

Review Questions

1. Explain how torque and power are inter-related.
2. With suitable examples, explain the two main types of torque.
3. Explain the working principle of a mechanical torsion meter with a neat sketch of construction.
4. Write the advantages and disadvantages of mechanical torsion meter.
5. Explain how an optical torsion meter measures the torque transmitted by an engine shaft, with a neat diagram.
6. How does an electrical torsion meter operate to determine the angular displacement of a shaft connected to an engine?

7. With a neat sketch, explain how a four-strain gauge system is used to measure the torque of an engine shaft.
8. Why does a strain gauge torsion meter require less maintenance compared to mechanical and electrical torsion meters?
9. Describe the construction details of the Heenan and Froude hydraulic dynamometer, with a neat sketch. Also explain its working principle.
10. Explain in detail the working principle of pony brake dynamometer, with a neat sketch.
11. Briefly explain the working principle of rope brake dynamometer, with a simple sketch.
12. How does the rope brake dynamometer differ from the pony brake dynamometer?
13. How does an eddy current dynamometer function? Explain with a neat sketch.
14. Why does eddy current dynamometer require less maintenance?
15. Briefly explain how a beam and strain gauge transmission dynamometer differs from other dynamometers. Also explain its working principle, with a neat sketch.
16. Calculate the power transmitted if dead weight on brake is 70 kg and the engine has a speed of 300 rpm. One end of the rope has a diameter of 5 cm attached to a spring balance, which has a reading of 150 N; the diameter of the drum is 1 m. **Ans:** 8.851 kW

8

Measurement of Level, Vibration and Translation

8.1 MEASUREMENT OF LEVEL

In industry, liquids such as water, chemicals and solvents are used for various processes, and hence, are stored in containers and vessels. The volume of these stored liquids can be determined by measuring the level of the liquid in the container. Liquid level refers to the position or height of a liquid surface above a datum line. The level affects not only the quantity delivered but also the pressure and the flow rate. The substance to be measured can be in a container or in its natural form, like in a river or lake. Level measurement can be continuous or based on points. Continuous level sensors measure the level to determine the exact amount of substance in a continuous manner, whereas point level sensors indicate if the substance is above or below the sensing point. Measurement of the liquid level can be performed by direct and indirect methods.

In the direct methods, the level is indicated directly by means of simple mechanical devices. Measurement is not affected by changes in the density of the material. For example, dipstick, sight glass, hooks, resistance tapes, floats and radar.

In the indirect methods, change in the liquid level is converted into a measurable signal, like resistive, capacitive, inductive or hydrostatic pressure, using a suitable transducer. The measurement is affected by changes in the material, and hence, a corrective factor must be used to recalibrate the instrument. For example, load cell, hydrostatic head methods and capacitance.

The following section explains the details of various level measuring techniques and instruments under direct methods.

8.2 DIRECT METHODS

8.2.1 Hook Level Indicator

When the liquid level in an open tank is measured directly on a scale, it is difficult to read the level accurately because of parallax error. In this case, a hook level indicator is used. It consists of a thin wire made of a corrosion resistive alloy (like stainless steel) which is bent into a U shape with one arm longer than the other, as shown in Fig. 8.1. The shorter arm is pointed and has a 60° taper, while the longer one is attached to a slider with a Vernier scale that moves over the main scale to indicate the level of the liquid.

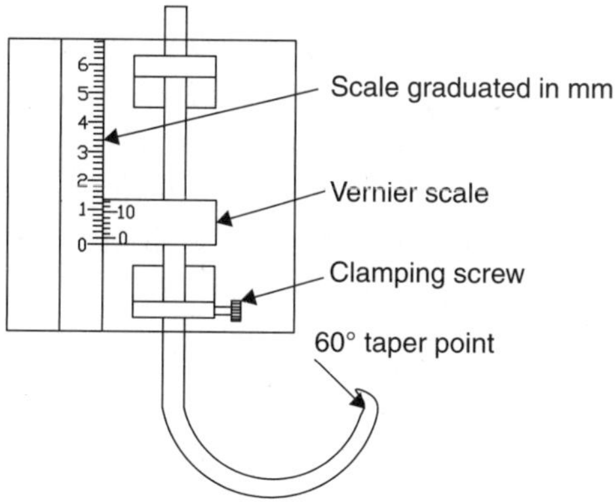

Figure 8.1 Hook level indicator

The hook is pushed below the surface of the liquid whose level is to be measured and gradually raised until the point is just about to touch the surface. It is then clamped and the level is read on the scale.

8.2.2 Sight Glass (Gauge Glass) Indicator

It is used for continuous indication of the liquid level within a tank or container. As shown in Fig. 8.2, it consists of a graduated tube of toughened glass connected to the bottom of the interior of the tank, which contains the liquid to be measured. This is a simple sight glass for an open tank in which the liquid level in the sight glass matches the liquid level in the tank. As the liquid level in the tank rises and falls, the level in the sight glass also rises and falls accordingly. Thus, by measuring

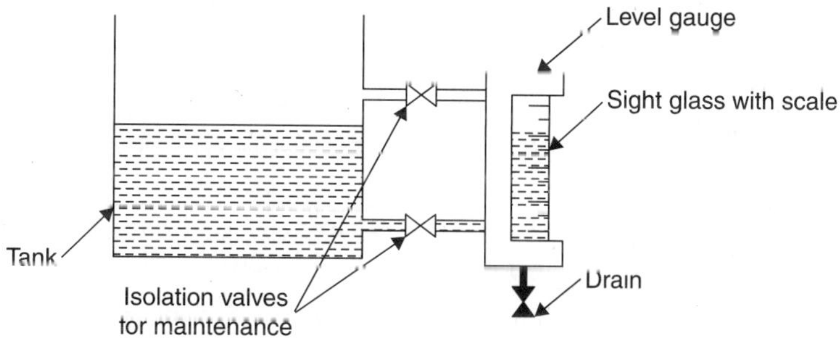

Figure 8.2 Sight glass indicator

the liquid level in the sight glass, the liquid level in the tank is measured. It is not necessary to use the same liquid as in the tank, and hence, coloured liquid can be used to aid visibility.

8.2.3 Float-operated Level Indicator

It is used to measure the liquid level in a tank in which a float rests on the surface of the liquid and follows the changing level of the liquid. The movement of the float is transmitted to a pointer through a suitable mechanism, which indicates the level on a calibrated scale. Various types of floats such as hollow mewl spheres, cylindrical floats and disc-shaped floats are used. A simple float level indicator is shown in Fig. 8.3. In this mechanism, the movement of the float is transmitted to a pointer by a stainless steel or phosphor-bronze flexible cable wound around a pulley, and the pointer indicates the liquid level in the tank. The float is also made up of corrosion resisting material, such as stainless steel, and rests on the liquid surface between two grids/guides to avoid error due to turbulence of the liquid surface.

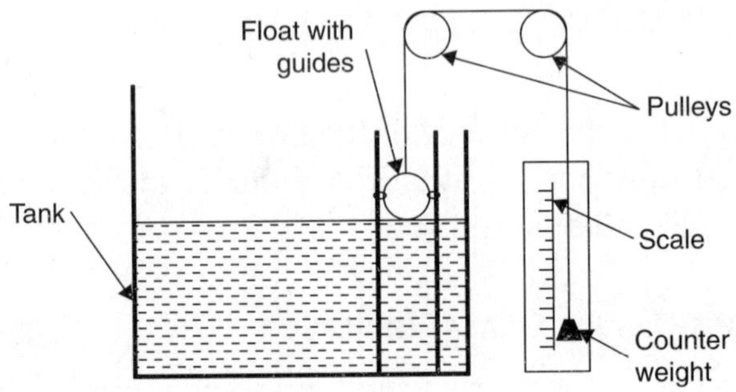

Figure 8.3 Float-operated level indicator

8.3 INDIRECT METHODS

This section describes the level measuring techniques and instruments used in indirect methods.

8.3.1 Capacitive Liquid Level Sensor

The principle of capacitive level measurement is based on change in capacitance. An insulated electrode acts as one plate of the capacitor and the tank wall (or reference electrode in a non-metallic vessel) acts as the other plate. Capacitance is based on the level of the liquid. An empty tank has lower capacitance than a filled tank.

As shown in Fig. 8.4, the sensor consists of an insulated capacitance probe (a metal electrode) fixed firmly near and parallel to the maul wall of the tank. If the liquid in the tank is non-conductive, the capacitance probe and the tank wall will form the parallel plates of the capacitor and the liquid in between will act as the dielectric medium. If the liquid in the tank is conductive, the capacitance probe and the liquid will form the parallel plates of the capacitor, and the insulation of the probe will act as the dielectric medium. A capacitance measuring device is connected to the probe and the tank wall, calibrated in terms of the liquid level in the tank.

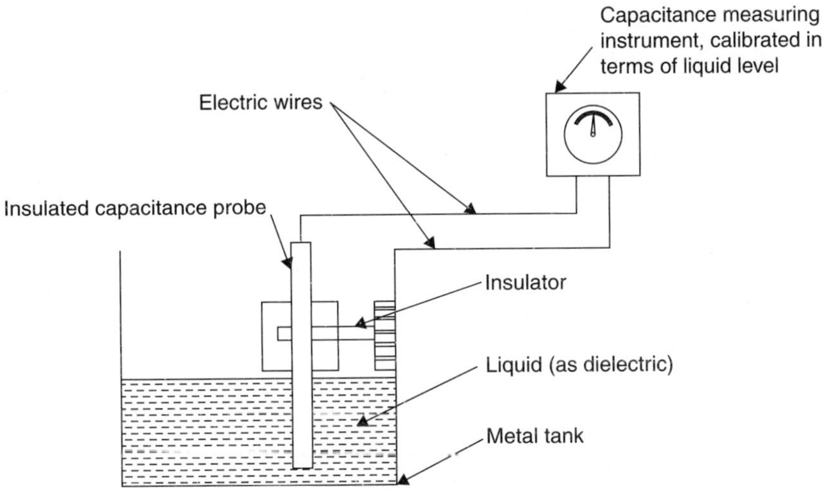

Figure 8.4 Capacitance liquid level indicator

The value of capacitance C is given by $C = E\ (K\ A/d)$, where C = capacitance in picofarads (pF); E = constant, absolute permittivity of free space; K = relative dielectric constant of the insulating material; A = effective area of the conductors; and d = distance between the conductors.

The change in capacitance can be measured using an AC bridge by applying an RF signal between the conductive probe and the vessel wall. The RF signal results in very low current flow through the dielectric process material in the tank, from the probe to the vessel wall. When the liquid level in the tank drops, the dielectric constant also drops, causing a fall in the capacitance reading and a minute drop in current flow. This change is detected by the level switch's internal circuitry and translated into a change in the relay state of the level switch, in the case of point level detection. In the case of continuous level detectors, the output is not a relay state, but a scaled analogue signal.

Level measurement can be divided into the following categories:

- **Measurement of non-conductive material:** A bare probe arrangement is used as liquid resistance which is sufficiently high to enable it to act as a dielectric medium. As the electrode and the tank are fixed in place, the distance d is constant, and hence, the capacitance is directly proportional to the level of liquid acting as dielectric medium.

- **Measurement of conductive material:** The probe plates are insulated using a thin coating of glass or plastic to avoid short-circuiting. The conductive liquid material acts as the ground plate of the capacitor.
- **Non-contact (Proximity) measurement**: In this type, the area of the capacitance plates is fixed, but the distance between the plates varies. This type of measurement does not produce a linear output and is used when the level varies by several inches.

Advantages

- Versatile, simple to use
- Relatively inexpensive
- No moving parts, and hence, requires minimal maintenance and is easy to clean
- Reliable and good range of measurement
- Can be designed for high temperature and pressure applications

Limitation: It is difficult to measure the level of low density materials and materials with particle sizes exceeding a diameter of 13 cm. This is because of their very low dielectric constants due to the large air space between the particles.

Applications: The capacitive liquid level sensor can be used to measure the level of:

- Liquid metals at very high temperature
- Powdered and granular solids
- Liquefied gases at very low temperature
- Corrosive materials like hydrofluoric acid

8.3.2 Ultrasonic Fuel Level Meter

This uses ultrasonic technology and is designed for monitoring the levels of water, milk, alcohol, acid, solvent, diesel, gasoline, liquefied gas, kerosene and other fuels kept in different types of tanks. It is widely used for fuel level measurement in vehicle fuel tanks in order to control the fuel consumption in real time.

It comprises an ultrasonic sensor (ultrasonic pulse source and echo receiver) and a data processing box (electronic processing unit and level indicator), as shown in Fig. 8.5. The sensor is installed directly on the external top part of the tank to detect the level of the liquid and is driven by the data processing box.

It works on the principle of reflection of sound waves from the surface of the liquid. The transmitter (pulse source) sends an ultrasonic wave towards the free surface of the liquid. This wave is reflected from the surface and is received by the

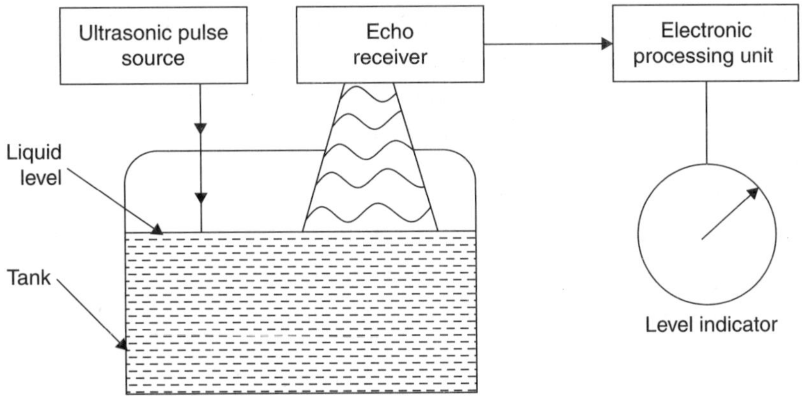

Figure 8.5 Ultrasonic fuel indicator

echo receiver. The time taken by the transmitted wave to travel to the surface of the liquid and then back to the receiver gives the liquid level, when it is calibrated. As the liquid level changes, the time taken by the wave to reach the surface and return to the receiver also changes. Thus, the change in time gives the change in the liquid level accurately.

Advantages

- Easy to install without tank drilling
- No contact with the fuel
- High measurement accuracy
- More safety from fire and explosions
- Compatible with any GPS/GPRS tracker
- Supports metal, plastic, glass or other tanks
- Supports tanks with a wall thickness of up to 6 mm

Drawbacks

- Experienced and skilled operators are required for measurement
- Very expensive

8.3.3 Magnetic Level Indicator

This is used to measure the level of liquids that contain corrosive and toxic materials. The construction details are given in Fig. 8.6. It consists of a float in which a magnet is placed, inside a stainless steel sealed chamber containing the liquid whose level is to be determined. The float moves up or down with increase or decrease in the liquid level. The float contains a magnetic shielding device with flux rings and a visual indicator housing containing small flags arranged in series and attached to the sealed chamber. The individual flags contain an alignment magnet that couples with the float magnet as the float moves up or down within the piping column. The flags are coated with luminous paint of different colours on both sides and are capable of rotating 180°. As the liquid level changes, the float moves up or down along with the magnet. Due to the movement of the magnet, the flags rotate in the opposite direction and change their colour, indicating the liquid level on the measuring scale.

Advantages

- Provides easy access to clean the sealed chamber
- Can be adapted to a variety of mounting styles

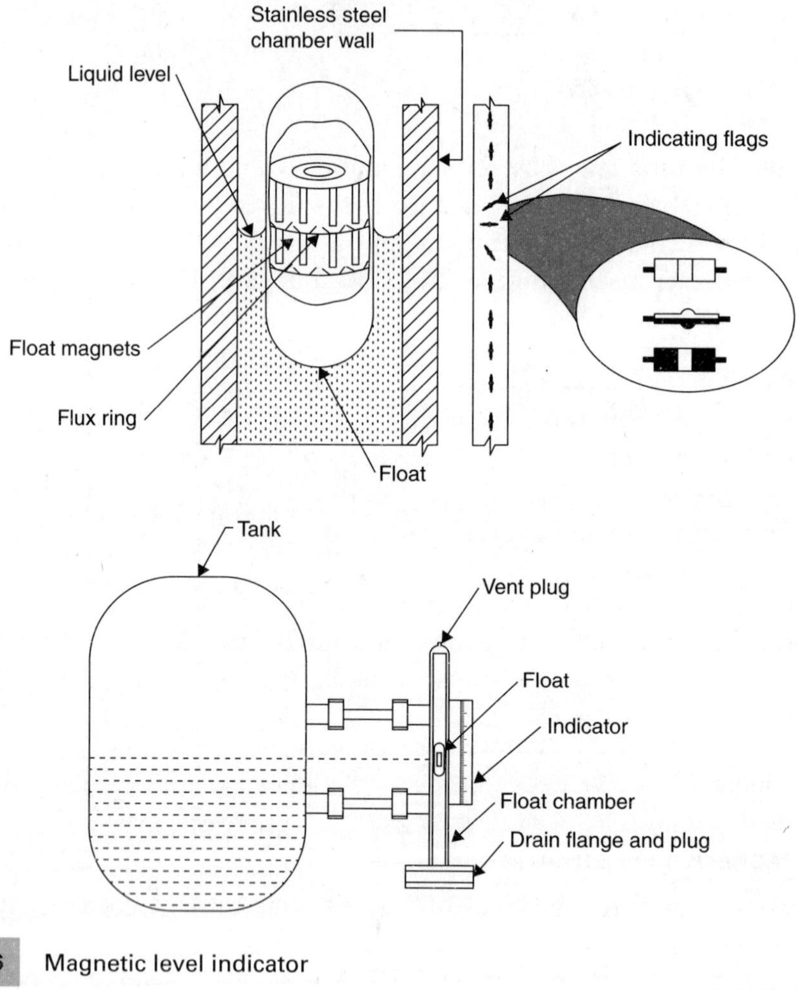

Figure 8.6 Magnetic level indicator

8.3.4 Cryogenic Fuel Level Indicator

Cryogenic fuels are in the gaseous state at normal temperature and change into the liquid state when cooled to very low temperature (usually below –150°C). The cryogenic fuel level indicator (Fig. 8.7) works on the principle that a superconductor exhibits zero resistance at a temperature below which the cryogenic fluid exists in liquid state. When the level of the cryogenic fuel changes, there is a change in the resistance level of the superconductor wire loop, and hence, a change in the conductivity level of the superconductor. This change in conductivity is measured using a high impedance ohmmeter. When the ohmmeter is calibrated for different liquid levels, the measurement directly gives the level of the cryogenic fuel in the storage tank. Cryogenic fuels need to be stored at extremely low temperatures in order to remain in the liquid state. These fuels are used in machinery that is operated in space (e.g., satellites, rockets) because ordinary fuel cannot be used in space due to the absence of oxygen in the environment, which supports combustion. Cryogenic fuels most often constitute liquefied gases such as liquid hydrogen. Liquefied natural gas (LNG) and liquid methane are commonly used commercial cryogenic fuels.

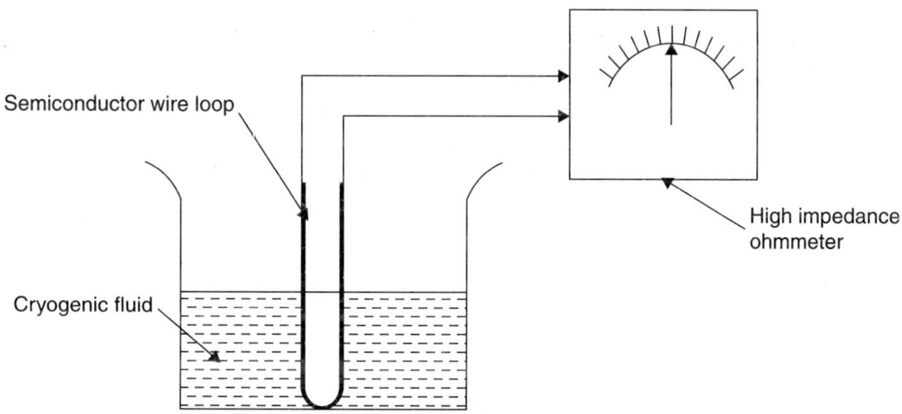

Figure 8.7 Cryogenic fuel level indicator

8.3.5 Bubbler (Purge) Level Indicator

Bubbler tubes provide a simple and inexpensive but less accurate (± 1–2%) fluid level measurement system for corrosive or slurry fluid containers and open or vented fluid containers, especially those in harsh environments such as cooling tower sumps, swimming pools, reservoirs, vented fuel tanks and drain sumps.

It is based on the principle that the liquid level is determined by measuring the pressure required to force a gas into a liquid at a point above the bottom surface of the container. The bubbler indicator uses compressed air or an inert gas introduced through a dip pipe called the bubble/sensing tube immersed at a fixed depth, usually at the bottom of the liquid container. The air flow restrictor (pressure regulator) reduces air flow to a very small amount, slightly higher than the hydrostatic pressure at the lower end of the bubbler tube. As the pressure builds, bubbles are released from the end of the bubble tube. The pressure inside the bubble tube is maintained as air bubbles escape through the liquid. Changes in the liquid level cause a change in the air pressure in the bubble, which can be detected by a pressure sensor (gauge) attached to the top of the bubble tube. Most bubble tubes use a small V-notch at the bottom of the tube to assist with the release of a constant stream of bubbles, because intermittent large bubbles will affect the consistency in measurement of pressure. To avoid blockages at the bottom of the bubble tube due to debris in the fluid, the tube tip is kept about 3 inches from the bottom of the tank. If the pressure gauge is calibrated properly in the unit of liquid level, the pressure inside the bubble tube will directly give the level of the liquid in the tank. A schematic diagram of a bubbler level indicator is given in Fig. 8.8.

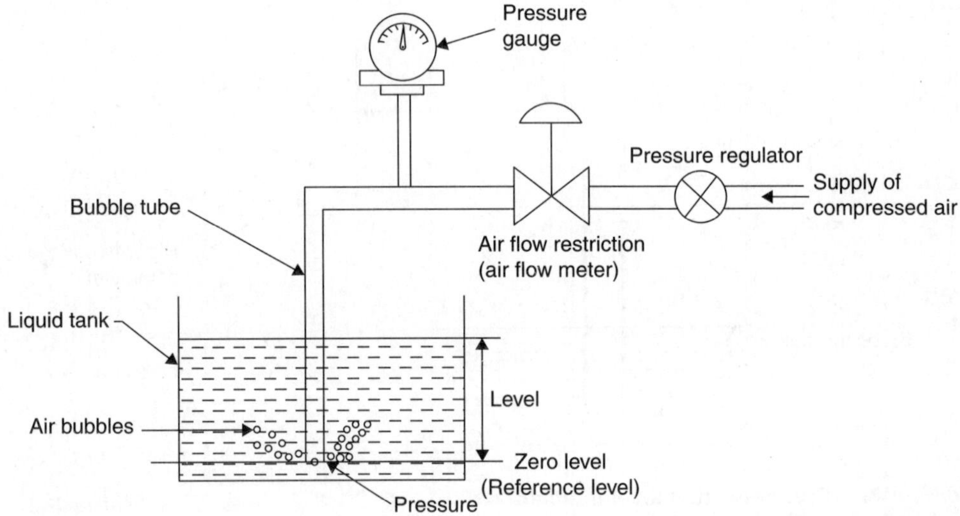

Figure 8.8 Bubbler level indicator

Advantages

- Simple and inexpensive
- Suitable for use with corrosive fluids
- Intrinsically safe
- Can be used for high-temperature applications

Drawbacks

- Requires compressed air and installation of air lines
- Debris/Dirt found at the bottom of the liquid may block the flow of air in the bubble tube

- Susceptible to density variations, freezing and coating by the process fluid
- Device must be able to withstand the maximum air pressure imposed if the pipe becomes blocked

8.4 MEASUREMENT OF VIBRATION

Vibration is a time-based displacement of an object around a central static position. Viewing vibration in the time domain reveals the instantaneous location of the vibrating surface at different moments in time. Vibration measurement is complex because it has many components, such as displacement, velocity, acceleration and frequency. Also, each component is measured in different ways, such as peak-to-peak, peak, average and RMS.

Vibration measurement is sometimes used as an indirect measure of some other parameter. Often, condition monitoring of wear, fatigue and failure requires vibration measurement. Other applications are concerned with the displacement of the object under study because unintended displacements affect the performance of the system. Machine tools and hard disk drives are examples of this type of positional or relative vibration.

Basic principle of vibrometry: Vibrometry is based on the principle of the Doppler effect – if a wave is reflected by a moving object and detected by a measurement system, the measured frequency shift of the wave can be given as $2v/\lambda$, where v is the velocity of the object and λ is the wavelength of the reflected wave.

8.4.1 Laser Doppler Vibrometer (LDV)

This is a scientific instrument used to make non-contact vibration measurements of a surface. The set-up of the LDV is shown in Fig. 8.9.

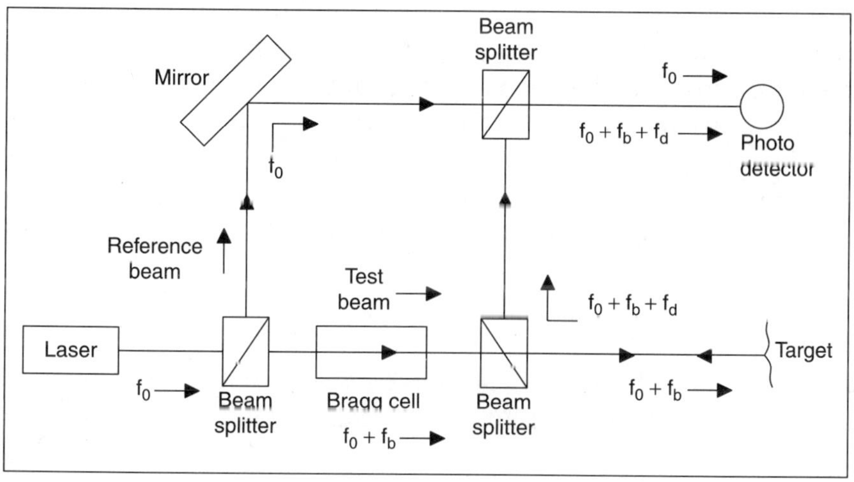

Figure 8.9 Laser Doppler vibrometer

A laser beam from the LDV is directed towards the surface of interest, and the amplitude and frequency of vibration of the surface is extracted from the Doppler shift of the reflected laser beam frequency due to vibration of the surface of interest. The output of an LDV is generally a continuous analogue voltage that is directly proportional to the target velocity component along the direction of the laser beam.

The LDV consists of a two-beam laser interferometer, which measures the frequency difference between an internal reference beam and a test beam. The test beam is directed at the target and the scattered laser beam from the target is collected and it then interferes with the reference beam on a photo detector, typically a photodiode. Most commercial vibrometers work in a heterodyne regime by adding a known frequency shift to one of the beams. This frequency shift is usually generated by a Bragg cell. From Fig. 8.6, it can be seen that the laser beam with a frequency f_0 is split into a reference beam and test beam using a beam splitter. The test beam passes through the Bragg cell to add a frequency shift f_b. This frequency shifted beam is then directed towards the surface of interest (target).

The motion of the target adds a Doppler shift to the beam, given by $f_d = 2v(t) \cos \alpha/\lambda$, where $v(t)$ is the velocity of the vibration of the surface (target) as a function of time, a is the angle between the laser beam and the velocity vector and λ is the wavelength of the laser beam. The laser beam scatters from the target in all directions. However, some portion of the beam is collected by the LDV and reflected by the beam splitter to the photo detector. This beam has a frequency equal to $f_o + f_b + f_d$. This scattered light beam is combined with the reference beam at the photo detector.

The initial frequency of the laser beam is very high, with a frequency of more than 10^{14} Hz, which is higher than the response of the photo detector. However, the photo detector responds to the beat frequency between the two beams, which is at $f_b + f_d$, in the frequency range of tens of MHz. The output of the photo detector is a standard frequency modulated (FM) signal, with the Bragg cell frequency as the carrier frequency and the Doppler shift as the modulation frequency. This signal can be demodulated to derive the velocity against time of the target.

A scanning vibrometer consists of two mirrors in the X and Y planes, enabling the laser beam to be moved across the surface of interest in both the planes. A three-dimensional vibrometer measures all three components of the velocity of the target by measuring the location, with three independent beams striking the target from three different directions.

Applications

- In aerospace, as tools in non-destructive inspection of aircraft components
- In acoustics, as standard tools for speaker design, and to diagnose the performance of musical instruments
- In architecture, for bridge and structure vibration tests

- In automotives, for brake diagnostics and structural dynamics
- For biological purposes, for eardrum diagnostics and insect communication
- In the detection of buried landmines
- In hard disk drive diagnostics, specifically in the area of head positioning
- In calibration of transducers

8.4.2 Seismic Instruments (Vibrometers and Accelerometers)

A vibrometer is a detector-transducer yielding an output, usually a voltage that is proportional to displacement or velocity. Displacement or velocity is determined primarily by the secondary transducing element. For example, if a differential transformer (LVDT) or a voltage dividing potentiometer is used, the output will be proportional to the displacement. Alternatively, if a variable reluctance element is used, the output will be a function of velocity. The term **accelerometer** is applied to devices whose outputs are functions of acceleration. There is a basic difference in design and application between vibrometers and accelerometers.

Vibrometer and accelerometers are usually of seismic mass form. A typical seismic motion measuring instrument is shown in Fig. 8.10. A spring-supported mass is mounted in a suitable housing, with a sensing element provided to sense the relative motion between the seismic mass and the housing. Suitable damping is provided and is shown as a dashpot mounted between the mass and the housing. The action of the seismic instrument is basically a function of acceleration through the inertia of the mass. The output is determined by the relative motion between the mass and the housing.

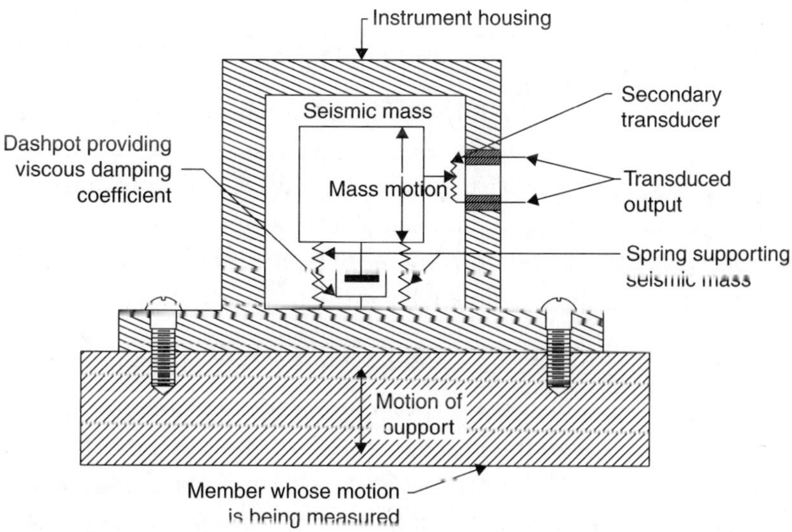

Figure 8.10 Seismic instrument

This principle results in two types of seismic mass instruments:. the vibrometer and the accelerometer. Vibrometers use a variable reluctance transducer, in which the relative motion between a coil and the magnetic flux field is used. Vibrometers are velocity sensitive because the output is proportional to the rate at which the lines of magnetic flux are cut. The same set-up can be designed so that the relative displacement between the seismic mass and the housing serves as a function of acceleration, with a proper selection of natural frequency and damping. The output in such a case is, therefore, calibrated in terms of acceleration.

The following sections discuss the principle of operation of vibrometers and accelerometers of seismic type to measure vibrations in terms of velocity and acceleration, respectively.

Vibrometer: Figure 8.10 shows the set-up of a seismic vibrometer. The instrument is screwed to a support placed on the vibrating member, whose vibrating motion is to be measured. When the member vibrates, the support and the instrument housing also vibrate. This causes a relative motion of the seismic mass inside the housing.

The following expression gives the relative displacement amplitude between the seismic mass M and the support.

$$S_1 = \frac{S_2(\Omega/\omega_n)^2}{\sqrt{[1-(\Omega/\omega_n)^2]+[2\varepsilon(\Omega/\omega_n)]^2}}$$

where S_1 is the relative displacement amplitude between the seismic mass and the supporting member; S_2 is the displacement amplitude of the instrument housing, and hence, of the supporting member to which it is attached; Ω is the exciting frequency; ε is the damping coefficient; ω_n is the undamped natural frequency of the system, equal to $\sqrt{k g_c/M}$; k is the deflection constant for the spring support; g_c is the dimensional constant; and M is the mass of the seismic element.

The secondary transducer (for example, LVDT) determines the relative displacement between the seismic mass and the support, and S_1 is its amplitude. Subsequent determination of S_2 from the above expression is used to find out the absolute displacement of the supporting member using the following relationship.

$$\text{Absolute displacement of the support} = S_2 \cos \Omega t$$

When the vibrometer is calibrated by subjecting it to steady-state harmonic motion of known amplitude and frequency, the output of the vibrometer is a sinusoidal voltage, measured by a cathode ray oscilloscope or a voltmeter. The vibrometer is then placed on the supporting vibrating member whose displacement needs to be measured, the output is measured through the secondary transducer and the corresponding sinusoidal voltage is obtained. This sinusoidal voltage is then compared with the calibrated values to determine the actual amplitude of vibration of the supporting member.

Piezoelectric accelerometer: This is a seismic motion-measuring instrument. It is a mechanical or electromechanical apparatus that can be used to measure the acceleration or deceleration of a moving or vibrating body. The sensors used in an accelerometer may be electromechanical servo, piezoelectric, liquid tilt, bulk micro-machined piezo-resistive or capacitive. Each has distinct characteristics pertaining to output signal, development cost and type of environment in which it functions best. Among them, the piezoelectric accelerometer has desirable features such as accuracy, durability, large dynamic range, ease of installation and long life span, though it is costlier than the other types.

The Greek word *piezein* means 'to squeeze or press'. When a physical force is exerted on the accelerometer, the seismic mass loads the piezoelectric element according to Newton's second law of motion. The force exerted on the piezoelectric material can be observed in the change in the electrostatic force or voltage generated by the piezoelectric material. The physical force exerted on the material can be of any type of bending and compression. Similarly, the piezoelectric materials also fall into one of two categories: single crystal (quartz) and ceramic materials. Crystals have a long life span in terms of sensitivity, while ceramic materials are more sensitive and less expensive. Ceramics use barium titanate, lead-zirconate-titanate and lead metaniobate. When low-sensitivity piezoelectric materials are used, two or more crystals can be connected for output multiplication. Nowadays, IC amplifiers/impedance convertors are generally packaged within the case of the accelerometer itself for amplified output. The inherent properties of piezoelectric accelerators – high frequency response and the ability to generate high resonant frequencies – make them a better alternative to strain gauges. The cross-sectional view of a piezoelectric accelerometer is shown in Fig. 8.11.

This instrument uses the piezoelectric effect of certain materials to measure dynamic changes in mechanical variables like acceleration, vibration and mechanical shock. Like transducers, it converts one form of energy into another and provides an electrical signal in response to the quantity, property or condition that is being measured. It acts upon a seismic mass restrained by a spring or suspended on a cantilever beam and converts a physical force or displacement into an equivalent electric signal. It uses an internal piezoelectric element coupled with a loading mass to form a single degree of freedom 'mass-spring' system.

The device is placed on the object whose vibration level is to be measured. The movements (vibrations) of the object exert a physical force (acceleration) on the housing of the accelerometer. An instantaneous change in the internal peizoelectric element produces a charge at the accelerometer's output terminals. This charge is proportional to the applied acceleration. For interfacing purposes, the piezoelectric accelerometer can be modelled as a voltage generator.

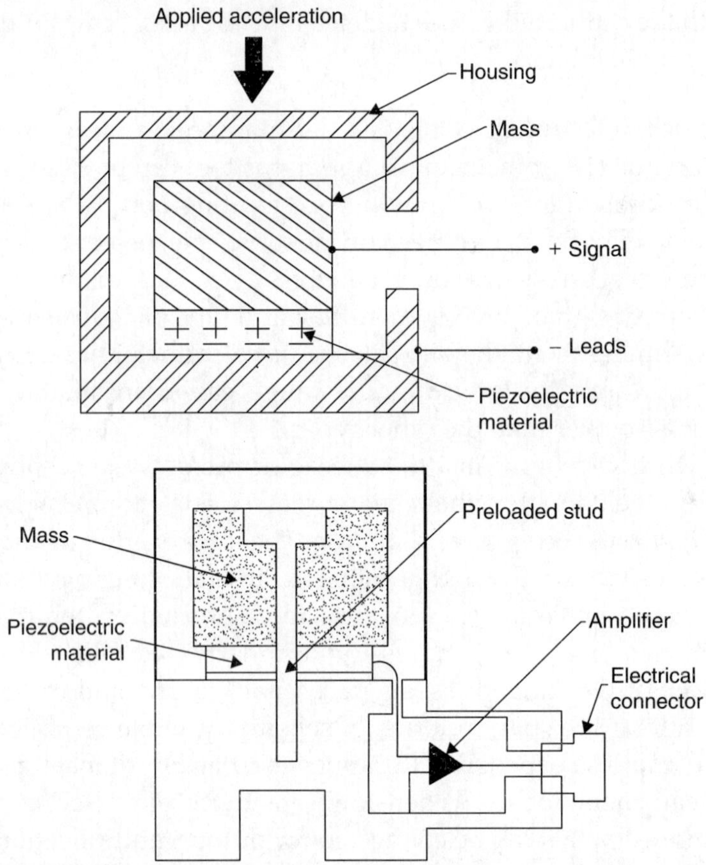

Figure 8.11 Piezoelectric accelerometer

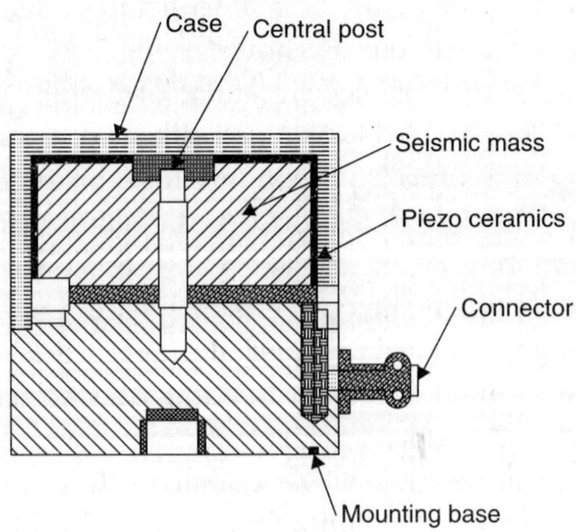

Figure 8.12 Piezoelectric accelerometer (compression design)

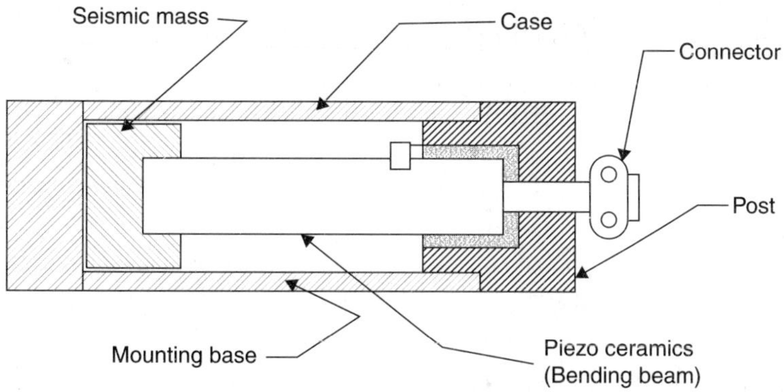

Figure 8.13 Piezoelectric accelerometer (bender design)

8.5 SOUND (DECIBAL) METERS

A sound level meter is an electronic device used for measuring acoustic waves (sound that travels through air). Sound travels in the form of pressure waves, increasing and decreasing as it moves through the air. The strength of the pressure waves determines the volume of the sound.

The meter is a handheld instrument with a microphone, the diaphragm of which responds to changes in air pressure caused by sound waves travelling through the air from a source of noise. The movement of the diaphragm, i.e., the sound pressure deviation (in pascals) is converted into an electrical signal (in volts). A microphone is distinguishable by the voltage value produced when it is calibrated with known sound pressures. As the measurement is based on the sound pressure level, the device is sometimes referred to as a sound pressure level (SPL) meter. As the reading from the sound level meter does not correlate well with human-perceived loudness, loudness is measured by another meter, called the loudness meter. Sound level meters are mainly used in construction sites and some manufacturing facilities that expose its workers to loud sounds over a period of time. Many sports stadiums and arenas are also fitted with sound level meters.

Devices that are used to measure the sound intensity in units called decibels are known as decibel meters. Decibel is a logarithmic scale first devised by the telephone pioneer, Alexander Graham Bell. Decibels do not increase in volume linearly, but logarithmically. On a ruler, a distance of 20 cm is two times the distance of 10 cm. However, the logarithmic decibel scale goes up by the power of ten. For example, every increase of 10 dB on the scale is equivalent to a 10-fold increase in sound intensity, which corresponds to a doubling in loudness. This means that a sound of 20 dB is 10 times more intense than a sound of 10 dB, and a sound of 30 dB is 100 times more intense. A sound of 100 dB is 10^9 times more intense than a sound of 10 dB, and not 10 times as intense. Table 8.1 provides a quick guide to some everyday sounds.

Table 8.1 A quick guide to some everyday sounds

Sound level in decibels	Activity	Times more intense	Times louder
10 dB	Falling leaves	1	1
20 dB	Watch ticking	10	2
30 dB	Birds flying by	100	4
40 dB	Quiet conversation	1,000	8
50 dB	Louder conversation	10,000	16
60 dB	Quiet traffic noise	1 lakhs	32
70 dB +	Louder traffic	10 lakhs	64
80 dB+	Loud highway noise at close range	1 crore	128
85 dB	Hearing damage after about 8 hours	–	–
100 dB	Pneumatic drill at close range; hearing damage after about 15 minutes	100 crores	512 (i.e., roughly 500 times louder than 10 dB)
110 dB+	Jet engine sound at about 100 m	1,000 crores	1024
120 db+	Threshold of pain. Hearing damage after very brief exposure	–	–

The sound level meter has a microphone, much like a recorder. It also has a screen or grill to protect a membrane (diaphragm) or movable cartridge inside. As sound pressure waves collide with the membrane inside the sound level meter, it causes the membrane to move back and forth, which creates an electric current. The more the membrane moves, the more electric current it creates. The strength of the electric current produced by the membrane is measured and converted into a numerical display.

Only when the ear is exposed to loud sounds for a long period is there long-term hearing damage. Hence, the sound level meter is used to record varying sound levels over a period of hours to days. Then, the average volume of a specific location is calculated. This can accurately determine if the given area experiences sound that is loud enough to damage a person's hearing if the person is exposed to it for a longer period of time.

A noise dosimeter is a specialised sound level meter intended specifically to measure the noise exposure of a person integrated over a period of time, usually to comply with Health and Safety regulations such as Occupational Safety and Health Administration (OSHA).

8.6 MEASUREMENT OF DISPLACEMENT

8.6.1 Linear Variable Differential Transformer

A linear variable differential transformer (LVDT), also called a linear variable displacement transducer, is a type of electrical transformer used for measuring linear displacement. As AC-operated LVDTs do not contain any electronics and are frictionless, they can be designed to operate at cryogenic temperatures or up to 650°C in harsh environments, under high vibration and shock levels. It works on the principle of mutual induction, whereby, it converts a position or linear displacement from a mechanical reference (zero position) into a proportional electrical signal with phase (for direction) and amplitude (for distance) information.

An LVDT contains three solenoid coils, placed end to end around a tube. As shown in Fig. 8.14, the centre coil is the primary winding and the two outer coils (at the sides) are the secondary windings. The number of turns in both the secondary windings is equal, but they are the opposite of each other; i.e., if the left secondary winding is in the clockwise direction, the right secondary winding will be in the anti-clockwise direction. Hence, the net output voltage will be the difference in voltages between the two secondary windings. A cylindrical ferromagnetic core, attached to the object whose position is to be measured, slides along the axis of the tube. An AC excitation voltage of 5–12 V drives the primary winding and causes a voltage to be induced in each secondary winding, proportional to the length of the core linking to the secondary.

As the core moves, the primary's linkage to the two secondary windings changes and causes the induced voltage to change. The coils are connected so that the output voltage is the difference between the left secondary voltage and the right secondary voltage. When the core is in the central position, equal distance is maintained between the two secondary windings, and hence, equal voltages are induced in the two secondary coils. However, the two signals (in different phases) cancel each other out, so that the output voltage becomes zero. When the core is displaced towards the left, the voltage in the left secondary winding increases, whereas the voltage in the right decreases. The resulting output voltage increases from zero. This voltage is in phase with the primary voltage. When the core moves in the other direction, the output voltage also increases from zero, but its phase is the opposite of that of the primary winding. The phase of the output voltage determines the direction of displacement (left or right) and amplitude indicates the amount of displacement. A synchronous detector can determine a signed output voltage that relates to the displacement.

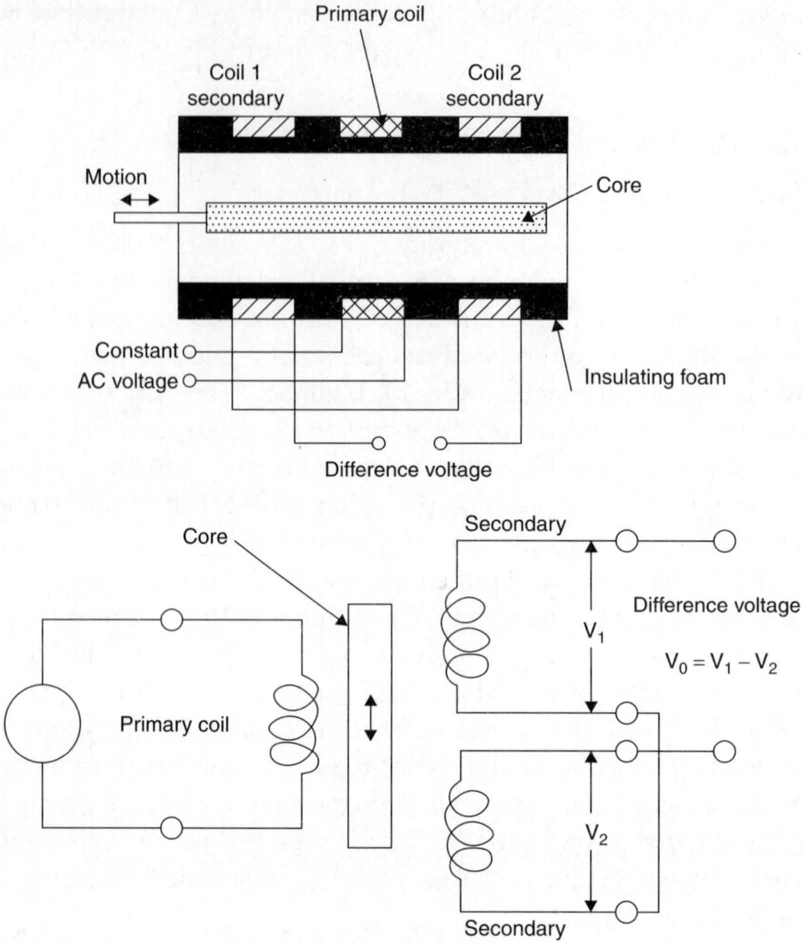

Figure 8.14 Linear variable differential transformer

LVDTs can be used as absolute position sensors. Even if power is switched off, on restarting, the LVDT shows the same measurement, and hence, no positional information is lost. Apart from the uni-axial linear movement of the core, any other movement, such as rotation of the core around the axis, will not affect the measurement. As the sliding core does not touch the inside of the tube, it can move without friction, making the LVDT a highly reliable measuring device. However, LVDTs need to be shielded from magnetic fields and their performance is affected by vibrations and temperature changes. LVDTs are commonly used for position feedback in servomechanisms and for automated measurement in machine tools.

8.6.2 Resistance Potentiometer

The potentiometer, also called pots, is an electrical device commonly used to measure the displacement of a body. The potentiometer is an electrical transducer or sensor and is of the resistive type because it works on the principle of change in resistance of a wire of uniform cross-section with its length. The resistance of the wire is directly proportional to the length of the wire.

A resistance potentiometer is an electric circuit in which the resistance can be changed manually by sliding contact (wiper), as shown in Fig. 8.15. In the given circuit, the voltage V_s is applied across the two points of the wire A and B. C is the variable contact point between A and B, and its position can be changed by sliding contact. The voltage V_o is measured between the points A and C.

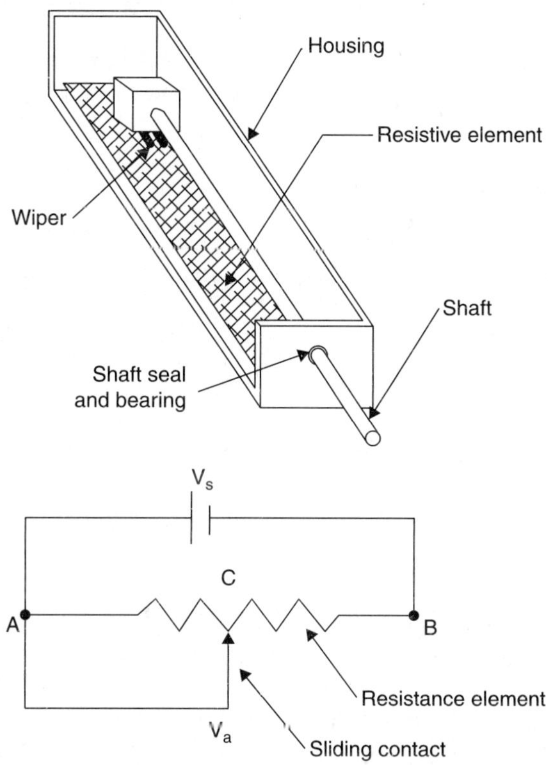

Figure 8.15 Resistance potentiometer

As per the resistance law of the conductor, the resistance of the conductor AC changes as the length of the wire AC changes. Accordingly, the output voltage V_o between A and C also changes. The point C is the slider whose position is changed by the operator or by the motion of the body whose displacement is to be measured.

The relationship between the length of the conductors and the voltage across the conductors can be stated as

$$\frac{V_o}{V_s} = \frac{AC}{AB}$$

To measure the displacement of a body, the moving body is connected to the sliding element of the potentiometer. As the body moves, the position of the slider located on the potentiometer also changes, and so, the resistance between the fixed point and the slider changes. Due to this change in the position of the slider, the voltage V_o across these points also changes. The change in the voltage or resistance is proportional to the change in the displacement of the body. Thus, the voltage change indicates displacement of the body.

Potentiometers are used not only for linear displacement, but also for rotary displacement. Rotary potentiometers are available with single or multi-turn abilities, commonly 3, 5 or 10 turns. A typical single-turn rotary pot is given in Fig. 8.16.

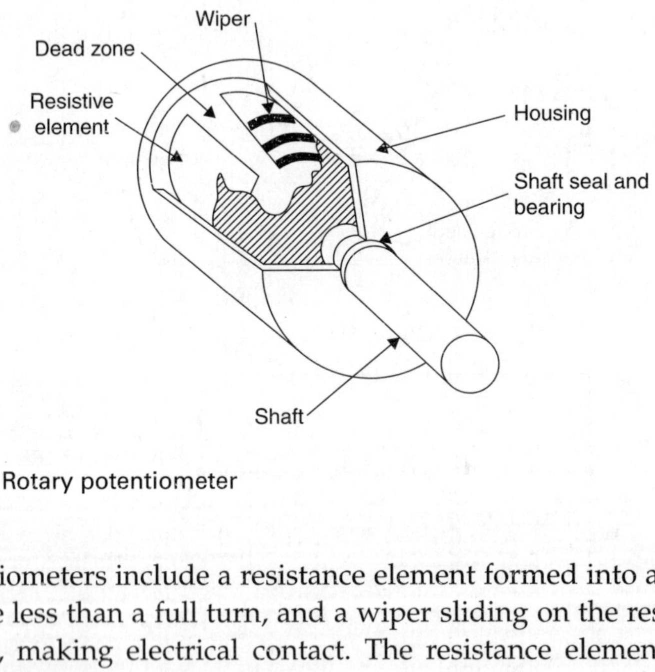

Figure 8.16 Rotary potentiometer

Rotary potentiometers include a resistance element formed into an arc of a circle, usually a little less than a full turn, and a wiper sliding on the resistance element when rotated, making electrical contact. The resistance element can be flat or angled. Each end of the resistive element is connected to a terminal on the case. The wiper is connected to a third terminal, usually between the other two.

Multi-turn potentiometers are also operated by rotating a shaft, but with several turns rather than less than a full turn. Some multi-turn potentiometers have a linear resistive element with a sliding contact moved by a lead screw. Some have a helical resistive element and a wiper that turns through 10, 20 or more complete revolutions, moving along the helix as it rotates.

A string potentiometer is a multi-turn potentiometer operated by an attached reel of wire turning against a spring, enabling it to convert a linear position to variable resistance.

Linear motion pots are available with maximum strokes ranging from roughly 5 mm to 4 m. String pots are available with maximum extensions exceeding 50 m. The resistive element of inexpensive potentiometers is often made of graphite. Other materials used include resistance wire, carbon particles in plastic and a ceramic/metal mixture called cermet. Conductive track potentiometers use conductive polymer resistor pastes that contain hard-wearing resins and polymers, solvents and lubricants, in addition to the carbon which provides the conductive properties.

8.7 MEASUREMENT OF SPEED

Speed is an expression of how fast an object moves. An object's speed is the rate at which it covers a distance in a given amount of time. The measuring units of speed include kilometres per hour (kph), miles per hour (mph) and metres per second (m/s). Instantaneous speed is the speed at some instant, or assumed constant during a very short period of time. By looking at a speedometer, one can read the instantaneous speed of a car. Average speed is the total distance covered divided by the time interval. Tangential speed is the linear speed of something moving along a circular path. For example, a point on the outside edge of a turntable travels a greater distance in one complete rotation than a point near the centre of the table. The speed along a circular path is known as tangential speed because the direction of motion is tangential to the circumference of the circle. Rotational speed involves the number of revolutions per unit of time. Rotational speed is expressed in rpm (revolutions per minute). Linear speed is determined by measuring the distance travelled divided by the amount of time taken. However, measurement of rotational speed requires some mechanism. This section describes the various methods used to measure rotational speed.

8.7.1 Mechanical Tachometer

The mechanical tachometer works on the principle that the centrifugal force developed by a rotating mass depends on the speed of rotation and can be used to stretch or compress a mechanical spring. A centrifugal mechanical tachometer is shown in Fig. 8.17.

The mechanical tachometer is usually handheld and is a linkage of shafts, gears and rotating flyweights. When the horizontal input shaft, as seen in Fig. 8.17, rotates, the vertical shaft in the tachometer also rotates, through a gear mechanism. The flyweights attached to the vertical shaft are hinged and free to move inwards and outwards, causing increase in compression of the spring as the speed increases.

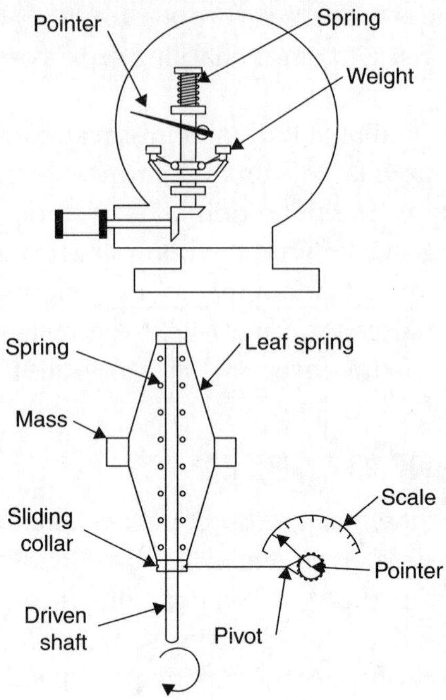

Figure 8.17 Mechanical tachometer

The movement of these flyweights causes the driving mechanism to operate a pointer through a rack-and-pinion mechanism over a scale that is calibrated to indicate the speed of the shaft in the desired unit, like rpm.

The scale is non-linear as the shaft speed is not directly proportional to the centrifugal force but to the square root of the centrifugal force. The tachometer indicates the instantaneous speed. If the driving cone of the tachometer is pushed too hard, it will cause additional mechanical loading on the driving shaft and slow down the shaft being measured, and hence, result in errors. This type of tachometer is capable of measuring a speed range of 0 to 50,000 rpm.

8.7.2 Electrical Tachometer

An electrical tachometer, otherwise called a tacho-generator, consists of an aluminium cup fitted on the driven shaft of the device, capable of rotating in a laminated iron electromagnet stator, as shown in Fig. 8.18. The stator has two separate windings at right angles to each other. The driven shaft of the device is pressed against the rotating shaft that needs to be measured. An AC power supply is given to one winding and eddy current is set up in the rotating aluminium cup. It results in an induced emf in the other stator winding, whose magnitude is proportional to the speed of rotation of the aluminium cup, and thus, the rotating shaft. The output voltage is measured on a voltmeter calibrated to read the units in rpm.

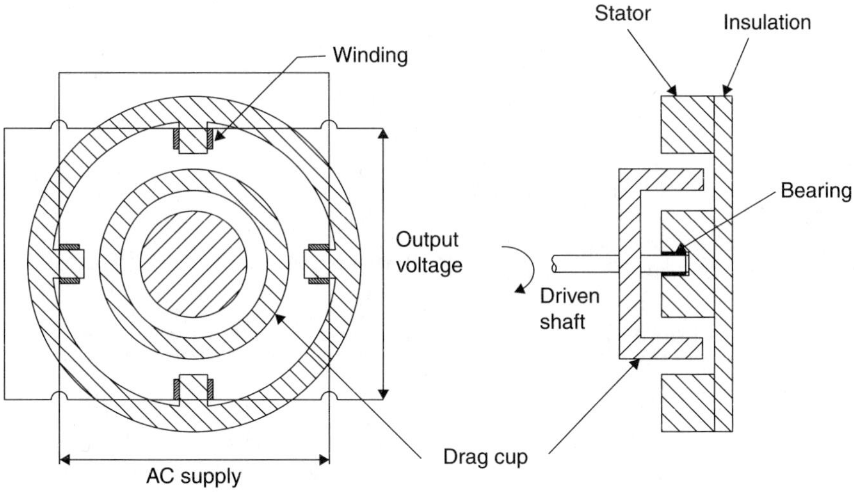

Figure 8.18 Electrical tachometer

As this device generates an emf proportional to the speed of the shaft, it is sometimes referred as tacho-generator. The emf produced may be of AC or DC, depending on the set-up.

8.7.3 DC Tachometer

It consists of a small armature that is coupled to the shaft whose speed is to be measured. This armature revolves in the field of a permanent magnet. The output shaft consists of a commutator and contact brushes, as in the case of a DC generator. The induced emf is proportional to the product of flux and the speed of the rotating shaft. As the flux of the permanent magnet is constant, the voltage generated is proportional to the speed of the shaft. The polarity of the output voltage indicates the direction of rotation. The voltage is measured with the help of a moving coil voltmeter having a uniform scale and calibrated directly in terms of speed (rpm). A typical DC tachometer is shown in Fig. 8.19. Series resistance is used in the circuit for the purpose of limiting the current from the generator in the event of a short circuit in the output circuit.

8.7.4 AC Tachometer

It is similar to a two-phase induction motor, in which two stator windings are placed in quadrature with each other and a rotor is short-circuited, as shown in Fig. 8.20.

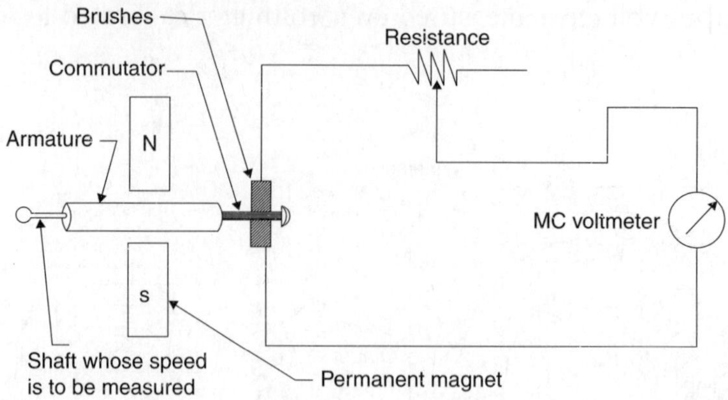

Figure 8.19 DC tachometer

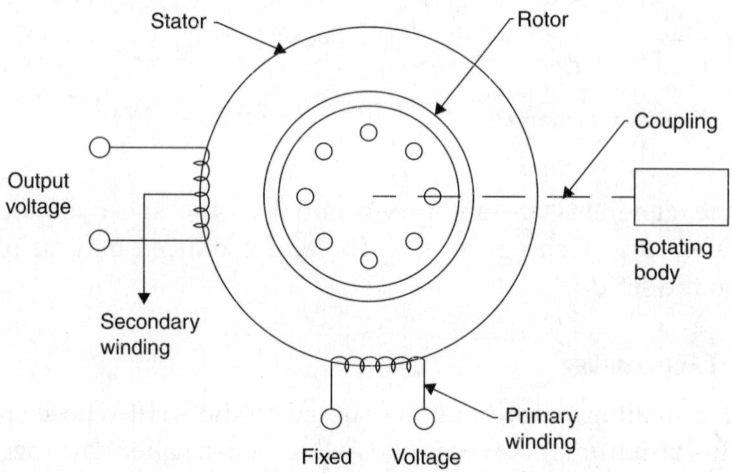

Figure 8.20 AC tachometer

An AC tachometer generator is a brushless alternator with a rotating multi-pole permanent magnet or electromagnet. The coil is wound on the stator, and hence, the problems associated with the commutator (as in the DC tachometer) are absent. The rotation of the magnet causes an emf to be induced in the stator coil, whose amplitude and frequency are proportional to the speed of rotation of the shaft. Thus, either the amplitude or the frequency of the induced voltage can be used as a measure of the rotational speed of the shaft being measured. Change in the direction of rotation of the shaft causes a phase shift. When the amplitude of the induced voltage is used as a measure of speed, the following circuit is used. The output of the AC tachometer is rectified and is measured with a permanent magnet moving coil instrument. The schematic diagram of a typical AC tachometer generator is given in Fig. 8.21.

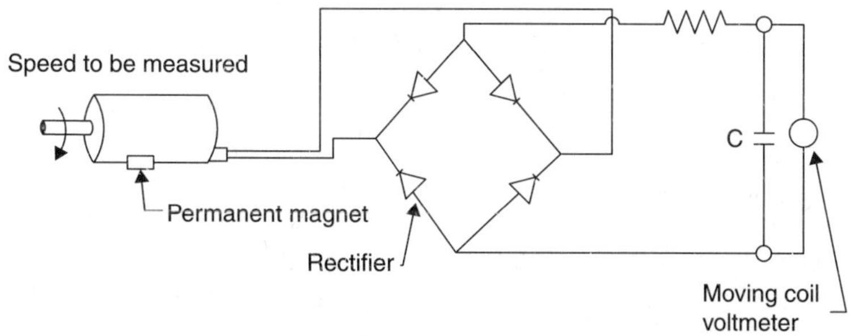

Figure 8.21 AC tachometer generator

8.7.5 Stroboscope

This is a non-contact type of tachometer. It is used to make a cyclically moving object appear to move slowly or to be stationary. It consists of a rotating disk with slots or holes, or a lamp such as a flashtube, which produces brief repetitive flashes of light. The rate of the stroboscope can be adjusted to different frequencies. When a rotating object is observed with the stroboscope at its vibration frequency, the rotating object appears to be stationary. In mechanical stroboscopes, the rotational speed of the rotating disk with slots or holes is adjusted so that it is synchronised with the speed of the rotating shaft being measured. This can be ensured by viewing the slot(s) of the rotating disk with the slot(s) or sticker(s) placed on the rotating shaft.

In electronic stroboscopes, the slotted disc is replaced by a lamp capable of emitting brief and rapid flashes of light. A gas discharge (neon) lamp or solid state (LED) lamp is used as they are capable of emitting light almost instantly when power is applied and of being extinguished as soon as the power is switched off. The frequency of the flash is adjusted so that it is equal to or a unit fraction of the speed of the rotating shaft, at which the rotating shaft is seen to be either stationary or moving slowly backward or forward, depending on the flash frequency. Figure 8.22 shows the principle of operation of a flash stroboscope.

8.7.6 Pick-up Tachometer

Various pick-up devices can be used in conjunction with a digital counter to give the direct reading of the speed of an object. In the inductive pick-up tachometer,

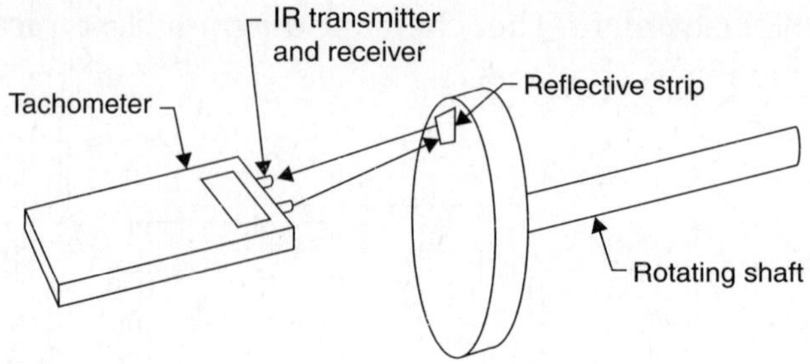

Figure 8.22 Stroboscope

an emf is induced, appropriately modified and then fed into a digital counter. In a capacitive pick-up tachometer, the rotating vane passes between the plates and a change in capacitance occurs in the form of a pulse. This pulse is modified and then fed into a digital counter to directly indicate the speed of the object. The schematic diagrams of inductive and capacitive tachometers are shown in Fig. 8.23.

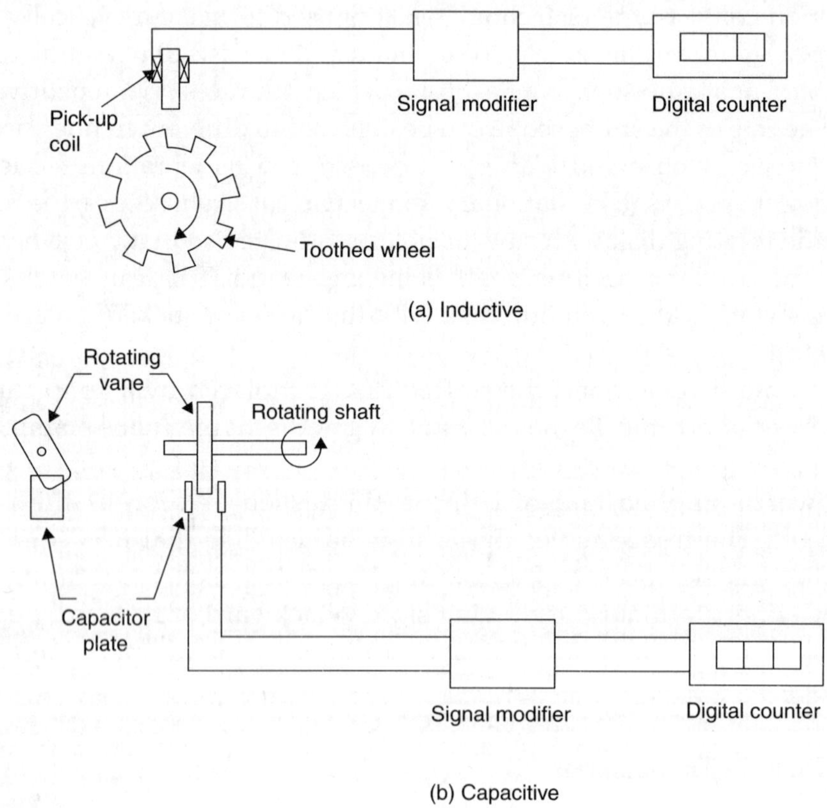

Figure 8.23 Pick-up tachometers

8.8 MEASUREMENT OF THE SPEED OF A MOVING OBJECT

Measurement of the speed of a moving object is of great significance in many fields. Measurement of the speed of automobiles in traffic surveys, the speed of balls in sports, the speed of aircraft, and so on, are a few examples. A radar gun is frequently used for this purpose. This radar gun works on the principle of the Doppler effect. In the Doppler effect, a frequency change occurs between the waves emitted by a source and the waves received by a detector or an observer. The basic operation of a radar gun includes sending radar waves and receiving them after the waves bounce off a targeted moving object. From the difference in the frequencies transmitted and received, the speed of the moving object can be calculated. As radar guns use radar waves and digital signal processors, they are very expensive, and the design of radar guns is also complex. Hence, ultrasonic waves and microprocessors are used in place of radar waves and digital signal processors, respectively.

If V_s is the speed of the moving object, V is the speed of the sound wave, V_o is the speed of the observer, F is the frequency of transmitted sound wave and F_o is the frequency of the received sound wave, then the relationship between these parameters can be written as

$$\frac{F_o}{F} = \left|\frac{V + V_o}{V \pm V_s}\right|$$

If the observer is stationary, then the expression for the speed of the moving object can be written as

$$V_s = \left[\frac{|F_o - F|}{F_o}\right] V$$

The term $|F_o - F|$ is identified as the Doppler shift because of the movement of the object, and is denoted as D. The expression for V_s can be simplified as

$$V_s = \left[\frac{D}{F_o}\right] V$$

The experimental set-up is shown in Fig. 8.24. The set-up consists of the following units:

- **Oscillator:** It is used to generate the sound wave with the desired frequency of 40 kHz. A microcontroller is used as an oscillator.
- **Transmitter:** It transmits the sound wave generated by the oscillator to hit the moving object. A piezoelectric transducer is used as transmitter.

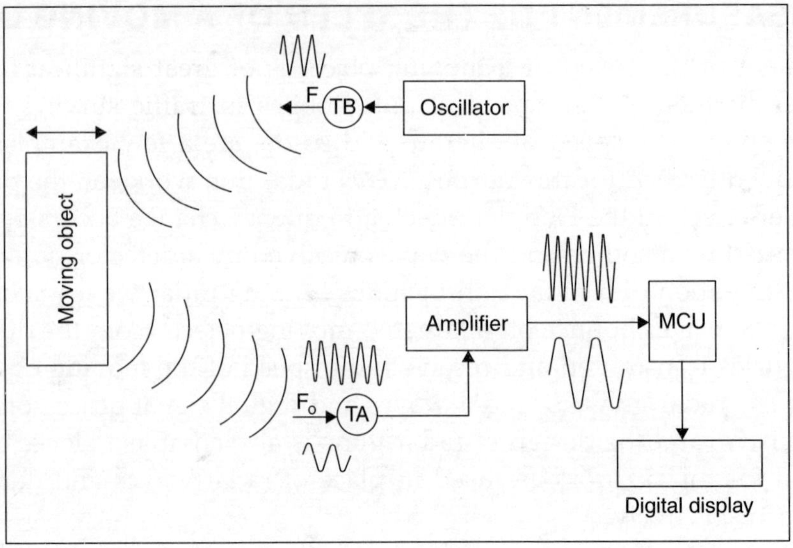

Figure 8.24 Speed measurement of a moving object

- **Receiver:** After it hits the moving object, the sound wave bounces back and is received by the receiver. Again, a piezoelectric transducer is used as receiver.
- **Amplifier:** It amplifies the received weak sound wave and feeds it into the microcontroller for amplification. An operational amplifier is used for this purpose.
- **Microcontroller:** It is programmed to calculate the change in frequency of the transmitted and received waves and the speed of the moving object.
- **Digital display:** It displays the speed of the moving object in the desired units.

SUMMARY

- Level measurement is either continuous or at a point. The methods used to measure level can be classified as direct or indirect. In the direct methods, the level is indicated directly by means of simple mechanical devices, whereas in the indirect methods, the change in liquid level is converted into a measurable signal. Measurement is not affected by changes in material density in the direct methods, whereas it is the opposite in the indirect methods.

- In the hook level indicator, a hook is pushed below the surface of the liquid whose level is to be measured and gradually raised until the point is just about to touch the surface.
- In the sight glass indicator, the liquid level in the sight glass matches the liquid level in the tank.
- In the float-operated level indicator, the float resting on the surface of the liquid, follows the changing level of the liquid and indicates the level on a calibrated scale.
- The capacitive liquid level sensor is based on the principle of change in capacitance based on the level of the liquid. An insulated electrode acts as one plate of the capacitor and the tank wall as the other plate. The change in capacitance is measured using an AC bridge by applying an RF signal between the conductive electrode and the tank wall. Various techniques are used to measure the level for conductive material, non-conductive material and non-contact vessels.
- An ultrasonic fuel level meter uses an ultrasonic sensor and a data processing unit. It works on the principle of reflection of sound waves from the surface of the liquid. It is used for monitoring the levels of alcohol, acid, solvent, diesel, gasoline, liquefied gas, kerosene and other fuels.
- In the magnetic level indicator, a magnet placed inside a float follows the change in the liquid level and is aligned with the respective flags; it then makes the flag rotate in the opposite direction. The change in colour indicates the liquid level on a measuring scale.
- The cryogenic fuel level indicator works on the principle that superconductors exhibit zero resistance at a temperature below which the cryogenic fluid exists in liquid state. The change in the conductivity of the conductor is measured and calibrated for various liquid levels of the cryogenic fluid kept in a storage tank in which the super conductor wire loop is inserted. This technique is used for measuring the level of liquid natural gas.
- The bubbler indicator works on the principle that liquid level is determined by measuring the pressure required to force a gas into a liquid at a point above the bottom surface of the container.
- The vibrometer is based on the principle of Doppler effect. Laser vibrometer measures the velocity of a moving object by finding the frequency shift of the laser beam.
- A piezoelectric accelerometer is used to measure the acceleration of a vibrating body. When a physical force is exerted on the accelerometer, the seismic mass loads the piezoelectric element. The force exerted on the piezoelectric material is observed in the change in the electrostatic

force or voltage generated by the material. The design of the piezoelectric accelerometer is based on the type of force exerted on the material, like bending, compression, twist, etc.

- Sound meters measure the volume of sound based on the principle that the strength of the pressure waves of sound determines the volume of the sound. In a sound meter, the change in sound pressure is converted into an equivalent electrical signal, which indicates the volume of the sound.
- LVDT works on the principle of mutual induction – it converts a position displacement from a mechanical reference into a proportional electrical signal with phase (for direction) and amplitude (for distance) information.
- A potentiometer works on the principle of change in resistance of the wire of uniform cross-section with its length. The resistance of the wire is directly proportional to the length of the wire. When a moving body is connected to the sliding element of the potentiometer, the resistance between the fixed point and the sliding element varies depending on the position of the moving body. The change in voltage or resistance measured is proportional to the change in the displacement of the body.
- A mechanical tachometer works on the principle that the centrifugal force developed by a rotating mass depends on the speed of rotation and can be used to stretch or compress a mechanical spring. The movement of the mechanical spring activates a point to move against a calibrated scale, which indicates the speed of the shaft.
- An electrical tachometer consists of a stator connected to two separate windings at right angles to each other. The rotor is connected to the shaft whose speed is to be measured. When an AC power supply is provided to one winding, an eddy current is set up in the rotor. It results in an induced emf in the other stator winding, whose magnitude is proportional to the speed of rotation of the rotor, and hence, the speed of the shaft being measured. Electrical tachometers are categorised as AC and DC tachometers, depending on the output emf mechanism.
- The stroboscope, which is a non-contact tachometer, makes a cyclically moving object appear to be stationary. In mechanical stroboscopes, the rotational speed of the rotating disk with slots is adjusted so that it is synchronised with the speed of the rotating body being measured. In electronic stroboscopes, the slotted disc is replaced by a lamp capable of emitting brief and rapid flashes of light. The frequency of the flashes is adjusted so that it is equal to the speed of the rotating body.
- A pick-up tachometer uses the technique of inductance and capacitance generated by the rotation of the object whose speed is being measured. The

strength of the inductance or capacitance varies with the speed of the rotating body.

❖ Radio waves or ultrasonic waves are used to measure the speed of a moving object using Doppler effect. Radar guns are used for this purpose.

EXERCISES

Multiple Choice Questions

1. The following is not a direct method of measurement:
   a) Dipstick　b) Load cell
   c) Radar　d) Sight glass

2. Dipsticks are used to measure
   a) Pressure　b) Flow
   c) Displacement　d) Level

3. The most common application of the float system is
   a) To monitor the fuel tank level in a motor vehicle
   b) To monitor the flow of solids
   c) To monitor the flow of liquids
   d) All of the above

4. Capacitive devices are used for level measurement of
   a) Only liquid　b) Solid in powdered form
   c) Both (a) and (b)　d) None of the above

5. In ultrasonic level gauge, the ultrasonic source is placed
   a) At the top of the liquid tank　b) In the middle of the liquid tank
   c) At the bottom of the liquid tank　d) Outside the liquid tank

6. Ultrasonic fuel indicator works on the principle of
   a) Change in timing of the sound wave received
   b) Change in strength of the sound wave received
   c) Change in direction of the sound wave received
   d) All of the above

7. The following is the disadvantage of a bubbler level indicator:
   a) It requires compressed air to function
   b) Dirt at the bottom of the liquid may block the bubble tube

c) Density variation in the liquid will affect the accuracy of measurement
d) All of the above

8. Magnetic level indicator is useful for level measurement of
a) Corrosive and toxic materials
b) Materials with magnetic property
c) Gaseous material
d) All of the above

9. Bubbler level indicator cannot be used for
a) Corrosive fluids
b) Gaseous material
c) Coloured fluid
d) None of the above

10. Piezoelectric material functions on the principle of
a) Newton's second law of motion
b) Conservation of energy
c) Thomson's effect
d) None of the above

11. The following is the sound level in decibels for louder traffic:
a) 10
b) 30
c) 70+
d) 120+

12. The following sound level in decibels will damage hearing:
a) 30+
b) 50+
c) 70+
d) 85+

13. In the radiation method, the detector device is located
a) At the top of the liquid tank
b) In the middle of the liquid tank
c) At the bottom of the liquid tank
d) Outside the liquid tank

14. If the zero point is at the top of a container, how would the level of a material in the container be measured?
a) From the bottom of the container to the surface
b) From the bottom of the container to the zero point
c) From the bottom of the container to the top
d) From the top of the container down to the surface

15. Vibration level sensor consists of
a) One piezoelectric oscillator
b) Two piezoelectric oscillators
c) Three piezoelectric oscillators
d) Four piezoelectric oscillators

16. Contact devices used for measurement of level are
a) Less reliable than devices that do not make contact with the material
b) More reliable than devices that do not make contact with the material
c) Not predictable
d) None of the above

17. Which of the following statements is true about stroboscopes?

a) It is a non-contact frequency instrument
b) It can measure frequency up to 5 Hz
c) It uses electromagnetic radiations to measure frequency
d) All of the above

18. Which type of instruments do not require a separate power source for measuring the vibratory response of a vibratory system?
a) Passive instruments
b) Active instruments
c) Both (a) and (b)
d) None of the above

19. Which of the following instruments measure the amplitude of a vibrating body?
a) Vibrometer
b) Seismometer
c) Both (a) and (b)
d) None of the above

20. Which type of monitoring system uses a stroboscope to measure the speed of the machine?
a) Basic condition monitoring system
b) Portable condition monitoring system
c) Computer-based condition monitoring system
d) None of the above

21. Sound is carried to our ears by
a) Pressure
b) Vibration
c) Humidity effect
d) All of the above

22. Sound moves in air because
a) Heat is passed on
b) Humidity is passed on
c) Pressure is passed on
d) Light is passed

23. Sound is a good example of
a) Transverse waves
b) Longitudinal waves
c) Both transverse and longitudinal waves
d) None of the above

24. LVDT is a kind of
a) Transformer
b) Transducer
c) Position sensor
d) All of the above

25. Which of the following statements is true for LVDT?
a) It can measure displacement and pressure
b) It is a mutual capacitive transducer
c) Presence of hysteresis gives precise readings
d) All of the above

26. The function of the transducer is to convert
    a) Electrical signal into mechanical quantity
    b) Electrical signal into non-electrical quantity
    c) Non-electrical quantity into electrical signal
    d) All of the above

27. Resistance potentiometer works on the principle of
    a) Change in resistance of the wire with its length
    b) Change in resistance of the wire with the applied voltage
    c) Change in resistance of the wire with the thickness
    d) All of the above

28. Strain gauge is a/an
    a) Passive device and converts electrical displacement into a change in resistance
    b) Active device and converts mechanical displacement into a change in resistance
    c) Passive device and converts mechanical displacement into a change in resistance
    d) Active device and converts electrical displacement into a change in resistance

29. LVDT is a/an
    a) Inductive transducer
    b) Non-inductive transducer
    c) Capacitive transducer
    d) Resistive transducer

30. If the displacement is measured with a strain gauge, then the number of strain gaugs normally required are
    a) One
    b) Two
    c) Three
    d) Four

Answers

1. b	2. d	3. a	4. c	5. c
6. a	7. d	8. a	9. b	10. a
11. c	12. d	13. d	14. d	15. b
16. a	17. a	18. a	19. c	20. a
21. b	22. c	23. b	24. d	25. a
26. c	27. a	28. c	29. a	30. d

Review Questions

1. Give some examples for direct and indirect level measuring instruments.
2. Explain the construction details and working of the following level measuring instruments, with neat diagrams. Also state their advantages, disadvantages and applications.
   i) Hook level indicator
   ii) Sight glass indicator
   iii) Float-operated level indictor
   iv) Capacitive liquid level sensor
   v) Ultrasonic fuel level meter
   vi) Magnetic level indicator
   vii) Cryogenic fuel level indicator
   viii) Bubbler level indicator
3. Why does a hook level indicator use a tapered pointer in the shorter arm of the U-bent wire?
4. Why is coloured liquid preferred in sight glass indicator for level measurement?
5. What is the purpose of using grids for holding the float in a float-operated level indicator?
6. State the acting parallel plates of a capacitor and dielectric medium when the liquid in the tank is a) conductive or b) non-conductive in the capacitive liquid level indicator.
7. Give some examples for non-contact level indicators.
8. Which instrument is most suitable for measuring the level of hazardous fuels like petrol, diesel and acids? State why.
9. How do the flags in a magnetic level indicator rotate and change their colours to indicate the level of the liquid?
10. State the principle of using superconductors in indicating the cryogenic fuel level in the storage tank.
11. Why is the pressure tube-tip not kept touching the bottom of the storage tank in a bubbler level indicator?
12. What is the purpose of using a small V-notch in the bubbler level indicator?
13. What is the purpose of the pressure regulator in the bubbler level indicator?
14. Define Doppler effect.
15. Explain in detail the working of the laser Doppler vibrometer, with a neat sketch.
16. Mention the advantages of scanner and 3D vibrometer over normal laser vibrometer.
17. What are the desirable characteristics of piezoelectric material to be used in accelerometers?
18. Explain in detail the construction and working principle of the piezoelectric accelerometer, with neat sketches for different types.
19. Why does the decibel meter use the logarithmic scale to measure the sound level?
20. Define the basic principle of operation of a sound meter.

21. Explain in detail the construction and working of an LVDT.
22. How is the direction of displacement identified in an LVDT?
23. Why is LVDT required to be shielded from the magnetic field during measurement of displacement of an object?
24. Explain in detail the principle of operation of a resistance potentiometer, with a neat circuit diagram and construction diagram.
25. What kind of speed of a moving object is measured by tachometers?
26. Describe the working principle of a centrifugal mechanical tachometer, with a neat sketch.
27. Why is the driving cone of the mechanical tachometer pushed gently against the driving shaft?
28. Mention a few differences between AC and DC electrical tachometers.
29. Why are electrical tachometers called tacho-generators?
30. Explain in detail the working principle of an AC electrical tachometer, with a neat sketch.
31. Mention a few differences between mechanical and electronic stroboscopes.
32. How does a mechanical stroboscope work?
33. What is the principle of working of an electronic stroboscope?
34. How do pick-up tachometers differ from conventional tachometers?
35. Explain the principle of operation of a radar gun, with a block diagram.

9

Measurement of Fluid Flow

9.1 MEASUREMENT OF FLOW

Measuring the quantity of fluid flow is one of the basic parameters required for metering and controlling the fluid flow in a process. Basically, flow is distinguished as open channel or closed conduit. Also, based on the molecular dispersion in the fluid, it can be classified as laminar or turbulent.

Flow can be measured directly by measuring the quantity of fluid flow in terms of mass or volume with respect to the time taken. This type of measurement is termed **primary or quantity measurement**. In other methods, flow is measured in an inferential way – fluid flow is inferred from related quantities such as pressure difference, temperature variation, force and velocity. This is termed **secondary measurement**.

Flow visualisation techniques are some special methods used for understanding the flow pattern in the flow field, such as regions of separation, recirculation and boundary layer.

The range of flow that a flow meter is capable of measuring with the required level of accuracy is called the **turn down ratio or range ability**. For example, if fluid flow varies between 200,000 m^3 per day and 2,000,000 m^3 per day in a particular application, the turn down ratio is 20:1. Therefore, the flow meter selected for the application should have a turn down ratio of 20:1.

Flow measurement methods or instruments are classified as follows:

- **Quantity meters:** Used for calibration of flow meters, in which the mass or volume of the fluid collected in a container is measured with respect to time and compared with the reading of the flow meter.
  - Weight tank (for gravimetric calibration of flow meters)
  - Volume tank (for volumetric calibration of flow meters)
- **Flow meters**
  - **Obstruction meters:** Flow rates are measured by observing the pressure differential in the flow.
    - Venturimeter
    - Orifice meters

- Flow nozzle
- Variable area meters (rotameters)

The venturimeter, orifice meters and flow nozzles are also called variable head meters.

➢ **Mass flow meters**
- Coriolis meters

➢ **Volume flow meters**
- Turbine flow meter
- Electromagnetic flow meters
- Ultrasonic flow meters

➢ **Velocity probes**
- Pitot static tube
- Hot wire anemometer
- Laser Doppler anemometer

➢ **Flow visualisation methods**
- Dye injection method
- Laser-induced fluorescence
- Hydrogen bubble technique
- Shadowgraph and Schlieren technique (observation of change in the refractive index of flow with respect to change in density)

9.2 OBSTRUCTION METERS

Obstruction or differential pressure meters work on the principle of Bernoulli's theorem. In this type of flow measurement, a pressure drop is created across the flow by positioning an obstruction device in a particular section.

Let p_1, p_2 and v_1, v_2 be the pressure and volume of a fluid upstream (at point 1) and downstream (at point 2) of the metered flow due to obstruction; z_1 and z_2 are the elevation of the points 1 and 2 from the datum. According to Bernoulli's theorem, the sum of pressure energy, kinetic energy and potential energy upstream and downstream is equal if losses and other energies are neglected.

Eq. 9.1 shows Bernoulli's equation.

$$\frac{p_1}{e_1} + \frac{v_1^2}{2} + z_1 \cdot g = \frac{p_2}{e_2} + \frac{v_2^2}{2} + z_2 g \tag{9.1}$$

For incompressible fluid, $\rho_1 = \rho_2$, and if the points 1 and 2 are in the same elevation, then $z_1 = z_2$. By the continuity equation, $\acute{m} = \rho_1 A_1 v_1 = \rho_2 A_2 v_2$, where A_1 and A_2 are the cross-sectional areas at points 1 and 2 and $\acute{m}$ is the mean flow rate.

$$\acute{m} = A_1 v_1 = A_2 v_2 \text{ (for incompressible fluid)}$$

$$v_1 = \left(\frac{A_2}{A_1}\right) v_2$$

Therefore, Eq. 9.1 becomes

$$\frac{p_1 - p_2}{\rho} = \frac{v_2^2 - v_1^2}{2} = \frac{v_2^2}{2}\left[1 - \left(\frac{A_2}{A_1}\right)^2\right] \tag{9.2}$$

$$v_2 = \frac{\sqrt{2(p_1 - p_2)/e}}{\sqrt{1 - \left(\frac{A_2}{A_1}\right)^2}} \tag{9.3}$$

The mass flow rate of the flow can be calculated by substituting v_2 in the continuity equation.

$$\acute{m}_{\text{theoretical}} = \rho A_2 v_2 \tag{9.4}$$

We know that the coefficient of discharge C_d is defined as the ratio of actual and theoretical mass flow rate. This coefficient is introduced to account for any losses that may occur in the different types of obstruction meters. Its values vary for different meters and are determined experimentally.

$$C_d = \frac{\text{Actual flow rate}}{\text{Theoretical flow rate}} \tag{9.5}$$

The types of obstruction meters available are venturimeter, orifice meter and flow nozzle.

9.2.1 Venturimeter

The schematic of a venturimeter is shown in Fig. 9.1. It consists of three parts: a converging section, a constant area throat and a diverging section. The converging section usually varies from 19° to 21° and the diverging section varies between 5° and 7°. The throat diameter is usually one-third to one-fourth that of the inlet diameter. The venturimeter is fitted in the flow line. As water enters the converging section, the flow converges and reaches the throat. At the throat, the pressure of the fluid decreases and velocity increases. A pressure difference thus exists between sections 1 and 2. This differential pressure is measured using a manometer, and the flow rate is calculated. The coefficient of discharge of the venturimeter varies between 0.98 and 0.99.

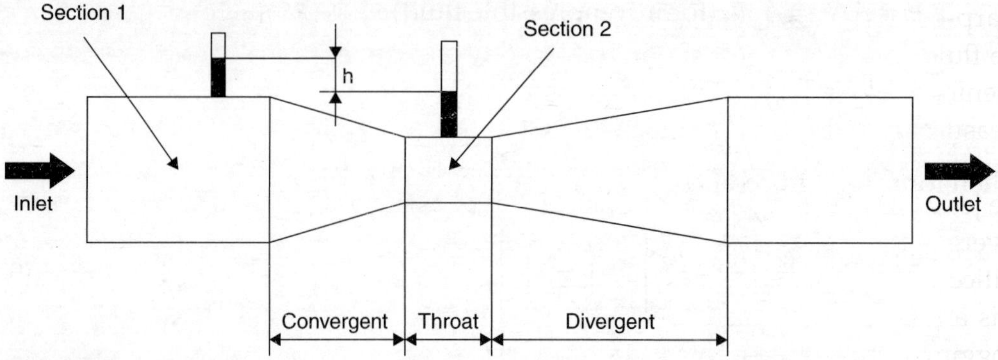

Figure 9.1 Venturimeter

$$Q = C_d \frac{a_1 a_2}{\sqrt{a_1^2 - a_2^2}} \times \sqrt{2g \frac{p_1 - p_2}{w}} \tag{9.6}$$

where Q = flow rate; C_d = coefficient of discharge; a_1 = area at the entrance of the venture; a_2 = area at the venture throat; p_1 = pressure at the entrance of the venture; p_2 = pressure at the venture throat; and w = specific weight of the fluid (= ρg).

9.2.2 Orifice Meter

The schematic of the orifice meter is shown in Fig. 9.2. The orifice meter is a circular, flat plate, usually made of stainless steel or phosphor bronze, with a small

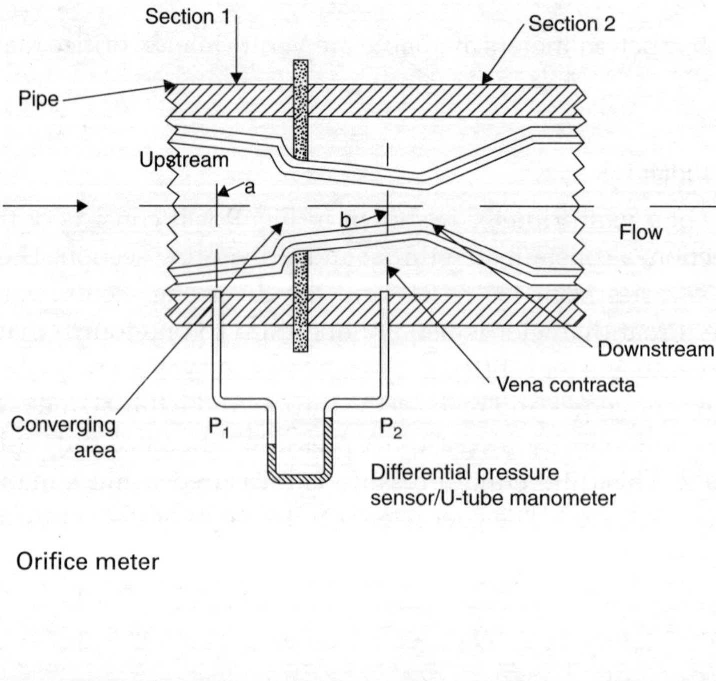

Figure 9.2 Orifice meter

sharp-edged circular hole at the centre. It is rigidly held by the pipe flange across the fluid flow conduit. It should be fitted in the straight portion of the pipe. Two openings are provided at sections 1 and 2; i.e., before and after the orifice meter, to measure the pressure difference by connecting a manometer. The fluid upstream converges before entering the restriction in the orifice meter and the diameter of the jet coming out of the orifice meter will be uniform for a small length before it diverges to the pipe diameter downstream. The portion of the jet that leaves the orifice meter with minimum and uniform diameter is called the vena contracta. It is an inexpensive method of measuring the fluid flow rate, but the chance of clogging is one of the main disadvantages.

9.2.3 Flow Nozzle

Figure 9.3 shows the schematic of a flow nozzle. It consists of a converging and a constant throat area section. It is held between the pipe flanges across the flow of

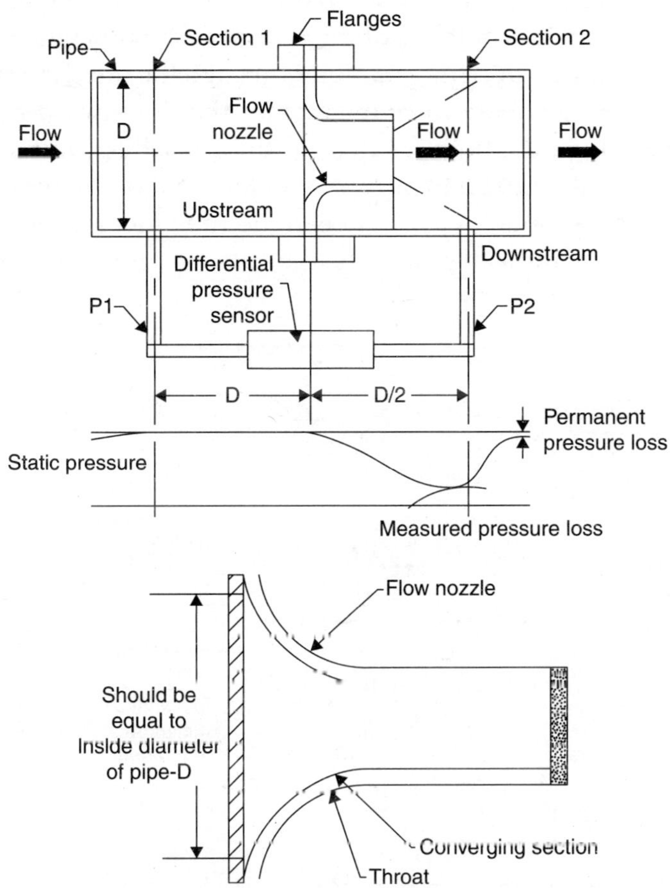

Figure 9.3 Flow nozzle

fluid, like in the orifice meter. Two small openings are provided at sections 1 and 2 to measure the pressure difference at those locations. As the fluid approaches the flow nozzle, its pressure continuously decreases and is maintained at minimum value at the throat cross-section. The differential pressure is measured by the sensors or the manometer connected to the arrangement.

9.2.4 Rotameter

A rotameter is a variable area meter used to measure the flow rate directly. It is made of a transparent and tapered glass tube. A float of conical shape is placed inside the glass tube and it is free to move between the ends of the tube in the axial direction. The diameter of the float is of a size that ensures that it completely closes the inlet of the glass tube when it is in the bottom position; when it is in any other position, an annulus is formed between the inside diameter of the glass tube and the float surface to allow the fluid to flow through the tube. The glass tube is graduated to indicate the fluid flow quantity. The rotameter should always be vertically positioned. The schematic of the rotameter is shown in Fig. 9.4.

When fluid is admitted into the rotameter through the inlet, the sum of pressure of the fluid and the buoyancy force overcomes the downward-acting weight of the float, and therefore, the float is lifted up till it reaches equilibrium. The total drag force acting on the float depends on the flow rate of the fluid and the cross-sectional area of the annular space between the tube inner surface and the float. The level of the float indicates directly the rate of fluid flow. The instrument

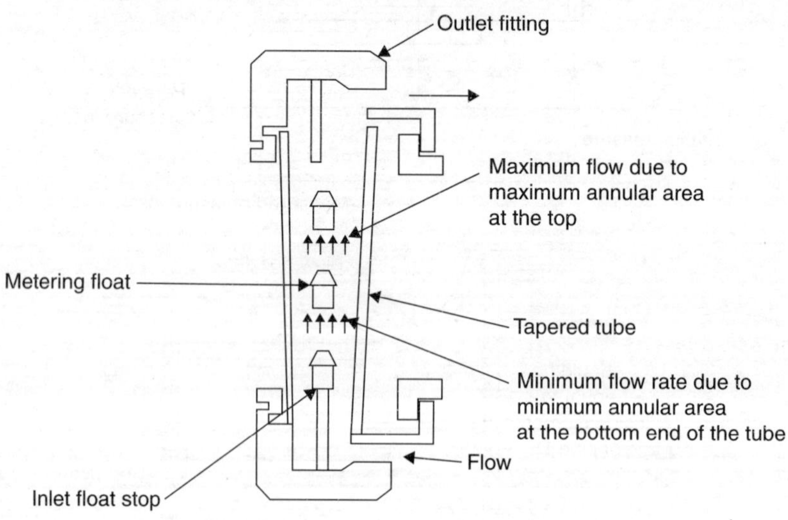

Figure 9.4 Schematic of a rotameter for fluid flow measurement

should be designed to minimise the effect of change in density and viscosity of the fluid.

9.3 CORIOLIS FLOW METER

The Coriolis flow meter works on the principle of the Coriolis effect. This effect is experienced by a mass that moves in a rotating system by a force (Coriolis force) acting perpendicular to the direction of flow and to the axis of rotation. This type of flow meter directly measures the mass flow rate and the density of the fluid. The schematic of the Coriolis flow meter is shown in Fig. 9.5.

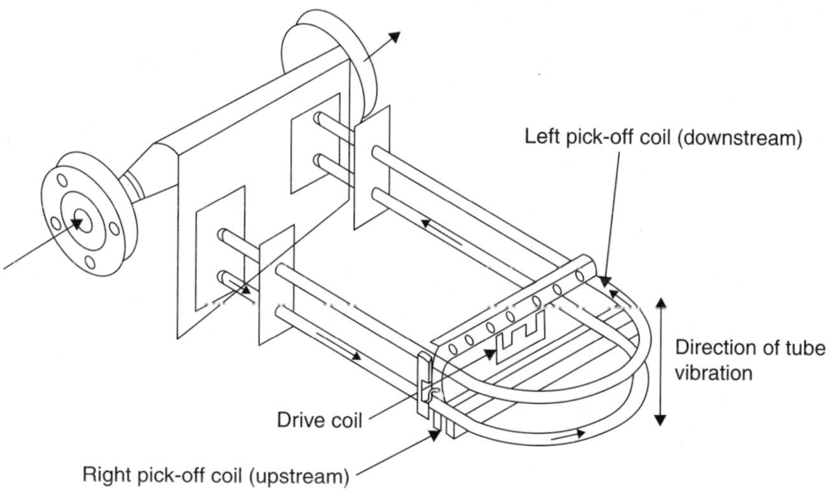

Figure 9.5 Coriolis flow meter

The basic dual tube Coriolis meter consists of two flexible tubes connected in parallel through which the fluid passes. An electromagnetic drive coil makes the tubes vibrate (or oscillate) towards and away from each other at the natural frequency. The tube frequency is measured by the stiffness and mass of the tube.

Two electromagnetic sensors or pick-off coils are placed upstream and downstream of the tube to measure the vibrations of the tube. When there is no fluid flow in the tubes, the two tubes move towards and away from each other in a uniform oscillation and the output signal observed through the upstream and downstream pick-off sensors will be in phase and equal.

As soon as fluid is admitted in the tubes, the Coriolis force comes into effect, and this will cause the downstream side of the tube to lead the upstream side, even though the earlier oscillation of the tube still continues. As the fluid direction upstream, i.e., at the entry side, and downstream, i.e., at the exit side, of the tube, are opposite in nature, the Coriolis force acts at the entry and exit sides of the tube, and are also opposite in nature; this causes the tube to twist. This twist is observed by the phase difference of the output signal of the two upstream and downstream pick-off sensors, and it is linearly proportional to the mass flow rate of the fluid.

The volume flow rate of the fluid varies with the changing pressure and temperature, but the mass flow rate remains the same. As the Coriolis flow meter measures the mass flow rate and density directly, the need for any compensation is not necessary to accommodate pressure and temperature variations. It measures the mass flow rate of a wide range of Newtonian and non-Newtonian fluids. To measure larger flow rates, two or more Coriolis meters should be connected in parallel.

9.4 TURBINE FLOW METER

The turbine meter possesses a magnetised turbine rotor introduced in the fluid flow, which will rotate in proportion to the rate of flow. The measure of rotor rotation is calibrated to be linearly proportional to the volume flow rate of the fluid. A variable reluctance pick-up coil is mounted close to the magnetic rotor.

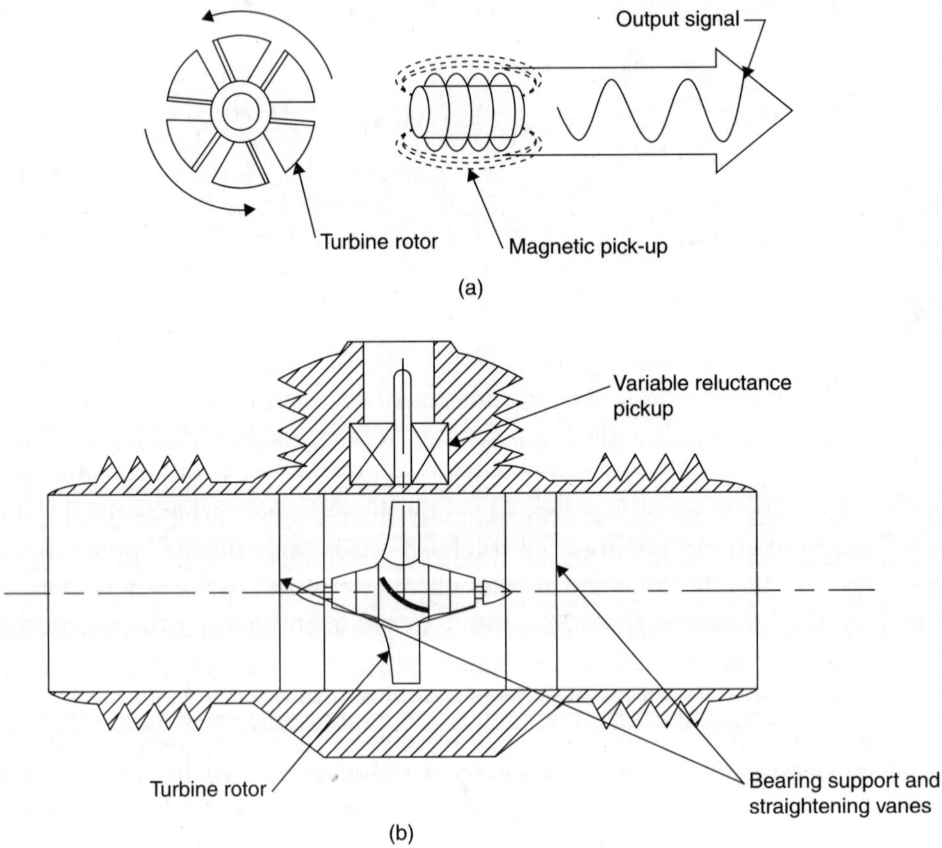

Figure 9.6 (a) Turbine flow meter (b) Working principle of turbine flow meter

When the rotor blade crosses the pick-up coil, a voltage pulse is generated in the coil due to the change in magnetic permeability.

This pulse is transferred as an output signal, which is processed by an electronic device to show the value of the measured flow rate. The turn down ratio of typical turbine flow meters are 30:1 and 10:1 for gases and liquids, respectively. Turbine meters are good for low viscosity fluids.

The wear and tear of the rotor bearing will reduce accuracy over a period. It is less accurate at low flow rates. The schematic of the turbine flow meter is shown in Fig. 9.6.

9.5 ELECTROMAGNETIC FLOW METER

The electromagnetic flow meter works on the principle of Faraday's law of electromagnetic induction (Fig. 9.7). According to Faraday, if the magnetic flux linkage between an electric conductor (a coil) and a magnetic field is disturbed by the relative motion between them, this change in flux will induce a voltage across the conductor (coil).

The magnitude of emf induced in the coil is proportional to the rate of change of flux in the coil. The induced voltage is measured by the following relationship:

$$e = k.B.d.v \qquad (9.6)$$

where e = emf or induced voltage; k = a constant; B = magnetic flux density (proportional to the number of coil turns); d = length of the conductor; and v = velocity of the conductor.

In the electromagnetic flow meter, the electric conductor described in Faraday's law is replaced by the fluid whose rate of flow needs to be measured. It consists of a pipe through which the fluid is allowed to flow.

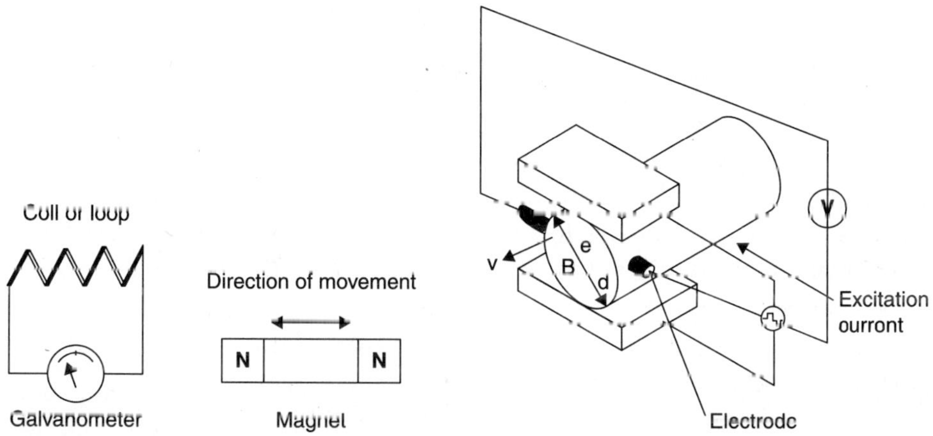

Figure 9.7 Faraday's law of electromagnetic induction

A current is passed through the wire coils mounted over the pipe, which creates an electromagnetic field. As the flowing fluid acts as a conductor and its velocity disturbs the magnetic field, an emf of voltage is induced in it. This emf is measured by electrodes placed on the pipe wall. The output is processed to provide the flow rate of the fluid. Electromagnetic flow meters are best suited for fluids with good electrical conductivity and which are corrosive and highly viscous because of the absence of moving parts, like in turbine flow meters. The accuracy range is between 0.5% and 2%.

9.6 ULTRASONIC FLOW METER

The ultrasonic flow meter is used to measure the mean velocity of fluid flow in a conduit using acoustic waves, called ultrasound waves. Ultrasound waves have a frequency above 20 kHz, which is not in the audible range of human hearing. The schematic of the ultrasonic flow meter is shown in Fig. 9.8.

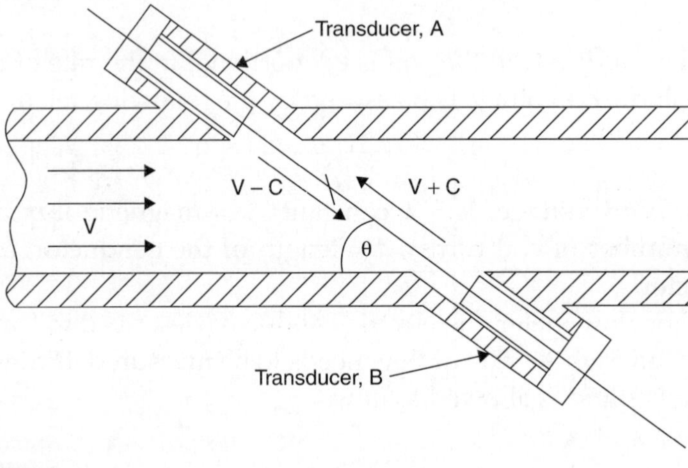

Figure 9.8 Ultrasonic flow meter

Two piezoelectric transducers are positioned outside the flow conduit in the direction shown in the figure, not necessarily in contact with the pipe. Both the transducers will act as transmitter and receiver, i.e., when transducer A sends the signal, transducer B will act as receiver and receive the signal, and vice versa. When the fluid inside the conduit is stationary or there is no fluid inside, then the time taken for the ultrasound waves to travel between the transducers A and B in both directions will be the same.

However, when flow takes place inside the conduit, the wave travelling from transducer A to transducer B will be aided by the velocity component of the fluid and will be opposed for the direction B to A. Let us assume the velocity of the fluid is v and the velocity of the ultrasound wave in static flow conditions is c. The velocity component v in the direction of ultrasound is $\pm v.\cos\theta$. Therefore, the effective velocity of the ultrasound travelling in the upstream to downstream direction is $c + v.\cos\theta$, and for the downstream to upstream direction is $c - v.\cos\theta$. This difference in the velocity of the wave is proportional to the mean velocity of fluid flow. The calculated velocity of the flow when multiplied with the cross-sectional area will provide the mean volume flow rate in the conduit. This flow meter does not obstruct the flow passage when there are moving or corroding parts. It is best suited for the measurement of fluid flow in clean and single-phase fluids because any interfering particles or surfaces will reduce the accuracy of the measurement.

9.7 PITOT STATIC TUBE

The Pitot tube was invented by the French scientist Henry Pitot in the 1700s to measure the flow of velocity in rivers. The Pitot static tube is used to measure the dynamic pressure of flowing fluid. It measures the static and stagnation pressure of the flowing fluid. Stagnation pressure is defined as the total pressure measured at a point to which the fluid is isentropically brought to rest or to zero velocity. It is the sum of static pressure and kinetic energy (dynamic head). Figure 9.9 shows the schematic of a Pitot static tube. It consists of L-shaped concentric tubes. The inner tube is used to measure the stagnation pressure and outer tube is used to measuring the static pressure. The inner tube is directed towards the flow, and fluid velocity becomes zero at point B. The outer body of the Pitot tube consists of ports at A, to sense the static pressure. Both the static pressure tube and the stagnation pressure tube are connected to manometers to measure the differential pressure, which is the dynamic pressure.

$$p_0 - p = \frac{\rho v^2}{2} = \rho g h \tag{9.7}$$

where p_0, p and v are the stagnation pressure, static pressure and velocity, respectively, at point A.

The Pitot tube may get clogged if the flowing fluid is not clean and is not free from dirt.

9.8 HOT-WIRE ANEMOMETER

This is a type of thermal anemometer. Here, fluid velocity is measured by measuring the heat convected away from a hot surface by the fluid. The instrument consists of a fine wire made of platinum or tungsten, generally 4 ~ 10 μm in diameter and around 1 mm in length. As the probe used is very small, it will be very fragile and

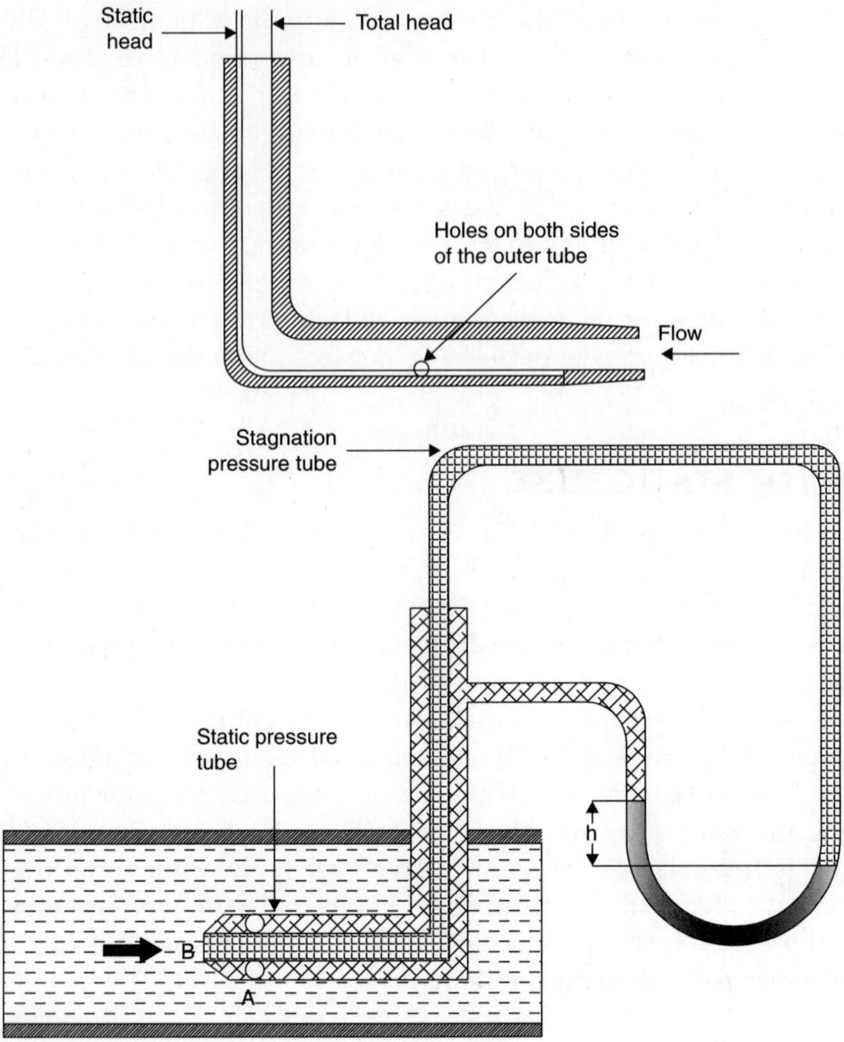

Figure 9.9 Schematic of a Pitot static tube

may break easily. Therefore, this type of anemometer is best suited for measuring clean gas, while the hot plate type is recommended for heavy and rugged fluid flow. A platinum film coated on quartz fibres of diameter 25–150 μm is used as probe in the hot plate type of anemometer. The hot wire is either heated by a constant current or maintained at a constant temperature, and the probe is controlled by a feedback bridge circuit. In both, the heat lost to fluid by convection is a function of the fluid velocity. The schematic of the hot-wire anemometer and the bridge circuit is shown in Figs 9.10 and 9.11.

In a constant temperature anemometer, a voltage-sensitive deflection bridge is used as a feedback circuit. One arm of the bridge circuit is connected to the

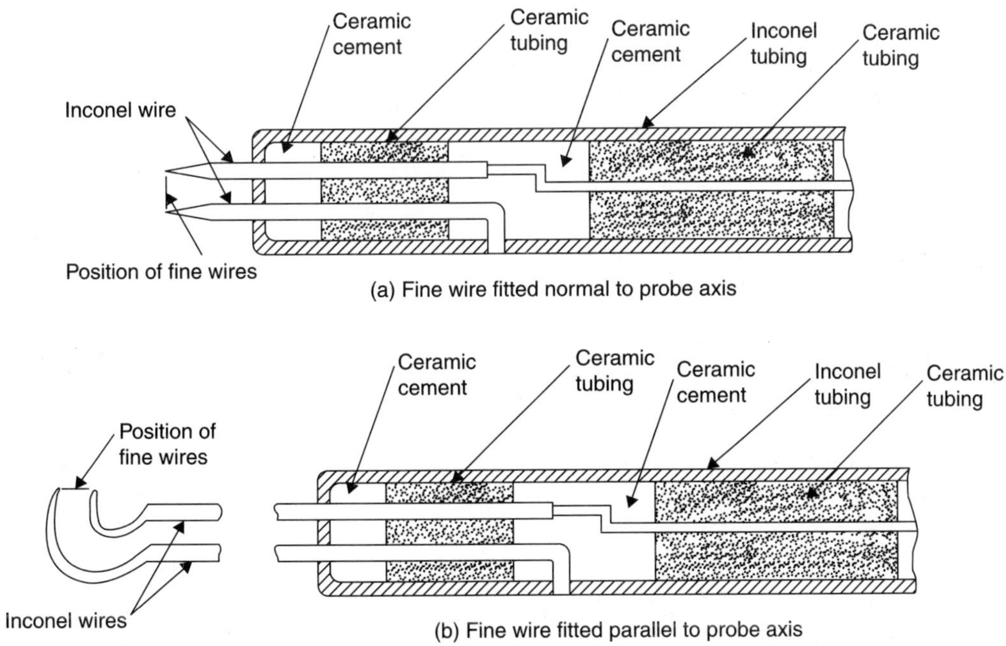

Figure 9.10 Schematic of hot-wire anemometer

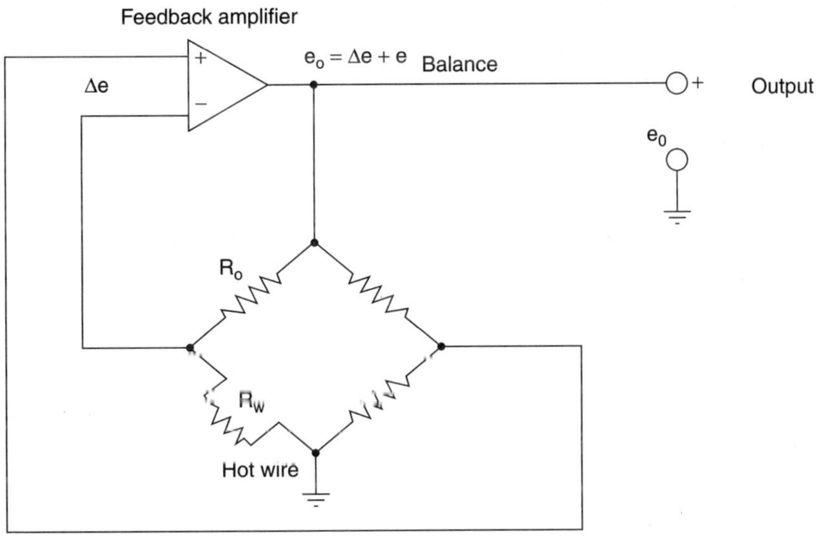

Figure 9.11 Schematic of constant temperature anemometer bridge circuit

hot-wire probe and current is passed through the bridge to heat the wire. The hot-wire probe is introduced to the flow field of the fluid. Due to removal of heat by convection from the hot wire, its resistance will tend to decrease, as it is a function of temperature. This action will create a voltage fluctuation, and the feedback amplifier will increase the power supply to the bridge. Therefore, the wire probe is heated to maintain a constant temperature; thereby, the bridge is balanced. The measure of the compensating voltage is proportional to the rate of fluid flow to be measured.

9.9 LASER DOPPLER ANEMOMETER

The laser Doppler anemometer works on the principle of Doppler shift (or effect). If a narrow beam of laser or ultrasonic wave of a particular frequency is projected on a moving object, the reflected beam will have a different frequency compared to that of the projected beam. This difference in frequency is known as Doppler shift. The measure of the frequency variation is proportional to the velocity of the moving object.

The schematic of the laser Doppler anemometer is shown in Fig. 9.12. It consists of an argon or helium-neon laser source that produces the laser beam.

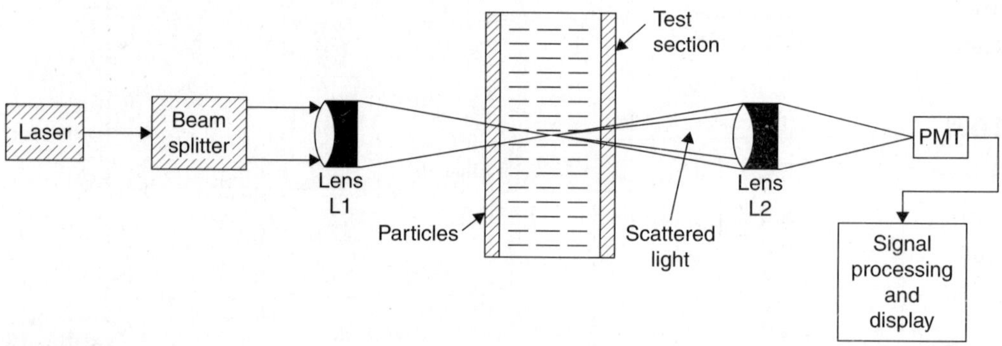

Figure 9.12 Schematic of a laser Doppler anemometer

The laser beam is split into two beams of equal intensity using a beam splitter. The two beams are focused by a lens on an intersection point in the flow field. The fluid particles that pass through the intersection point scatter in both the beams and the scattered beams are focused to a photo multiplier tube (PMT). This consists of a photo detector and a signal processing unit to measure the Doppler shift frequency of the laser beams. The magnitudes of the Doppler shift of both scattered beams are equal and opposite. The PMT is positioned in such a way that it collects both the direct and the scattered beam. The measure of the Doppler shift is proportional to the velocity of the fluid. LDA measurements are very accurate and can be used for the velocity measurement of both liquids and fluids. As particles are required to scatter the laser beams, LDA cannot be used for flow measurements of clear fluids.

$$V = A.v \tag{9.8}$$

where V = volume rate of fluid (m^3/s); A = cross-sectional area of the test section (m^2); and v = measured velocity (m/s).

9.10 FLOW VISUALISATION METHODS

In nature, fluid flows are very complex and experimental flow visualisation techniques are used to understand the flow pattern of a flowing fluid or fluid flow over a body and its surfaces. Flow visualisation techniques reveal the flow pattern of the entire flow field, unlike velocity probes, which provide information only at observed points or along a line. For unsteady and non-uniform flows, visualisation provides details of the flow field far more quickly than a compilation of point-by-point measurements could. Visualisation methods are used to understand the characteristics of flow past an object, such as a region of flow separation, eddy and pressure disturbances, and so on. Flow patterns can be visualised by a variety of methods – injecting dye particles in the flow and utilising variations in the refractive index of the fluid, with variation in density.

9.10.1 Dye Injection

A coloured dye is used to visualise the liquid flow by injecting it in the fluid flow through small holes on the flow surface in midstream. The path traced by the dye is recorded for study. The dye solution should be injected at the local speed of the fluid flow and should match the density of the fluid.

9.10.2 Laser-induced Fluorescence

In this method, an atom or molecule is excited to a higher energy level by the absorption of laser light, followed by spontaneous emission of light. Fluorescent dyes are injected in the fluid flow. A laser beam is projected on a particular plane of the flow so that the dyes in that particular plane alone emit the light.

9.10.3 Smoke Wire Visualisation

A very thin metal wire is placed in the fluid flow. The wire is completely coated with oil. When voltage is applied to the wire, it gets heated because of its resistivity to current flow. This creates thin lines of smoke which are carried away by the flow and help to visualise the flow path.

9.10.4 Hydrogen Bubble Visualisation

This method is used for water flow, where hydrogen bubbles are created by electrolysis and the movement of the bubbles in the flow region is used to observe the pattern of flow. A thin metallic wire of 10–100 μm diameter serves as the cathode and a thick carbon rod as the anode. They are used to electrolyse the water.

$$H_2O = H_2 + \frac{1}{2}O_2$$

9.10.5 Shadowgraph and Schlieren Technique

Variations in the refractive index deflect or phase shift light passing through the fluid. If a screen is placed opposite the light source, these effects create shadows on the screen, creating an image called a shadowgraph. The Schlieren method of flow visualisation is similar to the shadowgraph technique and follows the basic principle that light rays are refracted or bent whenever there is a change in the density of a fluid. The schematic of the Schlieren method is shown in Fig. 9.13. Collimated light rays from the light source are allowed to pass through the test section through which fluid flows. The rays emerging from the test section are focused by a lens on a screen that is placed at the conjugate focus of the lens L_2. A knife edge is placed at the focal point of the lens L_2. Now, if the parallel rays of light encounter a density gradient in the test section, the light is bent, or refracted. This is shown by the dashed line in the figure. This ray of light does not pass through the focal point, and is stopped by the knife edge. The resulting image recorded by the camera has darkened lines that occur where the density gradients are present. This technique provides better visualisation than the shadowgraph.

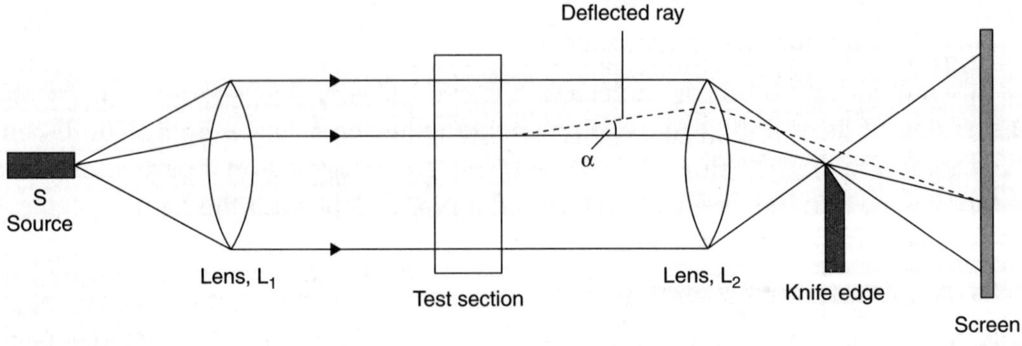

Figure 9.13 Schematic of Schlieren flow visualisation technique

The visualisation in the earlier versions of the Schlieren technique was as black and white images. If a prism is placed before the lens L_1, the white light is split into rays of different colours and the image would be colourful. This method is useful in the design of cars, aircraft and ships and to determine the effect of wind on large buildings.

SUMMARY

- Measurement of fluid flow is vital for metering and controlling fluids in any process application. Flow measurement methods are classified into quantity meters and flow meters. Quantity meters are mainly used for calibration of flow meters. Most flow meters measure the quantity of fluid flow by inferring from a measured quantity of related properties, such as pressure, temperature, density, force or velocity.
- Obstruction flow meters are the most commonly used and are a simple type of flow measurement device. In this type of meter, the quantity of variable head or the pressure difference created by the device arrangement is measured and the flow rate is calculated using Bernoulli's equation. The types of obstruction meters in use are venturimeter, orifice meters, flow nozzle and rotameters.
- The Coriolis flow meter is a device available for the measurement of the mass and density of fluid flow. It works on the principle of the Coriolis effect – an effect experienced by a mass moving in a rotating system by a force (Coriolis force) acting perpendicular to the direction of flow and to the axis of rotation.
- Turbine flow meters are similar to the anemometers used for measuring wind velocity. A turbine rotor with vanes is introduced directly in the flow passage. The rotation of the rotor is proportional to the rate of fluid flow, which is sensed by a pick-up coil arrangement. It is an accurate flow measuring device. Bearing wear and corrosion of turbine vanes may, however, create uncertainties.
- Electromagnetic flow meters work on the principle of Faraday's law of electromagnetic induction. The magnetic field across a flowing fluid which acts as a conductor is disturbed by the fluid motion, which produces an emf. The magnitude of emf induced is proportional to the rate of flow.
- Ultrasonic flow meters are used to measure the mean velocity of fluid flow in a conduit by using acoustic waves, called ultrasound waves, which have a frequency above 20 kHz – not in the audible range for human ears. The flow rate

of the fluid is the measured by observing the difference in velocity component of the reflected ultrasound waves beamed through the fluid passage.

- Pitot static tubes are used to measure the stagnation pressure or total pressure of the flowing fluid, which is the sum of static and dynamic pressure. The arrangement of the Pitot static tube helps to directly measure the static and stagnation pressure heads. The difference between the stagnation and static pressure heads will provide the dynamic or kinetic head of the flow, from which the velocity of the flow is calculated.
- The hot-wire anemometer is a type of thermal anemometer. Fluid velocity is measured by measuring the quantity of heat convected (or dissipated) by the fluid from a hot surface of known temperature.
- Laser Doppler anemometer (LDA) works on the principle of Doppler shift (or effect). If a narrow laser beam of a particular frequency is projected on a moving object, the reflected beam will have a different frequency compared to that of the projected beam. This difference in frequency is known as Doppler shift. The measure of the frequency variation is proportional to the velocity of the moving object.
- Flow visualisation techniques are experimental methods that provide the flow pattern of the entire flow field, unlike velocity probes, which provide information only at observed points. The flow pattern can be visualised by a variety of methods such as injecting dye particles in the flow, utilising variations in the refractive index of the fluid with variation in density, and so on. Some of the common methods are dye injection, laser-induced fluorescence, smoke-wire visualisation, hydrogen bubble visualisation, shadowgraph and the Schlieren technique.

EXERCISES

Multiple Choice Questions

1. Which of the following devices are flow obstruction meters?
   a) Orifice meter b) Venturi meter
   c) Flow nozzle d) All of these

2. Turbine meters are generally preferred for
   a) Low-viscosity and high-flow measurements
   b) High-viscosity and low-flow measurements

c) High-viscosity and high-flow measurements
d) Low-viscosity and low-flow measurements

3. Which one of the following is a mass flow meter?
a) Electromagnetic flow
b) Turbine flow meter
c) Coriolis meter
d) Ultrasonic flow meter

4. Which one of the following is a variable area flow meter?
a) Rotometer
b) Turbine flow meter
c) Electromagnetic flow meter
d) Ultrasonic flow meter

5. The coefficient of discharge of a venturimeter varies between
a) 0.5 and 0.575
b) 98 and 0.99
c) 0.25 and 0.3
d) None of the above

6. The ratio of the cross-sectional area of the fluid at the vena contracta to the cross-sectional area of the orifice is called
a) Coefficient of velocity
b) Coefficient of discharge
c) Coefficient of contraction
d) Coefficient of drag

7. Pitot tube is used to measure the
a) Average velocity
b) Velocity at any point
c) Differential pressure
d) Mass flow rate

8. Which of the following flow meters cannot be used for clean fluid?
a) Electromagnetic flow
b) Turbine flow meter
c) Coriolis meter
d) Laser Doppler anemometer

9. Which of the following flow meters is used to measure mean velocity?
a) Electromagnetic flow
b) Turbine flow meter
c) Ultrasonic flow meter
d) Laser Doppler anemometer

10. The turn down ratio of a typical turbine flow meter is
a) 30:1 for gases and 10:1 for liquids
b) 10:1 for gases and 30:1 for liquids
c) 10: 1 for both
d) 30:1 for both

11. Which of the following is not a flow visualisation technique?
a) Dye injection method
b) Hydrogen bubble visualisation
c) Shadowgraph and Schliren technique
d) Laser Doppler anemometer

12. The hot-wire anemometer is a
a) Variable inductance transducer
b) Variable resistance transducer
c) Variable frequency transducer
d) Variable capacitance transducer

13. In which of the following devices is a PMT used?
    a) Electromagnetic flow
    b) Turbine flow meter
    c) Ultrasonic flow meter
    d) Laser Doppler anemometer

14. Which of the following flow meters works on the principle of Bernoulli's theorem?
    a) Electromagnetic flow
    b) Venturi meter
    c) Coriolis meter
    d) Laser Doppler anemometer

15. Which of the following flow meters is not suitable for highly viscous and corrosive fluid flow?
    a) Electromagnetic flow
    b) Turbine flow meter
    c) Ultrasonic flow meter
    d) Laser Doppler anemometer

Answers				
1. d	2. a	3. c	4. a	5. b
6. c	7. b	8. d	9. c	10. a
11. d	12. b	13. d	14. b	15. b

Review Questions

1. What is the turn down ratio for a flow measuring device?
2. What are the various methods of fluid flow measurement?
3. What is an obstruction flow meter? Mention its types.
4. Explain the construction and working of the following:
    a) Venturimeter
    b) Orifice meter
    c) Flow nozzle
5. How does a rotameter work?
6. Derive Bernoulli's equation for the expression of flow rate for a venturimeter.
7. What is the coefficient of discharge?
8. Explain the principle and working of a turbine flow meter.
9. Explain how Faraday's law of electromagnetic induction is used for the flow measurement.
10. Explain the working of a Coriolis flow meter. What is the different between Coriolis flow meter and turbine flow meter?

11. Explain the following:
    a) Hot wire anemometer
    b) Laser Doppler anemometer
12. With a neat sketch, explain how flow is measured using an ultrasonic flow meter.
13. Explain how the dynamic pressure head can be calculated using Pitot static tube measurement.
14. Why is flow visualisation important? Name the methods.
15. Explain the following
    a) Dye injection method
    b) Smoke-wire visualisation
    c) Shadowgraph and Schlieren technique

10

Measurement of Temperature

10.1 INTRODUCTION

Temperature is the most frequently measured physical parameter. Temperature measurement in today's industrial environment encompasses a wide variety of needs and applications. For the safe and proper operation of any machinery and plant in an industry, it is mandatory that the temperature be properly measured and monitored. Temperature measurement ranges from simple monitoring of the cooling water temperature of an engine to complex measurements of the temperature of a welding. More difficult temperature measurements, such as the temperature of smoke stack gas from a power-generating station or blast furnace or the exhaust gas of a rocket, may also need to be monitored. As pressure, volume, electrical resistance and expansion coefficients are all related to temperature, through the fundamental molecular structure, they change with the temperature. These changes can be used to measure the temperature. This chapter comprehensively deals with the various aspects of temperature and its measurement.

10.2 BASICS OF TEMPERATURE

Before examining the different methods available for temperature measurement, let us look at a brief overview of the basics of temperature.

Temperature is a measure of heat. It is equivalent to potential in electricity and level in hydrostatics. Higher temperature would mean that heat would flow from it to a lower temperature, irrespective of the heat content of the body at the higher temperature. Therefore, temperature can be said to be equivalent to the heat level.

On the basis of kinetic theory, the temperature of a system (T) may be defined as

$$\frac{1}{2} mv^2 = \mathrm{k}T$$

where m is the mass of the molecule; v is its average velocity and k is Boltzmann's constant.

This equation holds good for gases. However, for liquid and solid systems, the temperature may again be defined by the intensity of molecular activity, but the relation is very complicated because of the presence of strong inter-molecular forces. Hence, temperature may be considered as a manifestation of the molecular kinetic energy of a body.

10.3 TEMPERATURE SCALES

Three temperature scales are commonly used: Fahrenheit, Celsius and Kelvin.

Fahrenheit scale: The Fahrenheit temperature scale is one that allots 32° for the freezing point of water and 212° for the boiling point of water; the interval between the two temperatures is divided into 180 parts. In the 18th century, Daniel Gabriel Fahrenheit, a German physicist, took for the zero of his scale, the temperature of an equal mixture of ice and salt, and selected the values of 30° for the freezing point of water and 90° for normal body temperature. This was later revised to 32° and 96°, respectively, but the final scale required an adjustment to 98.6° for 96°.

Celsius scale: The Celsius temperature scale is based on 0° for the freezing point of water and 100° for the boiling point of water. Invented in 1742 by the Swedish astronomer Anders Celsius, it is sometimes called the centigrade scale because of the 100° interval between the defined points. Figure 10.1 shows the relationship between the Fahrenheit and Celsius scales.

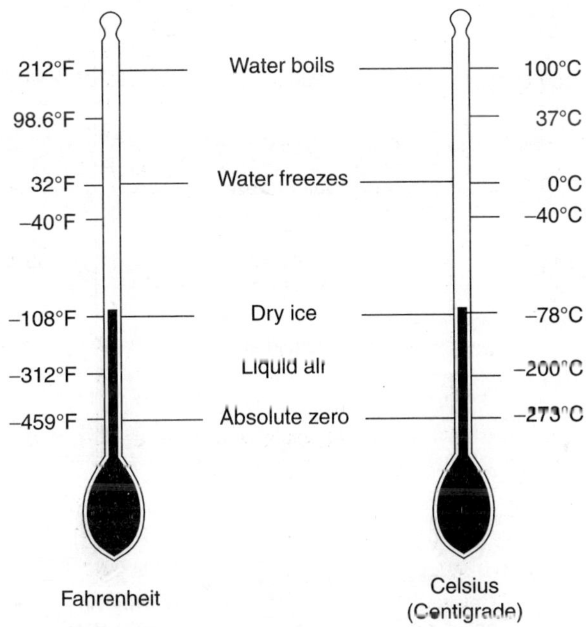

Figure 10.1 Fahrenheit and Celsius scales

Until the 1970s, the Fahrenheit temperature scale was in common use. The Celsius scale was used mostly for scientific purposes worldwide. Since that time, however, most countries have officially adopted the Celsius scale. It is in general use wherever metric units have been accepted, and it is used in scientific work throughout the world.

The relationship between the Celsius and Fahrenheit scales or the conversion formula for a temperature that is expressed on the Celsius (C) scale to its Fahrenheit (F) representation is given below:

$$F = \left(\frac{9}{5}\right)C + 32$$

Kelvin scale: The Kelvin temperature scale is the base unit of thermodynamic temperature measurement in the International System (SI) of measurement. It is defined as 1/273.16 of the triple point (equilibrium of the solid, liquid and gaseous phases) of pure water. The Kelvin (symbol K without the degree sign) is also the fundamental unit of the Kelvin scale, an absolute temperature scale named for the British physicist, William Thomson, 1st Baron Kelvin. The absolute Celsius scale is called the Kelvin scale. Such a scale has absolute zero, the theoretical temperature at which the molecules of a substance have the lowest energy, as its zero point.

Many physical laws and formulae can be expressed more simply when an absolute temperature scale is used. Accordingly, the Kelvin scale has been adopted

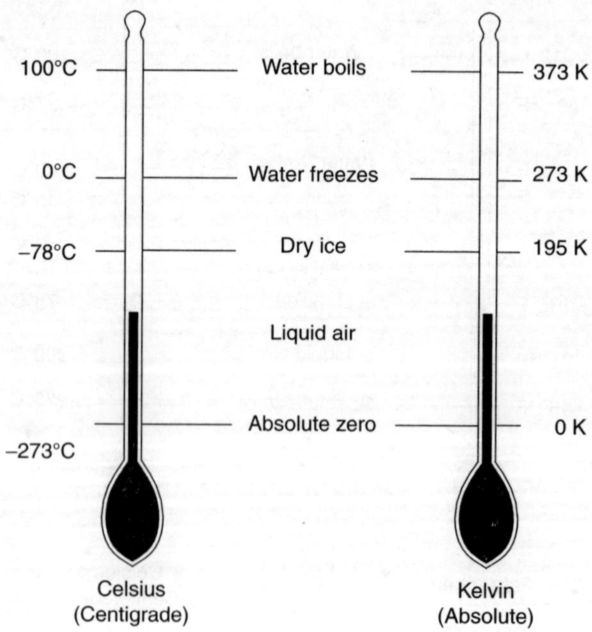

Figure 10.2 Celsius and Kelvin scales

as the international standard for scientific temperature measurement. It is related to the Celsius scale. The difference between the freezing and boiling points of water is 100° in each, so that the Kelvin has the same magnitude as the °C. Figure 10.2 shows the relationship between the Celsius and Kelvin scales.

10.4 METHODS OF TEMPERATURE MEASUREMENT

The methods available for temperature measurement may be broadly classified as contact method and non-contact method.

Contact method: In the contact method, the temperature-measuring device must make physical contact with the body whose temperature needs to be measured, and they must both establish a thermodynamic equilibrium between them. Examples of the temperature-measuring devices used in the contact method are thermometers and thermocouples.

Non-contact method: In this method, the temperature-measuring device and the body whose temperature needs to be measured do not make physical contact with each other. Even if they are separated by some distance, the temperature can be measured. The basic principle behind this method of measurement is the radiation mode of heat transfer. A radiation pyrometer is the best example of a temperature-measuring device used in this method.

10.5 PRINCIPLES OF TEMPERATURE MEASUREMENT

Behind every method of temperature measurement is a fundamental physical principle that governs the functioning of the temperature-measuring instrument. These principles are discussed below.

10.5.1 Expansion

This is one of the oldest principles of temperature measurement. On application of heat, the three states of any substance – solid, liquid and gas – expand. In other words, their volume increases on the application of heat. This increase in volume is almost proportional to the increase in temperature, and this makes this phenomenon of expansion one of the fundamental principles of temperature measurement.

The principle of expansion of solids when heated is used to measure temperature in the form of a bimetallic thermometer. It consists of two metal strips with different coefficients of thermal expansion. This means that the metals will expand at different rates on the application of the same quantity of heat or at the same temperature. These strips are brazed together so that they cannot move relative to each other, as shown in Fig. 10.3.

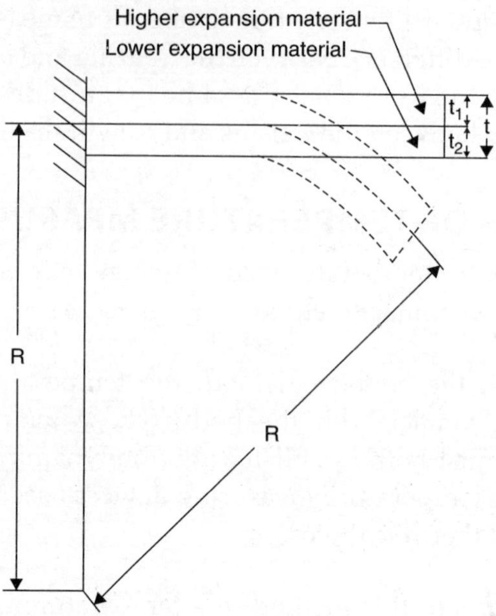

Figure 10.3 Bimetallic thermometer

When the two metals in the bimetallic strip are subjected to the same temperature change, they expand at different rates. As they are brazed together, the strips bend, as shown in Fig. 10.3. The radius of curvature of the bimetallic strip can be determined using the following relationship:

$$R = \frac{t\left[3(1+m)^2 + (1+mn)\left(m^2 + \frac{1}{mn}\right)\right]}{6(\alpha_H - \alpha_L)(T_2 - T_1)(1+m)^2}$$

$$m = \frac{t_1}{t_2} = \frac{\text{Thickness of lower expansion metal}}{\text{Thickness of higher expansion metal}}$$

$$n = \frac{\text{Modulus of elasticity of lower expansion metal}}{\text{Modulus of elasticity of higher expansion metal}}$$

where R = radius of curvature at temperature T_2; t = total thickness of the bimetallic strip $(t_1 + t_2)$; α_L = coefficient of expansion of the lower expansion metal; α_H = coefficient of expansion of the higher expansion metal; T_1 = initial temperature; and T_2 = final temperature.

Using the above formula, the temperature to which the bimetallic strip is subjected can be easily calculated. Bimetallic sensors are available in various forms, like the helix, spiral, cantilever, and so on.

Example: *Mercury Thermometer*

The commonly used mercury thermometer is also based on the principle of expansion. Figure 10.4 shows a typical mercury thermometer.

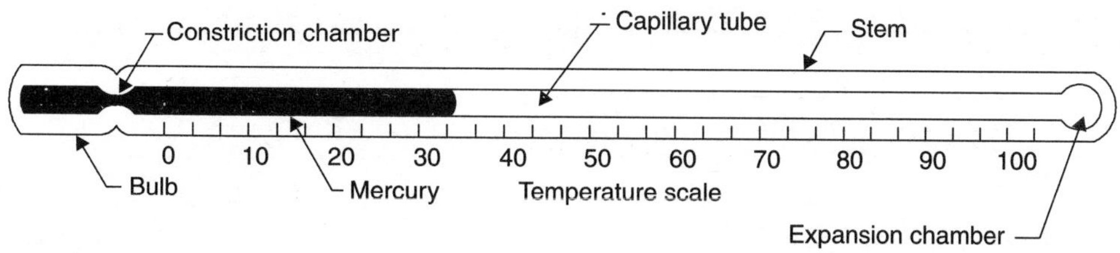

Figure 10.4 Mercury thermometer

When the mercury, a liquid metal, in the thermometer bulb is heated, it expands and thus, rises in the capillary tube, which represents the temperature on the scale marked on the tube.

The expansion of gases is governed by a universal gas law, which is given by the equation

$$|PV = \mathrm{R}T|$$

where P = absolute pressure; V = volume; R = universal gas constant and T = absolute temperature.

In this equation, if pressure is maintained at a constant, then volume is directly proportional to the temperature. This means that if the temperature increases, volume also increases, and vice versa. This is the basis of expansion, a principle of temperature measurement.

10.5.2 Change of State

Substances change their state from solid to liquid (fusion) or from liquid to gaseous (vapourisation) at a particular temperature; this temperature is called the melting (fusing) point or boiling point, respectively. This forms the relation between the state of a substance and its temperature.

Using this relation, temperature-measuring instruments such as pyrometric or Seger cones and vapour pressure thermometers have been devised.

10.5.3 Electrical Method

In this method, the change in temperature is converted into a change in either resistance or voltage, which would eventually result in a change in current or emf.

10.6 TEMPERATURE RANGES

The approximate ranges of various temperature-measuring devices within which they can be used effectively are given in Table 10.1.

Table 10.1 Temperature-measuring devices and ranges

Type of temperature measurement	Range in °C
Glass thermometers	–39 to 400
a) Mercury filled	
b) Alcohol filled	–70 to 65
Bimetallic thermometers	–74 to 540
Pressure gauge thermometers	
a) Vapour pressure	11 to 200
b) Liquid or gas filled	–150 to 600
Resistance thermometer	–240 to 980
Thermocouples	–185 to 1150
Thermistors	–100 to 260
Pyrometers	
a) Optical	760 and above
b) Radiation	540 and above

10.7 THERMOCOUPLES

Thermocouples are the most commonly used device for the measurement of temperature. The basic principle behind the operation of the thermocouple is the Seebeck effect or the principle of thermo-electricity, discovered by Seebeck in 1821.

If two dissimilar metal wires (A and B) are joined together to form a closed circuit, there will be two junctions where they meet. If one of these junctions is heated, an emf is produced, which establishes a flow of current in the circuit, and it can be detected by a galvanometer. The difference in temperature between the two junctions and the material used determines the amount of current, and is given by

$$E = \mathrm{a}\,\Delta T$$

where E = the emf produced; a = a constant that depends on the material selected and ΔT = the temperature difference between the two junctions.

Construction: The materials commonly used for the construction of thermocouples are copper–constantan, iron–constantan, chromel–alumel, platinum–rhodium and chromel–constantan. Any one of the above pairs of materials in the form of a wire is chosen, and they are joined to form a junction. This is covered by a protecting tube and head to prevent it from damage. This is called the measuring junction.

The other junction is the reference junction. The two junctions are connected to a voltmeter by means of extension wires, which are otherwise called compensating leads. Figure 10.5 shows a typical thermocouple.

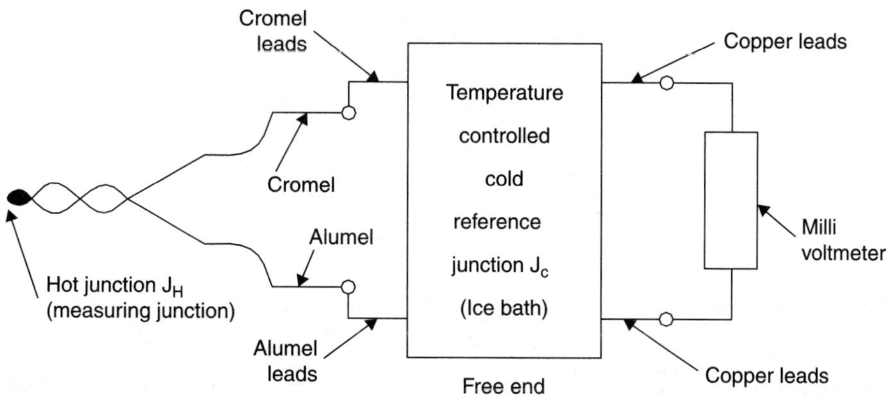

Figure 10.5 Thermocouple

It must be noted that the characteristics of the compensating leads and the materials chosen for the thermocouple must be similar so that the accuracy of the instrument is not affected.

Working principle: The measuring junction (hot junction) is introduced in the place where the temperature must be measured, and the reference junction (cold junction) is maintained at a constant known temperature. For ease of calculation, it is normally maintained at 0°C.

The reading from the voltmeter, which depends on the temperature difference between the junctions, is calibrated to give the temperature reading.

Laws of thermocouples: The operation of thermocouples is governed by the following laws:

- *Law of homogeneous circuit*: According to this law, "a thermo-electric current cannot be produced or sustained in a circuit of a single homogeneous material by the application of heat alone". So, to set a thermo-electric current in a circuit by means of temperature difference, we need at least two different materials
- *Law of intermediate temperatures*: This law states that, "if a single thermocouple, whose junctions are at temperatures T_1 and T_2, generates an emf E_1, and if it

generates an emf E_2 when its junctions are at T_2 and T_3, then the thermocouple will develop an emf $E = E_1 + E_2$ if its junctions are maintained at T_1 and T_3". This law is diagrammatically shown in Fig. 10.6.

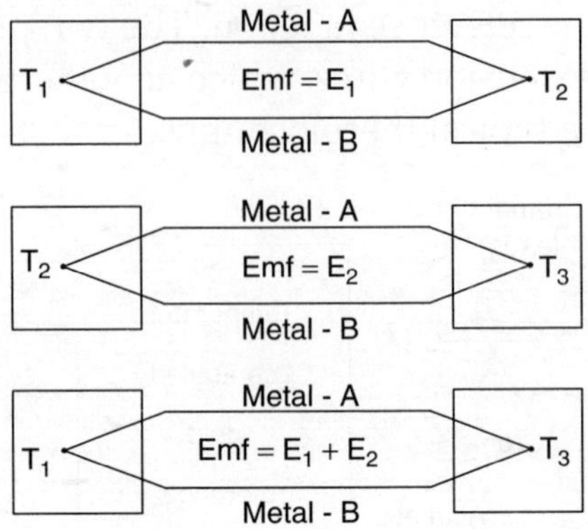

Figure 10.6 Law of intermediate temperatures

This law helps in making corrections to the thermocouple reading if the reference junction temperature is different from the one for which the thermocouple was calibrated.

- *Law of intermediate metals*: According to this law, when a third metal is introduced into a thermocouple circuit whose junctions are maintained at two different temperatures, the emf developed will not be affected, provided the two junctions introduced by the third metal are at the same temperature. This law is diagrammatically shown in Fig. 10.7.

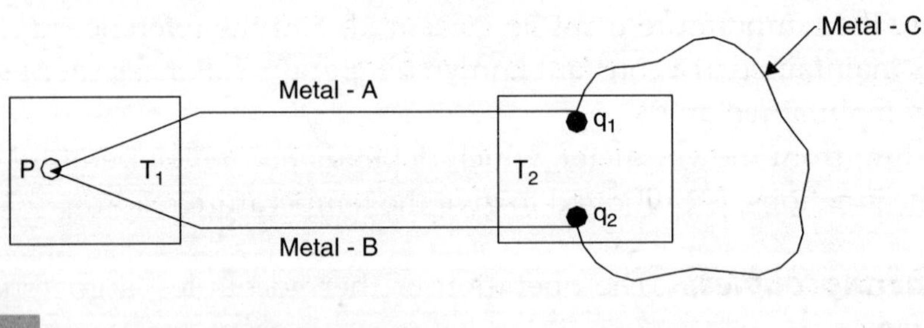

Figure 10.7 Law of intermediate metals

This law indicates the possibility of using extension wires of a metal different from that used in the thermocouple. Also, it facilitates soldering or brazing of the wires without any change in the performance.

- The algebraic sum of the emf produced in a circuit containing two or more thermocouples, all at the same temperature, is zero.

- The net emf of a circuit containing two thermocouples is unaffected by the addition of more thermocouples at same temperature as any of the others.

Advantages

- Inexpensive
- Wide temperature range of –270°C to 2800°C
- High response speed
- Good accuracy
- Good reproducibility
- Long transmission distance

Disadvantages

- Non-linearity of temperature–voltage relationship
- Reference junction temperature must be maintained at a constant, without any deviation
- Signal from the thermocouple needs amplification
- Accessories are expensive

10.8 THERMISTORS

Thermal resistors – shortened to thermistors – are temperature-sensitive variable resistors made of a ceramic-like semi-conducting material. They have negative coefficient of temperature resistance, which means that their resistance decreases with increase in temperature. They have about 10 times higher temperature coefficiency than that of copper. This enables thermistors to detect very small changes in temperature, which is not possible in thermocouples. Thermistors are used in precision temperature measurements.

Construction: Thermistors are made of metal oxides and their mixtures, namely, oxides of nickel, copper, manganese, cobalt, iron, etc. The powdered form of oxides and their mixtures are compressed to obtain the desired shape and size. Electrical contacts, in the form of lead wires, are embedded in the material and it is then heat-treated to form thermistors. The different shapes and sizes of thermistors available are shown in Fig. 10.8.

Thermistors are available in the form of beads 0.1 mm in diameter, discs 5 mm in diameter, and rods of length 0.8 mm to 50 mm.

For the measurement of temperature, thermistors can be pasted or hung in the place where the temperature needs to be measured. When the temperature changes, it produces a change in the resistance of the thermistor. This change in the relationship between resistance and temperature is given by

$$R_1 = R_o e^{\beta\left(\frac{1}{T_1} - \frac{1}{T_0}\right)}$$

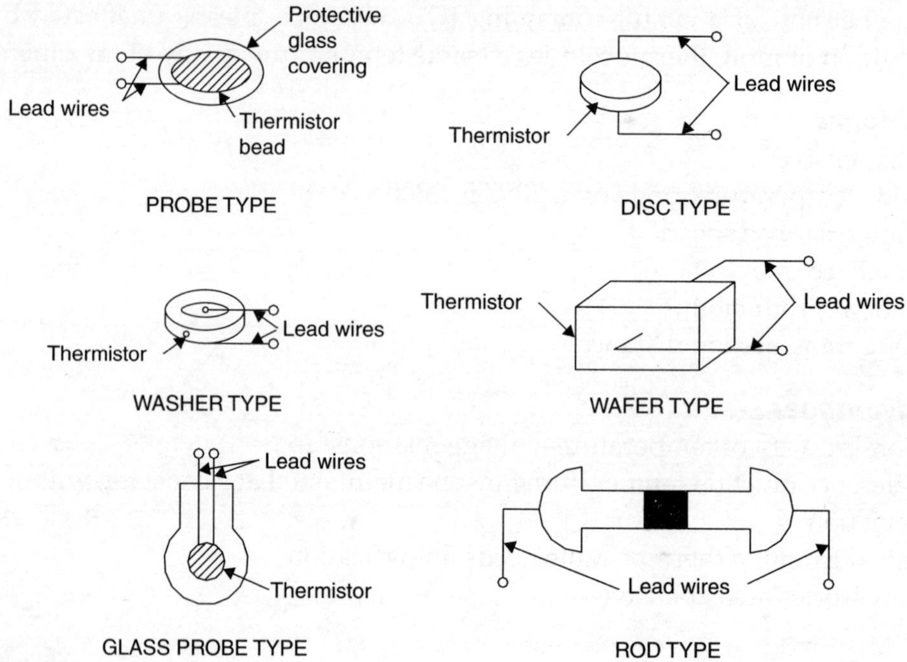

Figure 10.8 Thermistors

where R_1 is the resistance at temperature T_1K; R_0 is the resistance at temperature T_0K; and β is a constant varying from 3400K to 4600K based on temperature compensation.

Advantages

- ✦ Highly sensitive
- ✦ Precision is very high
- ✦ Wide temperature range from −60°C to 15°C
- ✦ Compact
- ✦ Inexpensive
- ✦ Good stability

Disadvantage

- ✦ High non-linear characteristic of resistance against temperature

Applications

- ✦ Used to control temperature
- ✦ Helps in temperature compensation
- ✦ Used to provide time delay
- ✦ Used in the measurement of thermal conductivity

10.9 PRESSURE THERMOMETERS

The principle behind the vapour pressure thermometer is that liquids and gases experience change in volume when they are subjected to a change in temperature – they expand when heated and contract when cooled. If a closed container is filled with fluid, the volume remains constant, but the pressure changes; i.e., if heated, pressure increases, and if cooled, pressure decreases. This forms the relationship between fluid pressure and temperature.

Construction: The construction of the pressure thermometer is given in Fig. 10.9. The pressure thermometer consists of a sensing bulb, a capillary tube, a Bourdon tube and a gauge. The Bourdon tube, which is a flexible tube with elliptical cross-section, is connected to the opening in the sensing bulb through the capillary tube. Both the bulb and the Bourdon tube, along with the capillary tube, are filled with liquid (mercury, acetone, ethyl alcohol) or gas (nitrogen, helium). Based on the filling medium, the pressure thermometer can be called liquid-filled or gas-filled, and sometimes, it can also be filled with liquid and gas.

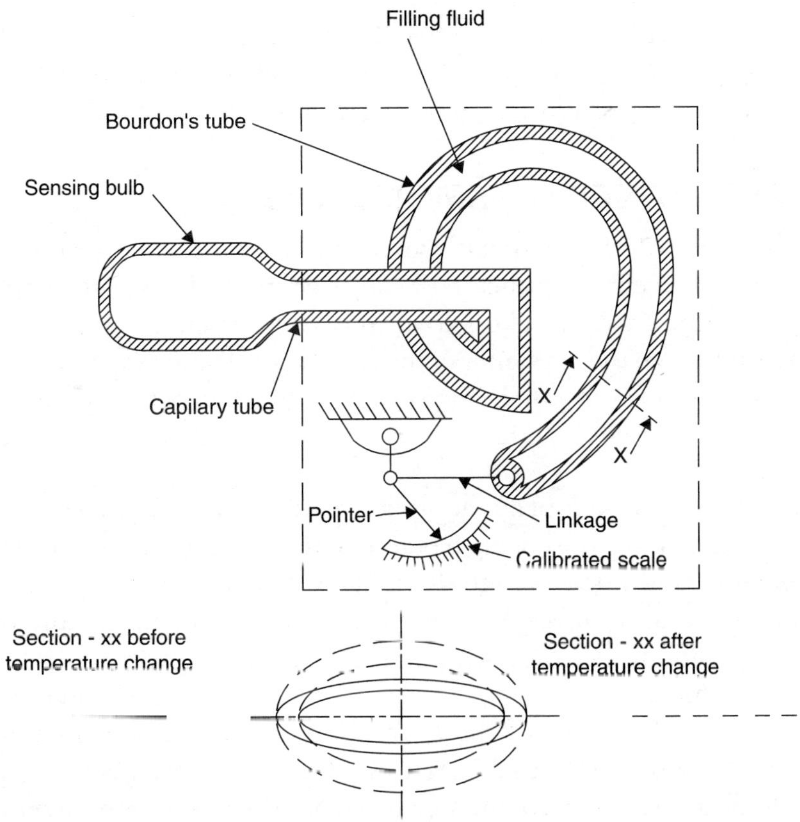

Figure 10.9 Pressure thermometer

The sensing bulb, capillary tube and Bourdon tube are sealed so that this system is leak-proof. The other end of the Bourdon tube is connected to a gauge with a pointer which sweeps over a temperature calibrated scale.

Working principle: The sensing bulb of the pressure thermometer is introduced in the place or medium whose temperature needs to be measured. Because of the change in temperature, the fluid inside the bulb undergoes change in its volume and pressure. This change is transmitted to the Bourdon tube, which in turn, undergoes a small displacement. This displacement of the Bourdon tube is sensed by the gauge, which points to the temperature on the calibrated scale.

Advantages

- Highly sensitive, and hence, has a quick response
- Cheap to construct

Disadvantages

- Decomposition of the fluid may affect the accuracy
- The tubes used are also temperature sensitive and may cause errors
- Temperature compensation must be carried out if a long capillary tube is used

10.10 RESISTANCE THERMOMETERS

Resistance thermometers are otherwise called resistance temperature detectors (RTDs). When an electric conductor is subjected to a change in temperature, it undergoes physical changes, like expansion or contraction, and these physical changes in turn change the resistance of the conductor. This change in resistance due to change in temperature leads to change in the relationship between the temperature and resistance to electricity, and hence, conduction of electricity takes place.

Construction: Wires made of materials like platinum, copper, nickel and tungsten are used as the temperature-sensing element in RTDs. The material selected for the sensing element should have a high thermal coefficient of resistivity, be corrosion resistant, and should not undergo phase changes due to temperature changes in the range of operation.

An insulating material made of glass or ceramic, which may be in the shape of a flat strip or a cylindrical rod, is wrapped around the wire. The insulating material wound with the wire is shielded by an evacuated or inert gas–filled protective sheath made of glass. The two ends of the wound-up wire are taken out of the protective sheath to form the two leads. Figure 10.10 shows the construction of the resistive thermometer.

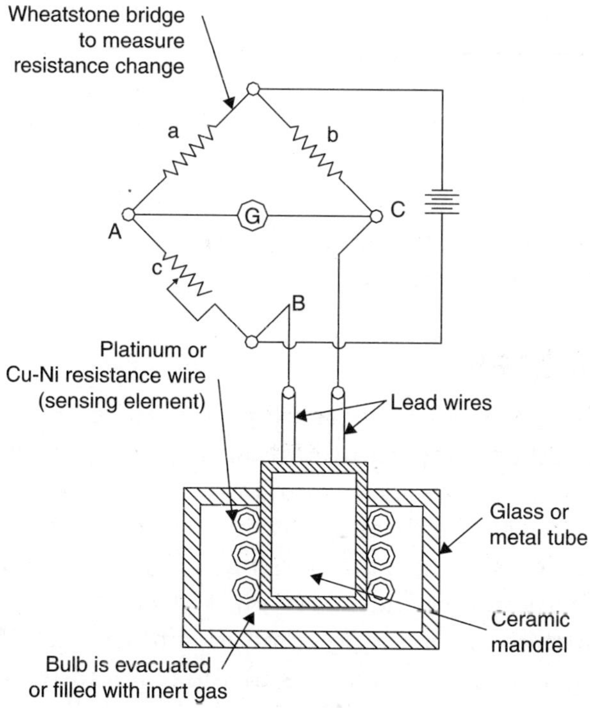

Figure 10.10 Resistive thermometer

Working principle: The two lead wires are connected to a Wheatstone bridge, as shown in the figure. When a known value of current is passed through the Wheatstone bridge, the initial resistance of the RTD is measured. When the RTD is introduced into the media whose temperature is to be measured, the resistance of the wire inside the RTD changes, and it causes a change in the flow of current. This can be detected using a galvanometer in the Wheatstone bridge. The change in the value of the current can be calibrated to measure the change in temperature.

The relation between the initial resistance and the resistance at the measured temperature is given by

$$R_t = R_o (1 + AT + BT^2)$$

where R_t is the resistance at temperature T; R_o is the resistance at the reference temperature or initial temperature; T is the measured temperature; and A and B are constants based on the material.

Advantages

- Accuracy is high
- The device is small, and hence, compact.
- Response is quick

- Good repeatability
- No temperature compensation is required
- High temperature range of –200°C to 650°C

Disadvantages

- High cost
- Requires Wheatstone bridge circuit
- Sensitive to vibration

10.11 PYROMETERS

Pyrometers are devices used to measure temperature above 500°C, which is not possible with other temperature-measuring devices like thermocouples and thermistors. Pyrometers are commonly used when the temperatures being measured are very high and physical contact with the object whose temperature is to be measured is not possible. They are also used in conditions where corrosive vapours or liquids would destroy thermocouples, RTDs and thermistors if they are made to come in contact with the medium to be measured. They can also be used to measure the temperature of moving bodies. Pyrometers use the principle of thermal radiation, and hence, are also called radiation pyrometers.

Radiation pyrometers measure the radiant heat emitted or reflected by a hot object. Thermal radiation is electromagnetic radiation emitted as a result of temperature. Thermal radiation lies in the wavelength region of about 0.1 μm to 100 μm. Practical radiation pyrometers are sensitive to a limited band of radiant energy, although the theory indicates that they should be sensitive to the entire spectrum of energy radiated by the object.

The operation of a thermal radiation pyrometer is based on the blackbody concept. The total thermal radiation emitted by a blackbody q_b is given by

$$q_b = \sigma T^4$$

where σ is Boltzmann's constant = 57.2 × 10^{-9} W/m^2 and T is the absolute temperature in K.

Pyrometers are of three types: total radiation, optical and infrared.

10.11.1 Total Radiation Pyrometers

The total radiation pyrometer receives virtually all the radiation from a particular area of a hot body and focuses it on a sensitive temperature transducer, such as a thermocouple or thermopile. The term total radiation includes both visible light and invisible infrared radiation.

Construction: The total radiation pyrometer consists of a radiation-receiving element and a measuring device to indicate the temperature. A mirror radiation receiver is shown in Fig. 10.11.

A diaphragm unit along with a mirror is used to focus the radiation on a radiant energy–sensing transducer. Because of the absence of a lens, the mirror arrangement has an advantage in the absence of absorption and reflection effects. The device is arranged in such a way that the image of the front diaphragm is focused on the thermocouple by the mirror. Therefore, the temperature measurements are independent of the distance of the target. The presence of any absorbing media between the target and the pyrometer reduces the radiation received and makes the pyrometer read low. Substances like smoke, dirt and gases absorb radiation, and therefore, cause negative static errors. On the other hand, the presence of heat sources like hot gases, high temperature particles and flames cause the meter to read high.

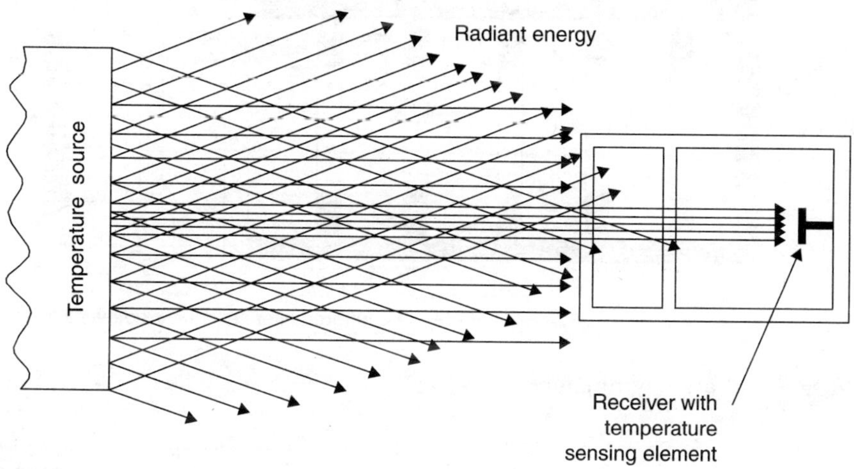

Figure 10.11 Total radiation pyrometer

Total radiation pyrometers are non-linear and exhibit poor sensitivity in the lower temperature ranges. Despite this lower sensitivity, they are preferred for use in low-temperature applications to partial radiation pyrometers, which provide better results in applications involving the measurement of high temperatures.

Total radiation pyrometers cannot be used for temperatures lower than 600°C, since at lower temperatures, errors may be introduced by the fact that the temperature of the pyrometer itself may not be negligible when compared with that of the hot body. Therefore, total radiation pyrometers are used for a temperature range of 1200°C to 3500°C.

10.11.2 Infrared Pyrometers

Infrared pyrometers are partial radiation pyrometers. Infrared energy is invisible to the human eye, but can be felt. There is a proportional increase in infrared energy as the temperature of the surface of the radiating body increases. Above 550°C, a surface starts to radiate visible light energy and, simultaneously, there is a proportionate increase in the infrared energy.

Different types of photoelectric transducers are used for infrared transducers. The most successful transducer for this purpose is the photovoltaic cell. High speed of response is an outstanding feature of the pyrometers that use photovoltaic cells. The infrared radiation is focused on to a photovoltaic cell, and therefore, it is necessary to ensure that the cell does not become over-heated. Figure 10.12 shows an infrared pyrometer that uses a photovoltaic cell.

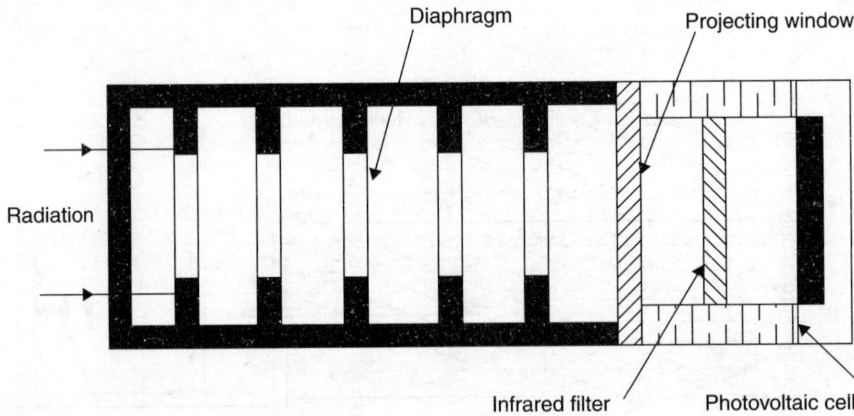

Figure 10.12 Infrared pyrometer

The cone of radiation passing to the photo cell is defined by the area of the first diaphragm. The protective window is made of thin glass and it serves to protect the cell and filter from any physical damage. The filter is normally used in the range of 1000°C to 1200°C in order to reduce the infrared radiation passing to the photo cell. This prevents the photo cell from getting over-heated.

Infrared pyrometers depend on the transmission of the infrared radiant energy emitted by a heated body to a detector in the measuring system through the surroundings. There is no physical contact with the surface whose temperature is being measured. The head of the sensor is focused on the object whose temperature is measured. The infrared rays falling on the detector change the resistance in the detector in proportion to the temperature. The change in resistance is then indicated on a meter or digital display, which in turn indicates the temperature.

10.11.3 Optical Pyrometers

The temperature of an object has a relation to its colour or brightness. Optical pyrometers use this principle as the base for temperature measurement. It is the most accurate of all the radiation pyrometers. The most commonly used type of optical pyrometer is the disappearing filament. A schematic representation of a disappearing filament optical pyrometer is shown in Fig. 10.13.

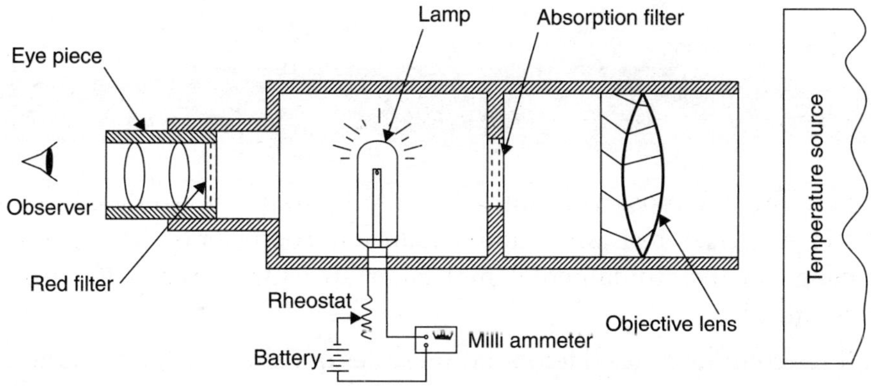

Figure 10.13 Disappearing filament optical pyrometer

It has a lens that makes an image of the target on a flat tungsten filament lamp. To increase the range of the tungsten filament, a calibrated absorption filter is inserted in front of it. An eyepiece is used to view the image of the target and the filament. The brightness of the filament changes depending on the current flowing through it.

Initially, the tungsten lamp is calibrated by comparing the brightness of the image of a blackbody whose temperature is known with that of the tungsten filament. The current flowing through the filament is controlled until the filament disappears in the superposed field of the target. When the filament disappears, the brightness of the tungsten filament and that of the target are equal. With this, a relation between the current flowing through the filament i and the temperature

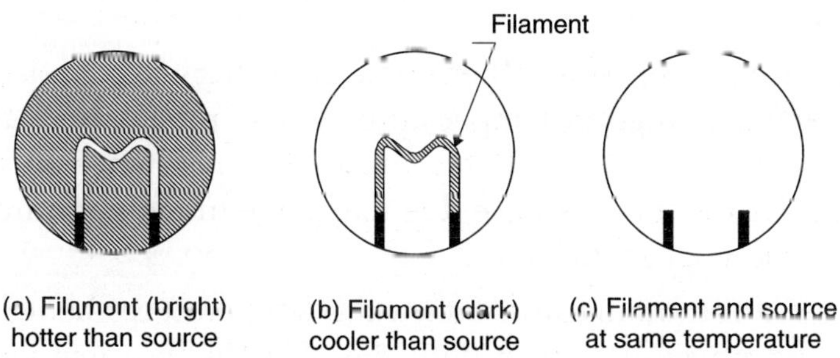

Figure 10.14 Filament conditions

of the blackbody T is arrived at. Figure 10.14 shows the appearance of the filament under different temperature conditions of the target.

Generally, a red filter, which enables a monochromatic condition, is placed between the filament and the eyepiece for easy matching. Even though there is no upper limit of temperature that can be measured by the optical pyrometer, considering the life and stability of the filament, the tungsten lamp will not normally be operated above a certain brightness or current.

SUMMARY

- Temperature, the most frequently measured parameter, is related to pressure, volume electrical resistance and expansion coefficients. The phenomenon of change in the fundamental molecular structure is used to measure the temperature.
- The three commonly used temperature scales of Fahrenheit, Celcius and Kelvin have different references and a fixed relationship.
- The fundamental principles of expansion, change of state and electrical resistance enable various techniques of temperature measurement. The principle of expansion of solids and liquids when heated is used to measure temperature using a bimetallic thermometer and a mercury thermometer, respectively.
- Thermocouples function based on the Seebeck effect. The operation of thermocouples is governed by three laws: law of homogeneous circuit, law of intermediate temperatures and law of intermediate metals.
- Thermistors are based on the phenomenon of decrease in resistance with increase in temperature.
- The principle behind the vapour pressure thermometer is that liquids and gases experience change in volume when they are subjected to a change in temperature.
- The basic principle of resistance thermometers is that when an electric conductor is subjected to a change in temperature, it undergoes physical changes like expansion or contraction, and this makes the resistance of the conductor change. This change in resistance due to change in temperature leads change in the conduction of electricity.
- Pyrometers are commonly used when the temperatures being measured are very high and physical contact with the object whose temperature is to be measured is not possible. They are also used in conditions where corrosive

vapours or liquids would destroy thermocouples. The operation of thermal radiation pyrometers is based on the blackbody concept.

EXERCISES

Multiple Choice Questions

1. Which law of thermodynamics forms the basis of temperature measurement?
   a) Zeroth law of thermodynamics
   b) First law of thermodynamics
   c) Second law of thermodynamics
   d) All of the above

2. The thermocouple circuit that is used to measure temperature works on
   a) Peltier effect
   b) Seebeck effect
   c) Thomson effect
   d) None of the above

3. Absolute zero on the Kelvin scale is equal to
   a) 0 K
   b) 273 K
   c) 373 K
   d) –273 K

4. The following value corresponds to –40°C:
   a) 0°F
   b) –4°F
   c) 273°F
   d) None of the above

5. Which of the following relationships between the Celsius and Fahrenheit scales hold good?
   a) $C = (9/5) F + 32$
   b) $C = (5/9) F + 32$
   c) $F = (9/5) C + 32$
   d) $F = (5/9) C + 32$

6. In electric resistance thermometers, the thermometric property is
   a) Electric current passing through a metal wire
   b) Voltage between two extreme end points of a metal wire
   c) Resistance of a metal wire
   d) None of the above

7. Which of the following is chosen as a standard thermometric substance?
   a) Mercury
   b) Gas
   c) Thermocouple
   d) Electric resistance

8. The materials commonly used for the construction of thermocouples is/are
   a) Copper–Constantan
   b) Iron–Constantan
   c) Chromel–Alumel
   d) All of the above

9. Thermocouples are
   a) Inexpensive
   b) Measure temperatures ranging –270°C to 2800°C
   c) Have good reproducibility
   d) Possess all of the above properties

10. Thermistors are
   a) Thermal resistors
   b) Made up of semi-conducting material
   c) Devices that possess negative coefficient of temperature
   d) Devices that possess all the above properties

11. Thermistors are made of oxides of
   a) Nickel
   b) Copper
   c) Manganese
   d) All of the above

12. The following characteristic of a material taken as an indication of change in temperature is known as
   a) Thermodynamics property
   b) Thermostatic property
   c) Thermometric property
   d) None of the above

13. In mercury thermometers, it is assumed that the length of liquid and temperature are
   a) Inversely proportional to each other
   b) Directly proportional to each other
   c) Independent of each other
   d) Weakly dependent on each other

14. In a resistance thermometer, a metal wire shows a resistance of 500 Ω at ice point and 550 Ω at hot point. Calculate the temperature that corresponds to a resistance of 535 Ω.
   a) 50°C
   b) 60°C
   c) 70°C
   d) 80°C

15. Every temperature-measuring instrument makes use of a physical property of a substance in order to measure temperature objectively. Which physical property is used by a mercury-in-glass thermometer?
   a) Resistance of a piece of metal
   b) Electromotive force
   c) Volume of a fixed mass of liquid
   d) Pressure of a fixed mass of gas at constant volume

16. Every temperature-measuring instrument makes use of a physical property of a substance in order to measure temperature objectively. Which physical property is used by a thermocouple?
a) Electromotive force
b) Volume of a fixed mass of liquid
c) Pressure of a fixed mass of gas at constant volume
d) Resistance of a piece of metal

17. Every temperature measuring instrument makes use of a physical property of a substance in order to measure temperature objectively. Which physical property is used by a resistance thermometer?
a) Electromotive force
b) Volume of a fixed mass of liquid
c) Pressure of a fixed mass of gas at constant volume
d) Resistance of a piece of metal

18. Which one of these thermometers is portable and simple to use?
a) Constant volume gas thermometer b) Resistance thermometer
c) Mercury-in-glass thermometer d) Thermocouple

19. Which one of these thermometers is best suited for measuring rapidly changing temperatures?
a) Thermocouple b) Mercury-in-glass thermometer
c) Resistance thermometer d) Constant volume gas thermometer

20. The sensitivity of a thermometer refers to
a) Amount of change in thermometric property for a unit change in temperature
b) How quickly the thermometer can register change in temperature
c) Minimum and maximum temperatures that the thermometer can measure
d) None of the above

21. The responsiveness of a thermometer refers to
a) Amount of change in thermometric property for a unit change in temperature
b) How quickly the thermometer can register change in temperature
c) Minimum and maximum temperatures that the thermometer can measure
d) None of the above

22. A thermocouple shows an emf of 5 mV with one of its junctions at 0°C and other at 100°C. If the 100°C junction is taken out of boiling water and placed in a glass of water of room temperature (25°C), the emf registered on thermocouple would be
a) 1.25 mV b) 5 mV
c) 10 mV d) 20 mV

23. Which feature of laboratory thermometers makes them responsive?
    a) Thin-walled glass bulb
    b) Small bulb containing a small amount of liquid
    c) Relatively small size of thermometer
    d) Fine as well as uniform bore of capillary tube

24. Which feature of laboratory thermometers makes them sensitive?
    a) Thin-walled glass bulb
    b) Small bulb containing a small amount of liquid
    c) Relatively small size of thermometer
    d) Fine as well as uniform bore of capillary tube

25. Pyrometers are used when
    a) Temperature to be measured is more than 500°C
    b) Physical contact with the measuring body is not possible
    c) Corrosive liquids would destroy thermocouples
    a) All of the above

Answers

1. a	2. b	3. a	4. b	5. c
6. c	7. b	8. d	9. d	10. d
11. d	12. c	13. b	14. c	15. c
16. a	17. d	18. c	19. a	20. a
21. b	22. a	23. b	24. d	25. a

Review Questions

1. List three areas where temperature measurement is important.
2. Define temperature.
3. Give the relationship between various temperature scales.
4. Name the parameters (principles) used for measurement of temperature.
5. Explain the basic principle on which the bimetallic thermometer works.
6. List five properties a material should have to be an element of a bimetallic strip.
7. List the common metals used in bimetallic strips.
8. List any four major disadvantages of using bimetallic strips for temperature measurement.

9. What is a pressure thermometer?
10. With a neat sketch, explain the construction and working principle of vapour pressure thermometer.
11. What are the limitations of a thermistor?
12. What is a thermocouple? State its applications.
13. What are the materials commonly used for thermocouples? What is a thermopile?
14. What is the purpose of connecting thermocouples in parallel?
15. Explain the working of helix and spiral bimetallic thermometers.
16. What are RTDs? On what basic principle do they work? Explain with a diagram of one of the RTDs.
17. List the advantages of temperature measurement by using the resistance thermometer.
18. What is a thermistor? How is it used for temperature measurement?
19. Explain how a thermocouple is used to measure temperature.
20. State and explain the three laws of thermocouples.
21. What is a pyrometer? What are its types?
22. Explain the working principle of any one pyrometer, with a neat sketch.

Quality Control

11

Statistical Quality Control

11.1 INTRODUCTION

The main objective of measurement is to ensure the existence of the required quality characteristics of a product, device or system at the required level. In other words, the purpose of measurement is to ensure quality. This chapter deals with the various techniques, methods and plans that use the statistics obtained from the process of measurement to check whether the manufacturing process will produce the components at the required quality level.

According to the ISO 9000 standards, quality is defined as, "a degree to which a set of inherent characteristics fulfils requirement". The term 'quality' can be used with adjectives such as 'poor', 'good' or 'excellent'. 'Inherent' means the characteristics that exist permanently in something.

The requirements can be generated by different interested parties such as customers, owners, suppliers and various stakeholders of the organisation. In general, the quality should reflect the expectations of the end user (consumer or customer), who is in need of a particular product or service, according to his/her desire. As perception varies with individuals and also with the type of product or service, it is difficult to quantify quality. The precision of machining expected from a CNC lathe machine cannot be expected from an ordinary lathe. Also, a high degree of precision is not required for all kinds of jobs.

11.2 QUALITY CONTROL AND QUALITY ASSURANCE

Quality control (QC) is the process of achieving quality in a product or service by implementing standards and procedures. According to ISO 9000:2005, quality control is considered as a part of the quality management system focused on fulfilling quality requirements.

The basic steps required to implement quality control are as follows:

1. Setting the standards and inspection methods.
2. Designing the product or service according to standards.

3. Implementation (manufacturing the product or creating infrastructure for the service).
4. Inspecting and reviewing the standards and procedures.

According to ISO 9000:2005, **quality assurance (QA)** is considered a part of the quality management system focused on providing confidence that the quality requirements will be fulfilled. Quality assurance aims at the framework of activities to be followed in order to assure a desired level of quality or output by preventing defects, whereas quality control focuses on identifying the defects in the product or service and helps to prevent it from reaching the customer. Therefore, quality control is product oriented while quality assurance is process oriented. Feedback from the quality control team would be helpful in improving the standards and procedures, which in turn would augment the quality assurance.

Table 11.1 states the differences between quality assurance and quality control.

Table 11.1 Quality assurance versus quality control

Quality Assurance	Quality Control
It is the framework of activities used to assure a desired level of quality of a product or service.	It is the process of achieving quality in a product or service.
It prevents defects.	It identifies and corrects the defects.
All the team members are responsible for QA.	The inspection team is responsible for QC.
The statistical technique used in QA is known as statistical process control (SPC).	The statistical technique used in QC is known as statistical quality control (SQC).
It is process oriented.	It is product oriented.
It is a failure prevention system.	It is a failure detection system.
It is performed before QC.	It is performed after QA.

11.3 CAUSES OF VARIATION

In a mass production process, the expected outcome would be instances of the product with identical parameters and without any variation. However, though very high precision equipment, measuring instruments, raw materials, proper ambience and human resources are deployed, there will be slight deviations (variations) in the outcome for various reasons that cannot be traced or understood. Even if the reasons could be predicted in some cases, it would not be economical to overcome them. However, such deviations would be acceptable if they were within the range of the upper and lower control limits of the process. If the variations are within these control limits, the process is said to be stable or in control. Such sources or causes of variation are inherent and inevitable according to the law of nature, and therefore, are called **chance** or **inherent causes**.

When the process outcome exceeds the control limits, it would be easy to identify the causes and control them. This process situation could be easily understood by unusual patterns in the control charts. This type is called **assignable causes**.

WE Deming, the quality control scientist, has depicted chance and assignable causes as common and special causes in his work. Table 11.2 states the differences between chance and assignable causes of variation.

Table 11.2 Chance cause versus assignable cause

Chance Cause	Assignable Cause
It consists of many individual causes.	It consists of one or a few individual causes.
Any one chance cause will result in only a minute amount of variation.	Any one assignable cause can result in a large amount of variation.
It cannot be easily detected.	It can be easily detected.
If detected, it can be easily eliminated from a process economically.	Even if detected, the action to eliminate the cause may need to be justified.
It follows the natural pattern of variation.	It follows an unnatural pattern of variation.
The effects are random.	The effects are systematic.
Random error occurs.	Systematic error occurs.
It occurs due to poor design, poor maintenance of machines, and poor working conditions like lighting, noise, dirt, temperature, etc.	It occurs due to poor adjustment of equipment or machines, faulty controllers, malfunctioning of machines, operator carelessness, poor batch of raw materials, power surges, etc.

Figure 11.1 shows a typical control chart that explains the behaviour of a process over time. The centre line implies the process average or mean, also called the **central limit**. Variations in the data plotted within the upper and lower control limits (UCL and LCL) are due to chance causes, and the process is considered to be stable or in the control limit. The data marked above the upper control limit is due to an assignable cause and indicates that the process exceeds the control limit.

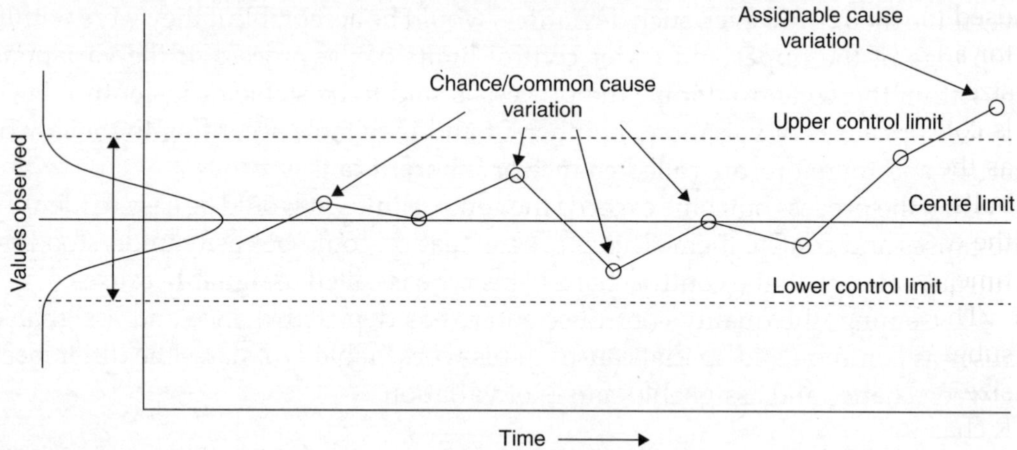

Figure 11.1 Chance and assignable causes of variation

11.4 CONTROL CHART AND ITS SELECTION

A control chart is an online process monitoring technique that utilises statistical calculations and graphical representations to help us understand the ongoing process in real time and to identify the process capability – whether the process is stable and predictable. Quality control charts are broadly classified into two groups, known as control charts for **attributes** and **variables.**

If the data available from a process is perceptive or intuitive in nature, and would reveal whether the quality of the parts produced are within the acceptable range or not (i.e., conforming or non-conforming), the analysis should be carried out using control charts for attributes. For example, when parts are inspected using the GO and NO GO gauges, or the quality of painting or a dent on a car body is inspected, the resulting data would indicate the number of parts or objects within the range of acceptability, out of the total sample size. In this type of data, the actual size of the individual part is unknown, unmeasurable or unimportant from the manufacturer's point of view.

When data such as the size and weight (for example, 12.05 mm or 250.25 kg) of a part is available, the analysis should be carried out using control charts for variables. It is important to note that the variable data available from a process can be converted into attribute data and analysis could then be carried out using control charts for attributes; however, it is not possible to convert attribute data into variable data.

The common types of control charts for variables are the average chart ($\overline{X}$ chart), moving range chart (mR chart), range chart (R chart) and standard deviation chart (S chart). The $\overline{X}$ chart is generally used to understand the central tendency, whereas the moving range (mR), range (R) and standard deviation charts (S) are used to understand the distribution of a process. The type of chart that is selected for a particular process depends on the number of subgroups and the subgroup size. The **subgroup** is determined based on the number of times the sample data is collected, while the number of samples collected for each subgroup is known as the **subgroup size**. For example, in a continuous production process for steel bolts, at every one-hour interval, exactly 10 samples are collected. Therefore, for the entire day, samples would have been collected 24 times and 10 samples each time. In this case, the number of subgroups is 24 and the subgroup size is 10.

The selection of a type of chart for a particular process depends on the nature and subgroup size of the data available for analysis. For variable data, if the subgroup size is 1, a combination of the $\overline{X}$ and mR charts is preferable. Whereas, the $\overline{X}$ and R charts are preferred if the subgroup size is between 2 and 10. In some cases, the subgroup size may vary in each subgroup and will not be constant. In such cases and if the subgroup size exceeds 10, the $\overline{X}$ and S charts are preferred. The common types of control charts for attributes are the P, nP, c and u charts.

The control charts for attributes are drawn based on two broad classifications of attribute data: **defective (non-conforming)** and **defect (non-conformity).** If an item or service deviates from the actual specifications expected from it, the situation is called a defect. It does not necessarily mention that the product or service is not fit for use. Suppose a particular product or service has several specifications and one particular specification is not met, but the product or service could still be used or improved. If an item or service is considered not fit for use, then the item is called defective. A defective item or service can have more than one defect.

The P and nP charts are used if data is available for the defective items of a process. If the volume of data available for each subgroup is constant, the nP chart should be used, whereas if the data available is in proportion to the total number of defectives in each subgroup, the P chart should be used. The c and u charts are used for data that is available in proportion to the number of defects. Again, if the volume of data available for a subgroup is constant, the u chart should be used, and if the volume of data in each group is in proportion to the total number of defects, the c chart should be used. The selection criteria for control charts are explained in Fig. 11.2.

The chart that is drawn before actually constructing the control chart, without the UCL, LCL and central mean, is called a **run chart**. The increasing or decreasing trend in the pattern of data plotted is called **runs**. The upward or downward moving trend of the data is called the **run up** and **run down,** respectively. A process may be considered as out of control when one or a few points are plotted outside the control limits. Even if the data is plotted at random within the control limits, the variations should be evenly distributed above and below the central line. In some cases, the points may be within the control limits and there will be a continuous run up or run down. Such conditions are also termed as out of control or unstable.

If a run chart is drawn for all the individual observations of each sample of a subgroup, it is called a **tier** or **tolerance chart**.

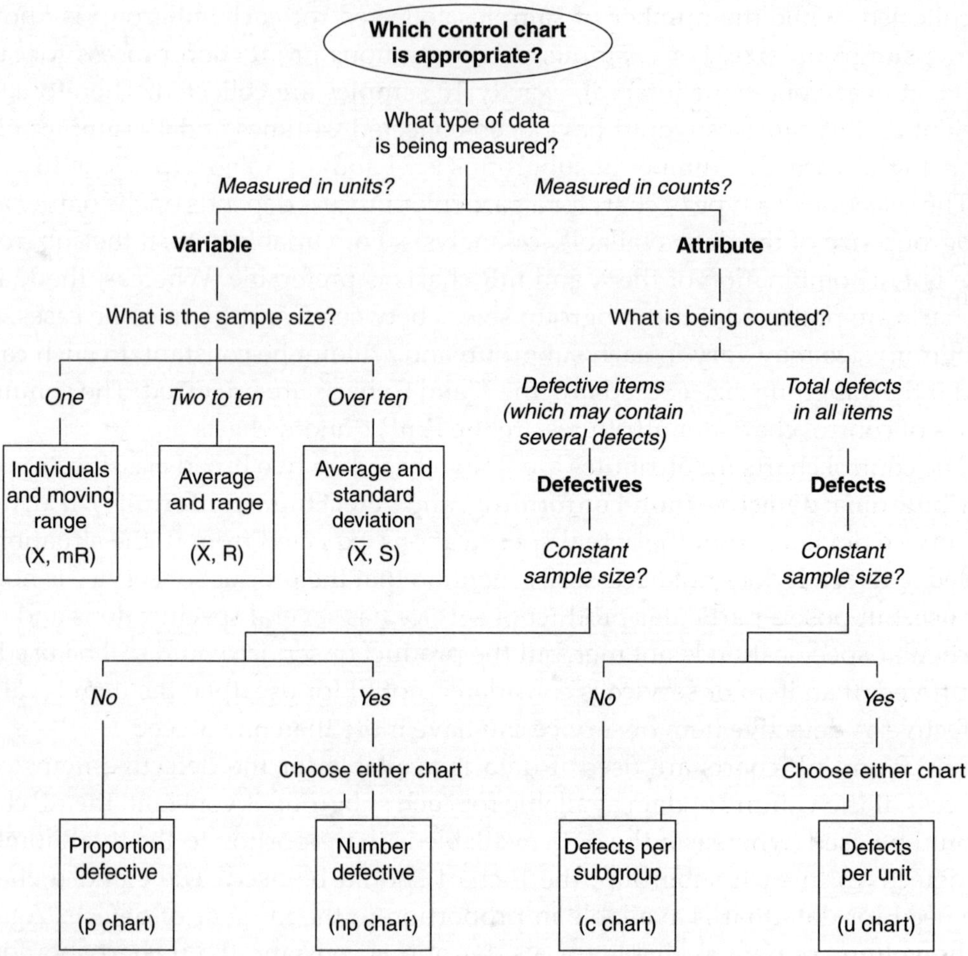

Figure 11.2 Flow chart for selection of control charts

Normally, the control charts are drawn based on the central line and the control limits are calculated from the available sets of data. As long as the trends of the runs are within the control limits and the random variations are in even dispersion around the central line, the process is said to be in control, and the same value limits could be used further. Suppose a single or few subgroup data sets are identified as out of control due to any assignable cause (can be eradicated), that subgroup data should be discarded and a revised control chart should be drawn with newly re-evaluated values of central and control limits.

11.5 PROCESS CAPABILITY

The control limits for a control chart are quantities derived from actual process data, which are calculated by using the average and standard deviation of data in statistical formulae. Therefore, the data plotted outside the control limits indicates that the process is out of control; however, the product size may be within the range of its tolerance limits or expected specifications (i.e., the upper specification level or lower specification level – USL and LSL). It is a proven fact that if the process data follows a normal distribution curve, 99.73% of the items will lie between $\mu + 3\sigma$ and $\mu - 3\sigma$, where μ and σ stand for mean and standard deviation, respectively. Therefore a **process spread** of 6σ (i.e., $\pm 3\sigma$) is usually considered for setting the control limits. Figure 11.3 shows the relationship between control limits and specification limits. This process spread is also called process capability.

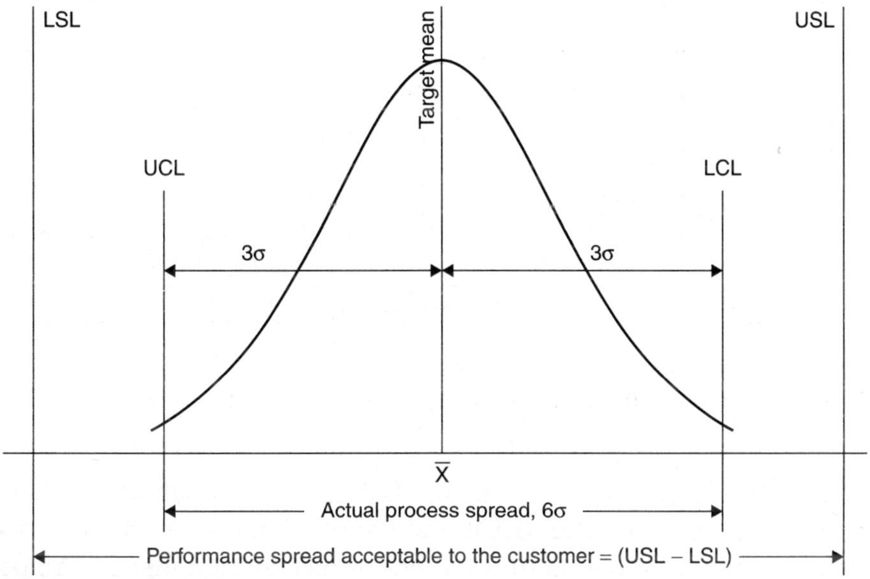

Figure 11.3 Relationship between specification and control limits

As discussed earlier, even if the process is within the control limits, we may not be able to know whether all the items produced are within the specification or tolerance limits. This could be determined using the **process capability ratio,** which is defined as the ratio of the difference between specification limits and process capability.

$$\text{Process capability ratio (Index), } C_p = \frac{\text{USL} - \text{LSL}}{6\sigma}$$

Case I: $C_p > 1$, which means $6\sigma <$ USL – LSL, process capability, 6σ is within the tolerance limit (Fig. 11.4). Even when the process average is shifted from the natural pattern of distribution, there will be a few out-of-control conditions, but the individual values of items produced are within their specification limits. This is the desirable condition.

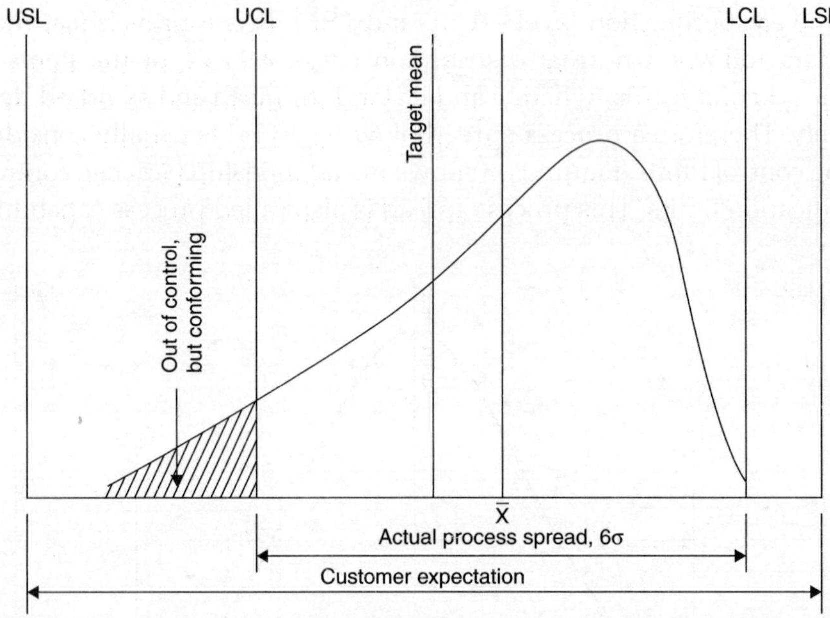

Figure 11.4 Process capability, $C_p > 1$

Case II: $C_p = 1$, which means $6\sigma =$ USL – LSL, process capability, 6σ is equal to tolerance limits (Fig. 11.5). The distribution shows that when there is a shift in the process average from the natural pattern of distribution, there is a possibility of a few non-conforming products which will not meet the specification limit. However, the products within the range of process control will conform.

Case III: $C_p < 1$, which means $6\sigma >$ USL – LSL, process capability, 6σ exceeds the tolerance limits (Fig. 11.6). In this case, even though the distribution follows the natural pattern of variation, a few individual values will fall outside the specification limits, though within the control limits. Therefore, this is the most undesirable condition.

Case IV: $C_p = 1$ but with the process average off-centred (Fig. 11.7). It means that though $6\sigma =$ USL – LSL, the process average is not centred with the target mean. In such cases, an adjustment is required to move the process mean back to the target.

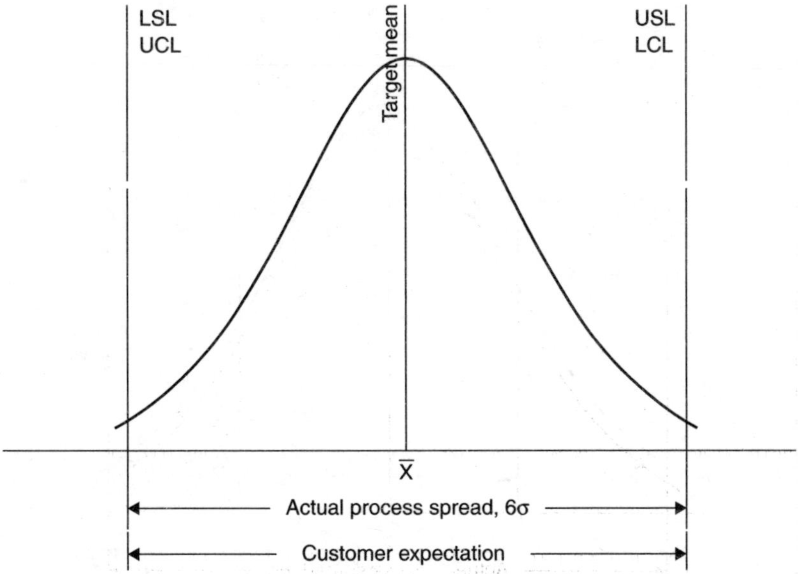

Figure 11.5 Process capability, $C_p = 1$

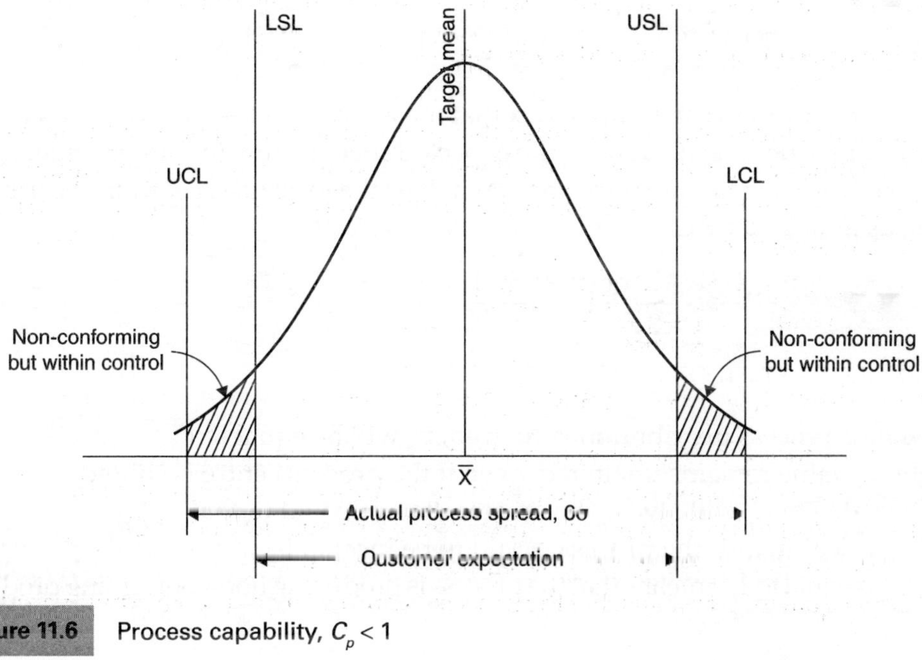

Figure 11.6 Process capability, $C_p < 1$

If no such action is taken, a considerable portion of the output will fall outside the specification limits (USL or LSL). Higher quantities of the output may need to be discarded even though the process may be in statistical control. Hence, this is also an undesirable case.

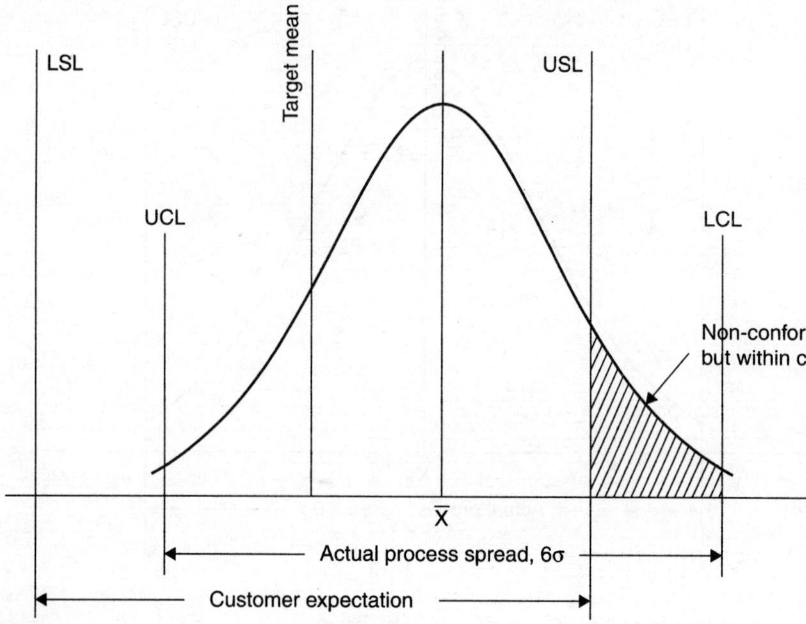

Figure 11.7 Process capability, $C_p = 1$ with process average off-centred

Note: The target mean line lies midway between the LSL and USL lines.

C_{pk} is another process capability index that measures the closeness of the measured quality or product specification to the expected specification (or target value) rather than just providing an inference about whether the process is centrally distributed or not, as indicated by C_p.

$$C_{pk} = \frac{\text{Min}\{(\text{USL} - \bar{x}) \text{ or } (\bar{x} - \text{LSL})\}}{3\sigma}$$

Some important features of C_p and C_{pk} are given below:

- When a process distribution is centred, C_p will be equal to C_{pk}.
- The C_p value remains unaltered, even if the process centre is shifted.
- The value of C_p will always be greater than or equal to that of C_{pk}.
- If C_p is less than 1, the process is said to be incapable.
- If C_{pk} is equal to 1, it means that the process is producing non-conforming products.

11.6 SAMPLING PLANS

For quality control and quality assurance, inspection is an inevitable process, in which decisions regarding the raw material, semi-finished goods and the final products are made at all stages of production, to verify whether they are conforming or non-conforming items. In some cases, based on the historical performance and quality of the product, the decision to accept an item can be made even without any inspection. In certain other cases, based on the precision requirement of the items (very low tolerance limits), a 100% inspection of items is necessary. However, this is a time-consuming, uneconomical and difficult method.

In statistical methods, a few samples from the entire output are inspected and if the required number of items is found to conform, then the entire lot is accepted. This method is called **acceptance sampling**. In acceptance sampling, the entire output of items (entire lot) is divided into a number of smaller lots (subgroups) and a few samples of accepted numbers (sample size) are selected at random for inspection. If the number of defective pieces is within the preset limit of some acceptance value, the entire lot of products is accepted as good, though it is well known that there would be a few defectives in the lot. In the SPC method, the process is continuously monitored and deviations in the product with respect to the actual standard are recorded and analysed using statistical tools and visual charts which will help determine whether the product is within the upper and lower control limits. This will also help identify the point at which the process has gone out of control and to incorporate remedial measures to bring the process back within control.

The sampling plan eliminates the method that requires all products to be inspected; it is a scientific method of quality control with sampling. Some of the other **benefits** of sampling plans are as follows:

- Reduced inspection cost
- Modern and scientific method of control
- Reduced wastage
- Quality assurance to all stakeholders
- Accurate understanding of process deviations and control factors
- Uniformity in production quality
- Better customer satisfaction

11.6.1 Classification of Sampling Plans

The broad classification of sampling plans is based on the quality characteristics being inspected (Fig. 11.8).

Acceptance sampling by attributes is further classified as follows:

- **Single sampling plan**: A decision to accept or reject a lot based on the evidence obtained from a single sample.
- **Double sampling plan**: A decision to inspect the first sample, leading to a decision to accept or reject that lot, or to draw a second sample for inspection which will lead to a decision to accept or reject the lot.
- **Multiple sampling plan**: An inspection in which after each sample, the decision is made to accept or reject or to draw another sample (up to the maximum

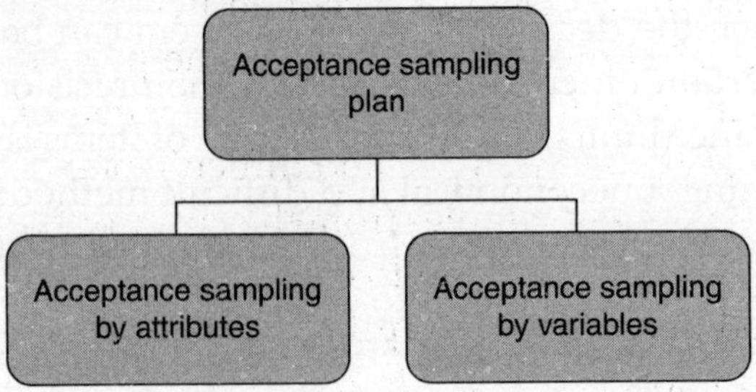

Figure 11.8 Classification of sampling plans

number of seven samples) by which a decision to accept or reject a lot will be reached.

- **Sequential sampling plan**: An inspection in which the items from a lot are inspected one by one (or group by group) and the decision to accept or reject a lot, or to continue inspection, is made on the basis of the evidence accumulated after each item is inspected.
- **Continuous sampling plan**: A plan in which the items are inspected in a continuous production or assembly line, to accept or reject a lot.
- **Multi-level sampling plan**: A plan which allows for any number of sampling levels, subject to the provision that transition can only happen between adjacent levels.
- **Lot-by-lot sampling plan**: A plan of evaluating a portion of the items in a lot for the purpose of accepting or rejecting the entire lot.

The following section deals with the first three popular types of sampling plans for which a few solved problems have been provided to help students understand the procedure of the sampling plans.

The procedure for the single sampling plan is given in Fig. 11.9.

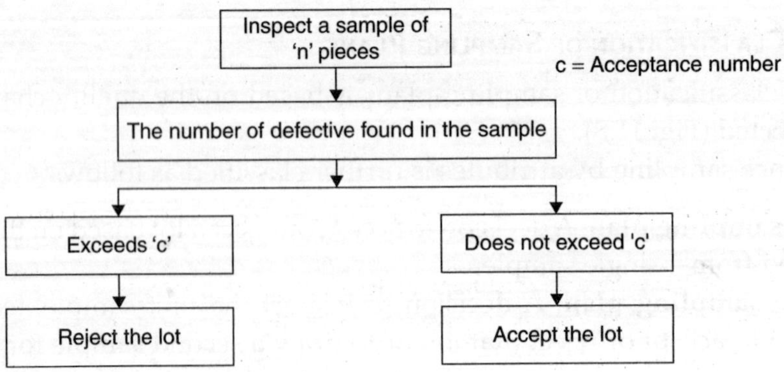

Figure 11.9 Single sampling plan

Figure 11.10 shows the procedure for the double sampling plan.

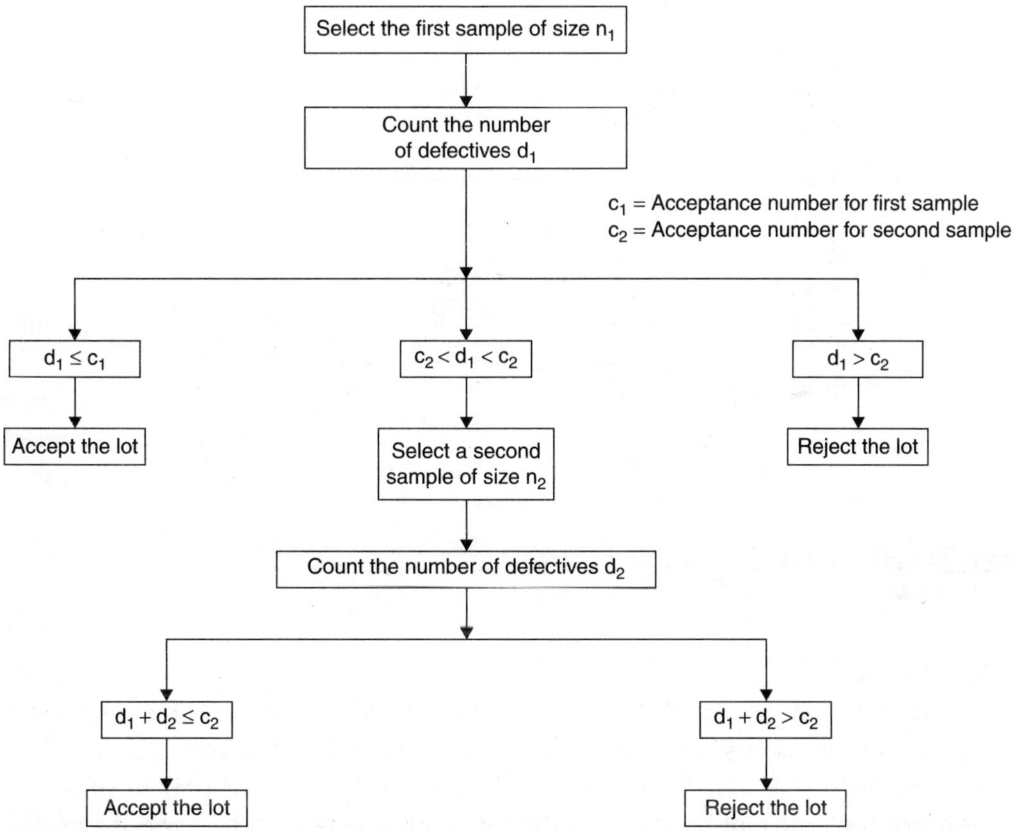

Figure 11.10 Double sampling plan

The procedure for the multiple sampling plan is similar to that of the double sampling plan. However, the sampling does not stop with two samples to determine the acceptance and rejection of the lot. Inspection of the sample continues till the sample is accepted or up to the seventh sample, whichever occurs first.

11.7 OC CURVE AND AOQ CURVE

The operation characteristic (OC) curve was developed along with SQC in the 1930s. It is used to differentiate between good and bad lots. The OC curve is an evaluation technique used to analyse the sampling plan and plot the fraction of defectives found in the sample against the probability of acceptance of the lot. The graph will indicate the probability that a particular lot with a certain percent of non-conforming items will be accepted or pass the inspection. The OC curve is roughly S shaped. A typical OC curve for the single sampling plan is shown in Fig. 11.11.

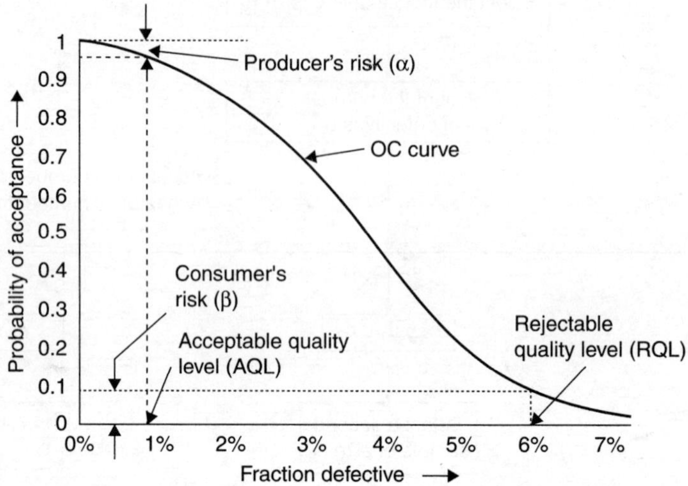

Figure 11.11 Typical OC curve for the single sampling plan

Salient features

- **Producer's risk (α)**: It represents the probability of rejection of a conforming (good) lot by the customer. It is otherwise called a Type-I error.
- **Consumer's risk (β)**: It represents the probability of acceptance of a non-conforming (bad) lot by the customer. It is otherwise called a Type-II error.
- **Acceptable quality level (AQL)**: It represents the maximum percent of defectives that will make a lot easily acceptable. It gives the fraction of defectives accepted without any serious effect on the quality and customer relations. Consumers are generally willing to accept lots that contain small percentages of defectives as 'good', especially if the cost related to a few defects is very low. This percentage is usually found in the range of 0.01–2.0%. AQL is also termed the producer's 'safe point'.
- **Rejectable quality level (RQL):** It represents the quality level that is unacceptable to the customer.
- **Lot tolerance percent defectives (LTPD)**: It represents the upper limit of the percentage of defective items that a consumer is willing to tolerate in an accepted lot. It represents the consumer's risk.
- **Acceptance number:** It is the largest number of non-conforming/defective items from a sample of size 'n' for which the lot should still be accepted.
- **Rejection number:** It is the minimum number of non-conforming/defective items from a sample of size 'n' for which the lot should be rejected.

- **Average outgoing quality limit (AOQL)**: It is the maximum value of AQL for accepting a given lot. It can be obtained from the AQL graph drawn with the fraction of defectives on the x axis and AQL values on the y axis.

A sampling plan and its OC curve have a one-to-one relationship. Determining the sample size and acceptance number is equivalent to determining the corresponding sampling plan's OC curve. The best way to determine a particular OC curve is to specify two points on it: producer's risk and consumer's risk. Many methods are used to select a suitable sampling plan. Among them, the Dodge and Romig (D-R) plan and the military standard (MIL-STD-105E) plan are popular.

AOQ: It is defined as the average proportion of the defectives (or average quality) of the outgoing products after inspection and includes all the accepted and rejected products which were identified and replaced by 100% inspection.

It can be computed as follows:

$$\text{AOQ} = \frac{\text{Outgoing number of defectives}}{\text{Outgoing number of items}}$$

If P_a is the probability of accepting a lot for a proportion of defectives, p; N is the number of items in a lot; and n is the number of items in the sample, then AOQ is given as follows.

Case 1: Defective items are not replaced.

$$\text{AOQ} = \frac{P_a(N-n)p}{N-np-p(1-P_a)(N-n)}$$

$$= \frac{P_a p}{1-p(1-P_a)} \quad \text{when } n \text{ is very small compared to } N$$

Case 2: Defective items are replaced.

$$\text{AOQ} = \frac{P_a(N-n)p}{N} = P_a p \quad \text{when } n \text{ is very small compared to } N$$

The following numerical example illustrates the concept of AOQ.

Let us assume a producer supplies 30 lots of 1000 items each, with 1% of non-conforming items in each lot. The sampling plan $n = 80$ and acceptance number $c = 1$ is considered for inspection. From the OC curve for the above sampling, the probability of acceptance is 0.808. Therefore, out of 30 lots, the number of lots accepted by the customer is 25 (i.e., $30 \times 0.808 = 24.24$). The number of rejected lots which are 100% inspected and supplied again to the consumer is 5.

Number of lots	Percentage of non-conforming items	Number of conforming items	Number of non-conforming items
25 lots (accepted)	1	25,000 (= 25 × 1000)	250 (= 25,000 × 0.01)
5 lots (rejected lot, then 100% inspected and 1% non-conforming discarded)	0	4,950 (= 5 × 1000 × 0.99)	0
Total		29,950	250

$$\text{AOQ (\% of non-conforming items)} = \frac{250}{29950} = 0.00835 = 0.835\%$$

A typical AOQ curve is given in Fig. 11.12.

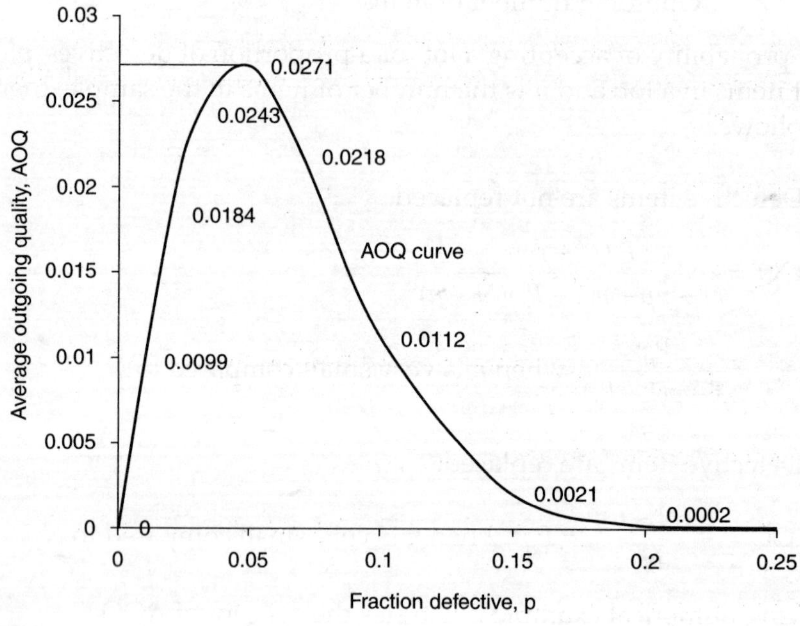

Figure 11.12 Typical AOQ curve

From the above graph, it can be seen that the AOQ value increases as p increases up to a limit called the average outgoing quality limit (AOQL), after which AOQ descends continuously towards 0.0. For the above case, AOQL is 0.0271, i.e., 2.71%.

Types of inspection plans

- **Normal inspection plan**: This is used at the beginning of any inspection and may be changed to tightened or reduced inspection based on the recent quality (AOQ) of the outgoing products.
- **Tightened inspection plan**: This is used when the producer's recent quality has reduced.
- **Reduced inspection plan**: This is used when the producer's recent quality has been exceptionally good.

11.8 SOLVED PROBLEMS

Example Problem 1: Plot the three-day moving average ($m\overline{X}$) and moving range (mR) control charts for the following data on the sales of a branded newspaper for a two-week period. Check if the sales process is in control.

Day	1	2	3	4	5	6	7	8	9	10	11	12	13	14
Sales in thousands	24	13	11	19	16	17	13	17	10	14	16	14	17	15

Solution: The averages of today's sales and the previous two days' sales and the range of values between today's sales and that of the previous two days are calculated separately from the third day onwards and tabulated as follows.

Day	Daily sales (in thousands), X	Three-day moving total	Three-day moving average, $m\_\overline{X}$	Three-day moving range, m_R
1	24	–	–	–
2	13	–	–	–
3	11	48	16.0	13
4	19	43	14.3	8
5	16	46	15.3	8
6	17	52	17.3	3
7	13	46	15.3	4
8	17	47	15.7	4
9	10	40	13.3	7
10	14	41	13.7	7
11	16	40	13.3	6
12	14	44	14.7	2
13	17	47	15.7	3
14	15	46	15.3	3
	Total		179.9	68

Sample size, $n = 3$

$$\text{Mean value of moving average, } m\_\overline{\overline{X}} = \frac{\Sigma m\_\overline{X}}{14-2} = \frac{179.9}{12} = 14.99$$

$$\text{Mean value of moving range, } m\_\overline{R} = \frac{\Sigma m\_R}{14-2} = \frac{68}{12} = 5.67$$

$$\sigma = \frac{\overline{R}}{d_2} = \frac{5.67}{1.693} \text{ .... } (d_2 = 1.693 \text{ for } n = 3 \text{ from Appendix 1}) = 3.35$$

Control limits: $UCL_{\overline{X}} = m\_\overline{\overline{X}} + 3\sigma = 14.99 + 3\,(3.35) = 25.04$

$LCL_{\overline{X}} = m\_\overline{\overline{X}} - 3\sigma = 14.99 - 3\,(3.35) = 4.94$

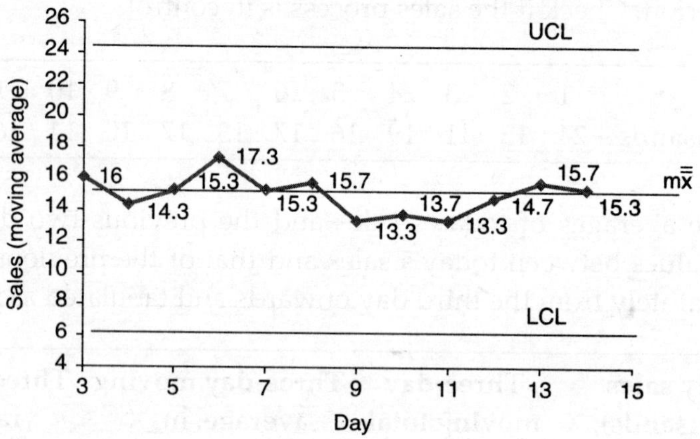

Figure 11.13 Moving average chart

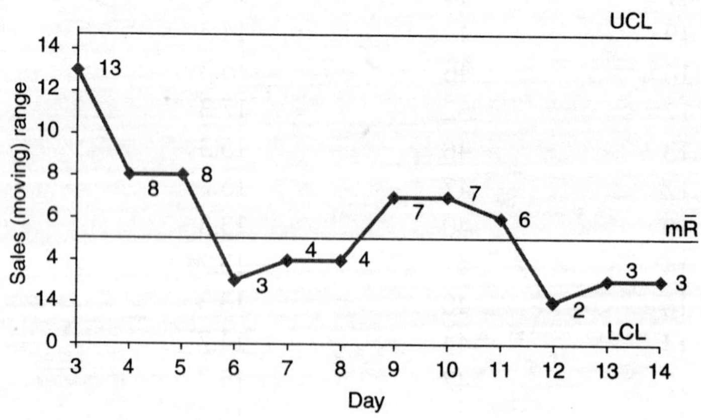

Figure 11.14 Moving range chart

$$UCL_R = D_4\, m\_\overline{R} = 2.574\ (5.67)\ (D_4 = 2.574 \text{ for } n = 3 \text{ from Appendix 1})$$

$$= 14.6$$

$$LCL_R = D_3\, m\_\overline{R} = 0\ (D_3 = 0 \text{ for } n = 3 \text{ from Appendix 1})$$

The m_$\overline{X}$ and m_R charts are shown in Figs 11.13 and 11.14, respectively.

From the m_$\overline{X}$ chart, it is clearly seen that all the points are within the control limits, and hence, the sales process is in control.

Example Problem 2: The following are the $\overline{X}$ and R values for 20 subgroups of 10 readings. The specification of the dimension to be measured is 4.037 ± 0.01. The values given are the last three digits of the dimension reading, i.e., X, 30.9 should be read as 4.0309 and R, 2 should be read as 0.002

Subgroup	1	2	3	4	5	6	7	8	9	10
$\overline{X}$	30.9	33.8	33.2	38.4	33.2	34.2	34.2	32.8	30.4	33.8
R	2	3	3	4	5	5	6	9	4	6

Subgroup	11	12	13	14	15	16	17	18	19	20
$\overline{X}$	34.4	34.0	32.8	34.4	33.2	33.2	33.8	37.8	37.0	34.0
R	4	4	4	5	2	5	9	6	4	4

a) Determine the control limits for the $\overline{X}$ and R charts for future use, eliminating all the out-of-control points.
b) Will the process be able to meet the specifications? Is shifting of the process centre required?

Solution: Given $n = 20$

$$\overline{\overline{X}} = \frac{\Sigma\overline{X}}{n} = \frac{30.9 + 33.8 + \cdots + 34.0}{20} = \frac{679.5}{20} = 34.0 \text{ (to be read as 4.034)}$$

$$\overline{R} = \frac{\Sigma R}{n} = \frac{94}{20} = 4.7 \text{ (to be read as 0.0047)}$$

Trial control limits on the X and R charts:

$$UCL_{\overline{X}} = \overline{\overline{X}} + A_2\,\overline{R}\ (A_2 \text{ value is obtained from Appendix 1 for } n = 20, \text{ as } 0.180)$$

$$= 34.0 + (0.180 \times 4.7) = 34.8$$

$$LCL_{\bar{X}} = \bar{\bar{X}} - A_2\,\bar{R} = 34.0 - (0.180 \times 4.7) = 33.2$$

Similarly,

$$UCL_R = D_4\,\bar{R} \text{ ...(} D_4 \text{ value is obtained from Appendix 1 for } n = 20\text{, as 1.585)}$$

$$= 1.585 \times 4.7 = 7.4$$

$$LCL_R = D_3\,\bar{R} \text{ ... (} D_3 \text{ value is obtained from Appendix 1 for } n = 20\text{, as 0.415)}$$

$$= 0.415 \times 4.7 = 1.9$$

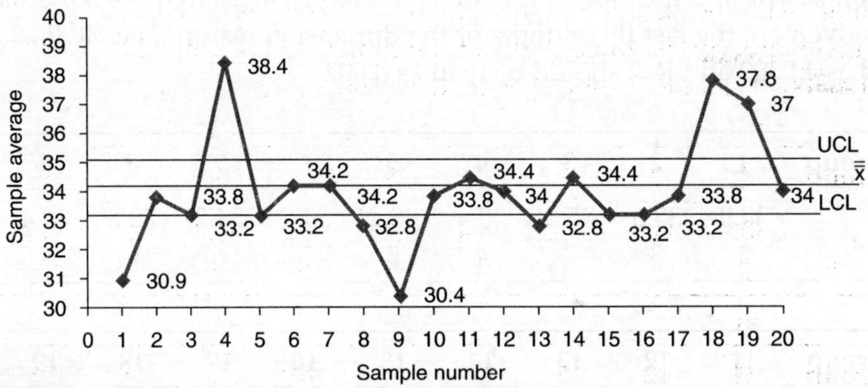

Figure 11.15 The $\bar{X}$ chart

On plotting the $\bar{X}$ chart (Fig. 11.15), it is noticed that the seven subgroups (samples) 1, 4, 8, 9, 13, 18 and 19 are beyond the control limits. Hence, the process is out of control. If assignable causes are known for these out-of control points, they can be eliminated, and revised values can be calculated as

$$\bar{\bar{X}}_{revised} = \frac{679.5 - (30.9 + 38.4 + 32.8 + 30.4 + 32.8 + 37.8 + 37)}{20 - 7}$$

$$= \frac{439.4}{13} = 33.8 \text{ (to be read as 4.0338)}$$

On plotting the R chart (Fig. 11.16), it is noticed that the samples 8 and 17 fall above the upper control limit, and no sample falls below the lower control limit. If assignable causes are known for these points, they can be eliminated and the revised $\bar{R}$ value can be calculated as

$$\bar{R}_{revised} = \frac{94 - (9 + 9)}{20 - 2} = \frac{76}{18} = 4.22 \text{ (to be read as 0.00422)}$$

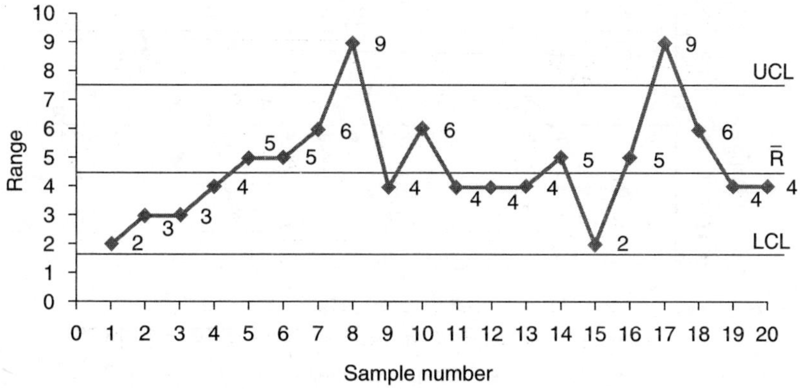

Figure 11.16 The R chart

Hence, the new control limits (first trial limits) are

$$UCL_{\bar{X}} = \bar{\bar{X}}_{revised} + A_2\,\bar{R}_{revised} = 33.8 + (0.180 \times 4.22) = 34.6$$

$$LCL_{\bar{X}} = \bar{\bar{X}}_{revised} - A_2\,\bar{R}_{revised} = 33.8 - (0.180 \times 4.22) = 33.0$$

Similarly,

$$UCL_R = D_4\,\bar{R}_{revised} = 1.585 \times 4.22 = 6.7$$

$$LCL_R = D_3\,\bar{R}_{revised} = 0.415 \times 4.22 = 1.8$$

Note: In all of the above calculations for revised values, the same constant values of A_2, D_3 and D_4 are used for the original sample size, though a few samples are ignored.

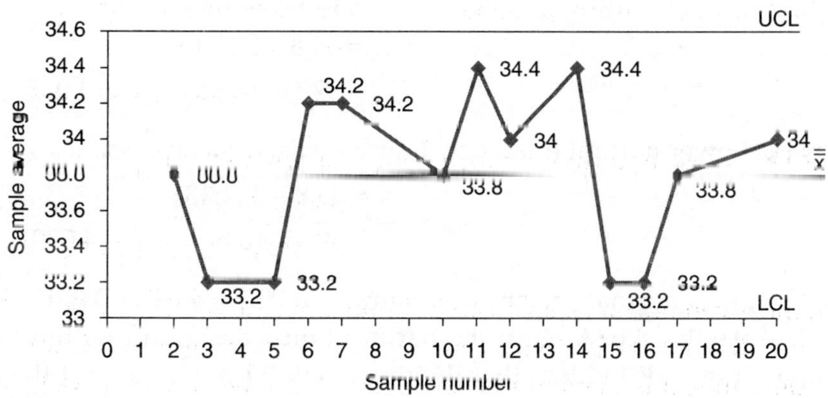

Figure 11.17 Revised $\bar{X}$ chart

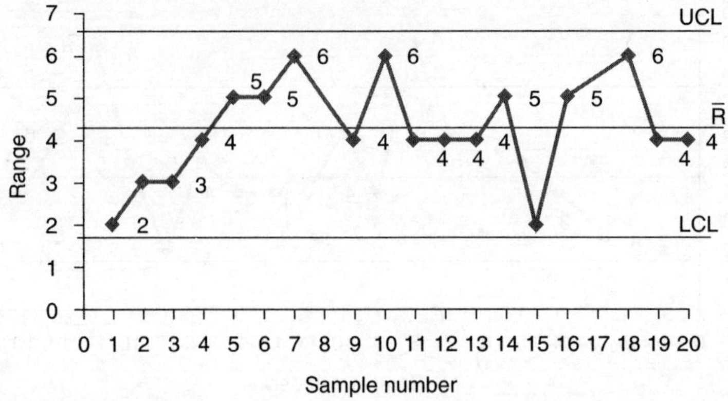

Figure 11.18 Revised R chart

On plotting the $\overline{X}$ and R charts (Figs 11.17 and 11.18), it can be seen that all the points are found within the control limits, and hence, the process is said to be in control.

(a) Hence, the first trial limits are considered as the final control limits for future use, if any.

(b) Population standard deviation, i.e., the estimate of the current value of the process ($\hat{\sigma}$) is given as

$$\hat{\sigma} = \frac{\overline{R}_{revised}}{d_2} = \frac{4.22}{3.735} \quad (d_2 \text{ value is obtained from Appendix 1})$$

$= 1.13$ (to be read as 0.00113)

3σ limits for revised values:

UNTL, Upper natural tolerance limit = Revised mean + $3\hat{\sigma}$

$= 33.8 + 3\ (1.13)$

$= 37.19$ (to be read as 4.037)

LNTL, Lower natural tolerance limit = Revised mean – $3\hat{\sigma}$

$= 33.8 - 3\ (1.13)$

$= 30.41$ (to be read as 4.030)

However, the given normal specification limits are USL = 4.037 + 0.001 = 4.047 and LSL = 4.037 – 0.001 = 4.027. Since the natural limits are found within the actual specification limits, it is evident that the process will be able to meet the required specifications, and hence, it is not advisable to shift the process centre.

Example Problem 3: The diameter of a stud is required to be within 40 ± 0.1 mm. The control charts $\bar{X}$ and R are initiated. After 30 samples of 5 studs each are examined, the results obtained are $\Sigma\bar{X} = 1203$ mm and $\Sigma R = 90$ mm.

(a) Estimate the mean and standard deviation of the process, assuming that it is in statistical control.
(b) Determine RPI control limits.
(c) Determine whether a stud with a diameter less than 38 mm must be scrapped and a stud with diameter greater than 42 mm must be reworked.
   i) Determine the percentage of rework and scrap, assuming normal distribution of the process.
   ii) What should be the centred value if scrap is to be maintained below 1%?
   iii) If a review committee recommends that the stud with diameter up to 43 mm can be used successfully, what will be the probability of rework?

Solution:

(a) $\bar{\bar{X}} = \dfrac{\Sigma\bar{X}}{n} = \dfrac{1203}{30} = 40.1$

$$\bar{R} = \frac{\Sigma R}{n} = \frac{90}{30} = 3$$

$$\sigma = \frac{\bar{R}}{d_2} = \frac{3}{2.326} = 1.289 \quad (d_2 \text{ is } 2.326 \text{ for } n = 5, \text{ from Appendix 1})$$

USL = 40 + 0.1 = 40.1; LSL = 40 – 0.1 = 39.9

(b) $\text{RPI} = \dfrac{\text{USL} - \text{LSL}}{\bar{R}} = \dfrac{40.1 - 39.9}{3} = 0.067$

$$UCL_{\bar{X}} = \bar{\bar{X}} + A_2\,\bar{R}$$

$$= 40.1 + 0.577\ (5)\ (A_2 \text{ is } 0.577 \text{ for } n = 5, \text{ from Appendix 1}) = 42.99$$

$$LCL_{\bar{X}} - \bar{\bar{X}} - A_2\,\bar{R} = 40.1 \quad 0.577(5) - 37.23$$

(c) With the process centred at m = 40.1, the percentage of scrap is calculated as follows.

   i) Value of Z corresponding to $\dfrac{X - \mu}{\sigma}$, i.e., $\dfrac{43 - 40.1}{1.289}$, i.e., 2.25

   = 0.9878 (right tailed) (from Appendix 2) = 1 – 0.9878 = 0.0122 = 1.22%

   Percentage of scrap = 1.22%

   Similarly, the percentage of rework expected is given as follows.

Value of Z corresponding to $\dfrac{X-\mu}{\sigma}$, i.e., $\dfrac{38-40.1}{1.289}$, i.e., −1.63
= 0.0516 (left tailed) (from Appendix 2) = 5.16%

Percentage of rework = 5.16%

ii) If the percentage of scrap is to be maintained below 1%, i.e., ≤ 0.0099, then the value of Z should be −2.33 (from Appendix 2). Then

$$-2.33 = \frac{38-\mu}{1.289};\ \mu = 41$$

Hence, the mean value of the stud should be maintained at 41 mm in order to have a scrap percentage of less than 1%.

iii) With the process centred between the newly suggested upper value of 43 and the existing lower value of 38, the mean value will be (43 + 38)/2, i.e., m = 40.5, probability of scrap is calculated as:

Value of Z corresponding to $\dfrac{X-\mu}{\sigma}$, i.e., $\dfrac{43-40.5}{1.289}$, i.e., 1.94
= 0.9738 (right tailed) (from Appendix 2) = 1− 0.9738 = 0.0262 = 2.62%

Probability of scrap = 2.62%

Similarly, the probability of rework is given as follows.

Value of Z corresponding to $\dfrac{X-\mu}{\sigma}$, i.e., $\dfrac{38-40.5}{1.289}$, i.e., −1.94
= 0.0262 (left tailed) (from Appendix 2) = 2.62%

Probability of rework = 2.62%

Example Problem 4: A packing machine is set to pack a given weight (in grams) of 10 samples, the sample size being 5. The data on the mean and standard deviation of the samples is obtained as follows. The specification of the packing is 100 ± 5 g.

Sample no.	1	2	3	4	5	6	7	8	9	10
Mean weight	101	102	100	106	103	107	100	101	104	105
S.D	2	3	1	4	2	5	6	3	4	2

(a) Construct the $\overline{X}$ and S charts.
(b) Determine the control limits for future use.
(c) Assuming the weight of the packing is to be normally distributed, what proportion of the products will not exceed the specifications?
(d) Find the process capability ratio.

Solution:

(a) $\bar{\bar{X}} = \frac{\Sigma\bar{X}}{n} = \frac{101 + 102 + 100 + \cdots + 105}{10} = 102.9$

$\bar{S} = \frac{\Sigma S}{n} = \frac{32}{10} = 3.2$

Control limits:

$UCL_{\bar{X}} = \bar{\bar{X}} + A_3\bar{S}$

$= 102.9 + 1.427\ (3.2)\ldots\ (A_3 = 1.427 \text{ for } n = 5, \text{ from Appendix 1}) = 107.466$

$LCL_{\bar{X}} = \bar{\bar{X}} - A_3\bar{S} = 102.9 - 1.427(3.2) = 98.334$

Figure 11.19 shows the $\bar{X}$ chart.

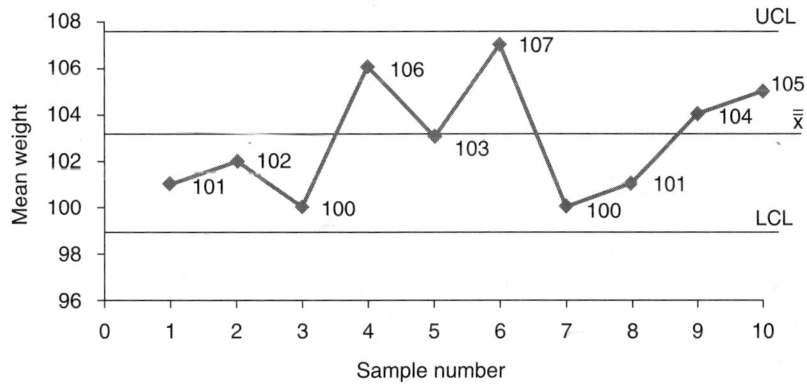

Figure 11.19 The $\bar{X}$ chart

From the above chart, it can be seen that all the sample averages fall within the control limits. Similarly, the control limits for the S chart are given below.

$UCL_s = B_4\bar{S} = 2.089\ (3.2) = 6.68\ \ldots..(B_4 = 2.089 \text{ for } n = 5, \text{ from Appendix 1})$

$LCL_s = B_3S = 0\ (B_3 = 0 \text{ for } n = 5, \text{ from Appendix 1})$

Fig. 11.20 shows the S chart.

From this chart, it is clear that all the sample averages are within the control limits.

(b) The control limits for the future remain unchanged from the trial limits, as all the sample averages are within the control limits of both the charts.

(c) To find the proportion of the output that does not meet the specifications, the standard normal value at the upper and lower specification limits must be determined first.

USL = 100 + 5 = 105; LSL = 100 – 5 = 95 and the σ value is given by

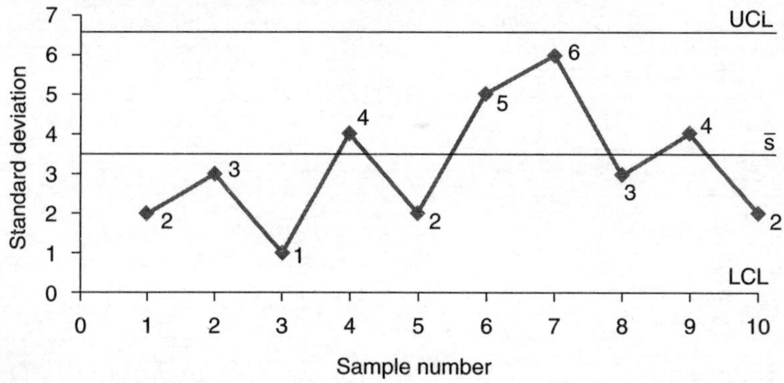

Figure 11.20 The S chart

$$\sigma = \frac{\bar{s}}{c_4} = \frac{3.2}{0.94} \text{ (from Appendix 1, } c_4 = 0.94 \text{ for } n = 5) = 3.404$$

At the lower specification limit, we get

$$z_1 = \frac{X - \mu}{\sigma} = \frac{95 - 102.9}{3.404} = -2.32$$

Value of Z at – 2.32 = 0.0102 (from Appendix 2) = 1.02%

Similarly, at the higher specification limit, we get

$$z_1 = \frac{X - \mu}{\sigma} = \frac{105 - 102.9}{3.404} = 0.617$$

Value of Z at 0.617 = 0.7486 (right tailed) (from Appendix 2)

= 1 – 0.7486 = 0.2514 = 25.14%

Figure 11.21 shows the process capability.

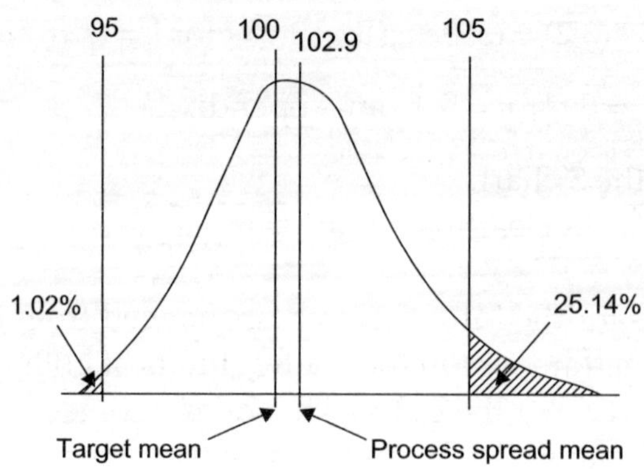

Figure 11.21 Process capability

From the above figure, it is seen that the total proportion of the end product of studs not meeting the specifications is given by

$$= 1.02\% + 25.46\% = 27.78\%$$

(d) $\text{PCR} = \dfrac{\text{USL} - \text{LSL}}{6\sigma} = \dfrac{105 - 95}{6(3.404)} = 0.489$

PCR < 1 indicates that the process for the actual required values is incapable, and hence, there are some rejections, as shown in Fig. 11.21.

Example Problem 5: Ten samples of parts were taken from a production line for inspection. Each sample contains 50 parts. The following number of non-conforming parts (defectives) was found in each sample: 2 4 6 5 3 4 5 3 3 4

(a) Calculate the fraction rejected ($\overline{\text{p}}$).
(b) Draw the distribution of defectives.
(c) Calculate the three-sigma control limits and the two-sigma warning limits.
(d) What are the future control limits?

Solution:

(a) $\text{N} = 10 \times 50 = 500$

$$\overline{\text{p}} = \frac{(2 + 4 + 6 + \cdots + 4)}{500} = \frac{40}{500} = 0.08$$

(b) The tally chart is first drawn in order to plot the distribution chart.
The frequency table is shown below.

Sample no.	No. of defectives (d)	Fraction defective (p)
1	2	0.04
2	4	0.08
3	6	0.12
4	5	0.1
5	3	0.06
6	4	0.08
7	5	0.1
8	3	0.06
9	3	0.06
10	4	0.08

The tally table is shown as follows:

p	Tally mark	frequency
0.04	/	1
0.06	///	3
0.08	///	3
0.10	//	2
0.12	/	1

Figure 11.22 shows the distribution of the defective items.

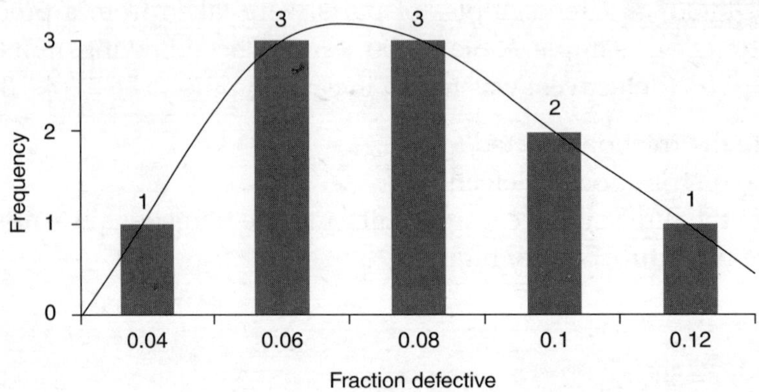

Figure 11.22 Distribution of defective items

From Fig. 11.22, it is clearly seen that the defectives are almost normally distributed.

(c) $\sigma_p = \sqrt{\dfrac{\bar{p}(1-\bar{p})}{N}} = \sqrt{\dfrac{0.08(1-0.08)}{N}} = 0.0121$

$2\,\sigma_p = 2\,(0.0121) = 0.0242;\ 3\,\sigma_p = 3\,(0.0121) = 0.0363$

Control limits:

$UCL_p = \bar{p} + 3\sigma_p = 0.08 + 0.0363 = 0.116$

$LCL_p = \bar{p} - 3\sigma_p = 0.08 - 0.0363 = 0.044$

Warning limits:

$UWL_p = \bar{p} + 2\sigma_p = 0.08 + 0.0242 = 0.104$

$LWL_p = \bar{p} - 2\sigma_p = 0.08 - 0.0242 = 0.056$

Figure 11.23 shows the p chart.

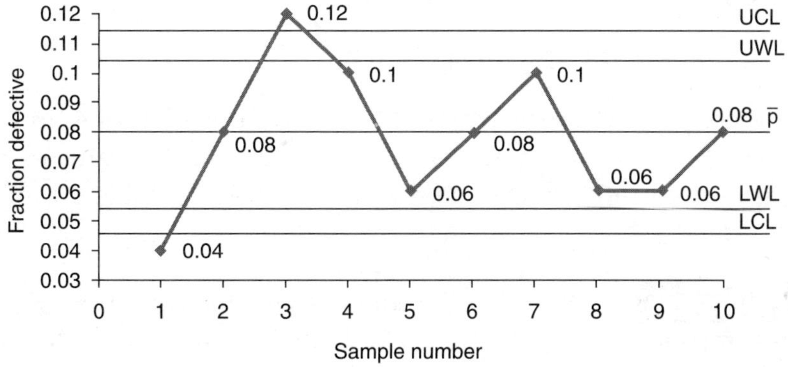

Figure 11.23 The p chart

d) Since the samples 1 and 3 are found to be beyond the control limits, they are ignored, and revised values are calculated as follows:

$$\bar{p}_{\text{revised}} = \frac{(4+5+3+\cdots+4)}{8\times 50} = 0.078$$

$$\sigma_p = \sqrt{\frac{0.078(1-0.078)}{400}} = 0.0134$$

Revised control limits:

$$UCL_p = \bar{p} + 3\sigma_p = 0.078 + 3\ (0.0134) = 0.118$$

$$LCL_p = \bar{p} - 3\sigma_p = 0.078 - 3\ (0.0134) = 0.038$$

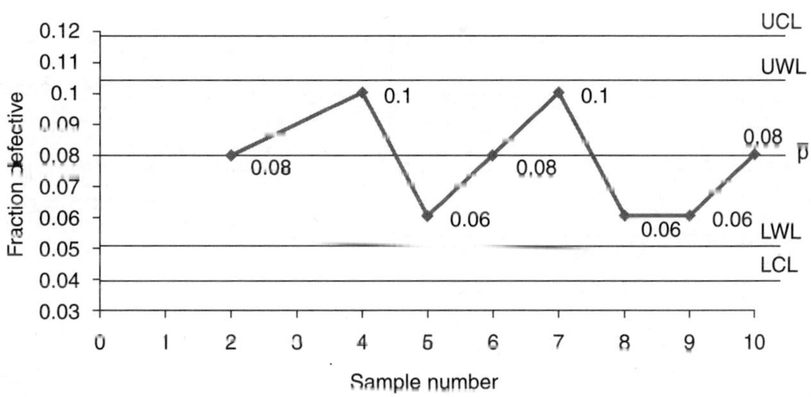

Figure 11.24 Revised p chart

Warning limits:

$$UWL_p = \bar{p} + 2\sigma_p = 0.078 + 2\,(0.0134) = 0.105$$

$$LWL_p = \bar{p} - 2\sigma_p = 0.078 - 2\,(0.0134) = 0.051$$

The revised p chart is drawn with revised control limits and is shown in Fig. 11.24. From this chart, it can be seen that all the samples are within the control limits, and hence, the revised limits are considered for all future inspections.

Example Problem 6: The following data gives the number of rejects for consecutive samples. The sample size is 200. Find the control limits for 100 p or the percentage rejected control chart and draw the control chart.

3 0 2 4 3 5 4 1 3 5

Solution: $N = 10 \times 200$

$$\bar{p} = \frac{(3+0+2+\cdots+5)}{2000} = \frac{30}{2000} = 0.015$$

$$100\,\bar{p}(\text{or})\ \%\,\bar{p} = 0.015 \times 100 = 1.5$$

$$3\sigma_p = \sqrt{\frac{\bar{p}(1-\bar{p})}{N}} = 3\sqrt{\frac{0.015(1-0.015)}{2000}} = 0.0027$$

Control limits:

$$UCL_p = \bar{p} + 3\sigma_p = 0.015 + 3\,(0.0027) = 0.0231$$

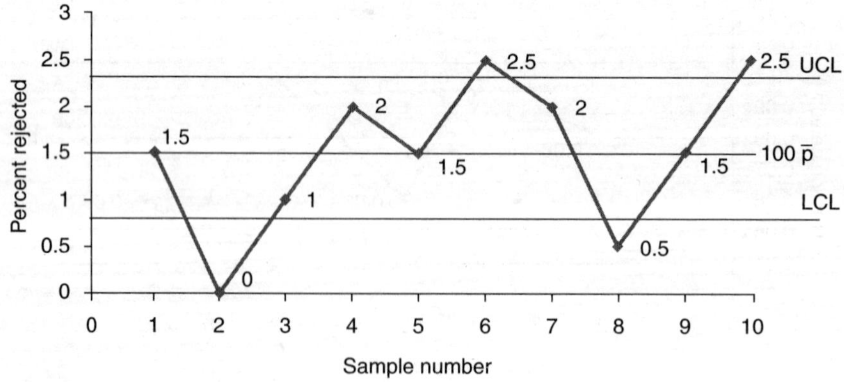

Figure 11.25 Percent rejected control chart

$$UCL_{100p} = 0.0231 \times 100 = 2.31$$

$$LCL_p = \bar{p} - 3\sigma_p = 0.015 - 3\ (0.0027) = 0.0069$$

$$LCL_{100p} = 0.0069 \times 100 = 0.69$$

To obtain the percentage rejected values for all samples, the number of rejects is divided by the sample size and multiplied by 100; thus, the revised values are

1.5 0 1 2 1.5 2.5 2 0.5 1.5 2.5

Figure 11.25 shows the control chart for 100p.

The samples 2, 6, 8 and 9 fall outside the control limits, and hence, they should be ignored to find out the revised control limits for future use.

Example Problem 7: The details of the inspection of 10 consecutive batches are given below:

Batch no.	1	2	3	4	5	6	7	8	9	10
No. of products in batch	162	150	175	213	165	221	203	201	181	209
No. of defectives in batch	4	4	5	7	4	9	8	6	5	8

Construct a control chart showing the varying 3σ upper control limit and 2σ upper warning limit.

Solution: Total number of products = 162 + 150 + ⋯ + 209 = 1880

Total number of defectives = 4 + 4 + ⋯ + 8 = 60

$$\bar{p} = \frac{60}{1880} = 0.032$$

Since the batch size varies, the control limit for each batch will be different. Hence, the control and warning limits are calculated for a convenient batch size that covers all the given batches, i.e., the smallest given batch size of 150 and the largest batch size of 221.

Since σ_p involves $\sqrt{n}$, the values for which the square root can be easily found, i.e., $\sqrt{121}, \sqrt{144}, \ldots \sqrt{256}$ are considered to cover all the given batch sizes.

$$3\sqrt{\bar{p}(1-\bar{p})} - 3\sqrt{0.032\,(1-0.032)} = 0.528$$

$$2\sqrt{\bar{p}(1-\bar{p})} = 2\sqrt{0.032(1-0.032)} = 0.352$$

The following table gives the values of upper warning and control limits for batch sizes ranging from 121 to 256.

Batch size (n)	UWL_p	UCL_p
121	0.064	0.080
144	0.061	0.076
169	0.059	0.073
196	0.057	0.070
225	0.055	0.067
256	0.054	0.065

For example, for value 121,

$$UWL_p = \bar{p} + 2\sqrt{\frac{\bar{p}(1-\bar{p})}{n}} = 0.032 + \left[\frac{0.352}{\sqrt{121}}\right] = 0.032 + \left[\frac{0.352}{11}\right] = 0.064$$

Similarly, $$UCL_p = \bar{p} + 3\sqrt{\frac{\bar{p}(1-\bar{p})}{n}} = 0.032 + \left[\frac{0.528}{\sqrt{121}}\right] = 0.032 + \left[\frac{0.528}{11}\right] = 0.08$$

The values for other batch sizes are calculated and tabulated as follows:

Fraction rejected p values for 10 batches are calculated using the formula

$$p = \frac{\text{Number of defectives}}{\text{Number of products in the batch}}$$

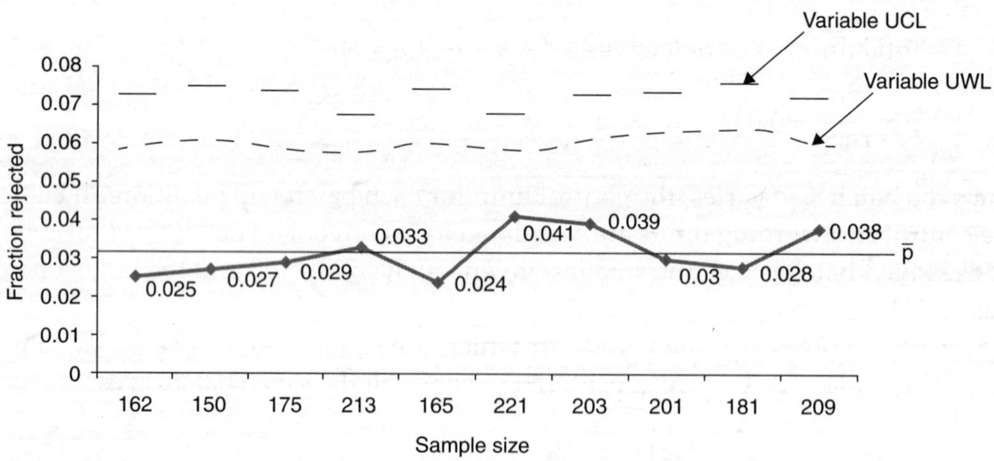

Figure 11.26 P chart for variable sample size (x axis = sample size)

Sample no.	1	2	3	4	5	6	7	8	9	10
n	162	150	175	213	165	221	203	201	181	209
p	0.025	0.027	0.029	0.033	0.024	0.041	0.039	0.030	0.028	0.038

Figure 11.26 shows the upper warning and control limits and the points plotted on the p chart with batch size on the *x* axis. This chart also can be drawn with the sample number marked on the *x* axis, as shown in Fig. 11.27.

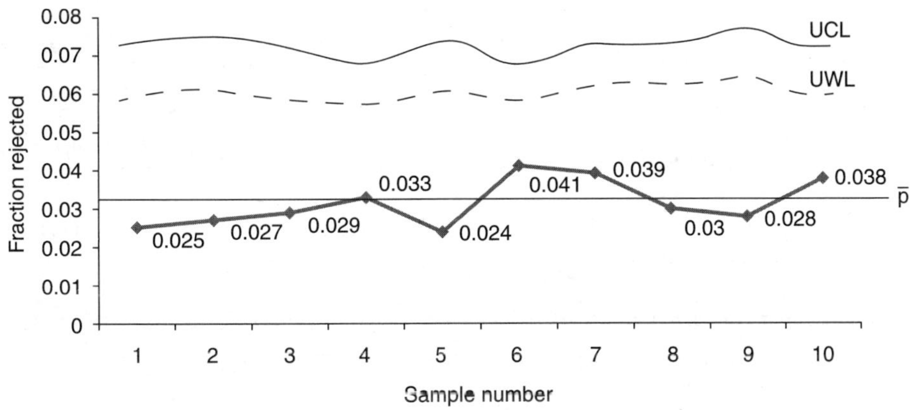

Figure 11.27 P chart for variable sample size (*x* axis = sample number)

Example Problem 8: A component is manufactured in lots of 200 each, and subjected to 100% inspection. The percentage of defective components is found to be 1.5% for the first 20 lots.

(a) Determine the trial control limits on an np chart.
(b) What is the probability that the twenty-first lot will contain exactly 7 defective components, if the defective average remains unchanged?
(c) What is the probability that the twenty-first lot will have 7 or more defective components?

Solution:

(a) $\bar{p} = 1.5\% = 0.015;\ n = 200;\ n\bar{p} = 0.015\ (200) = 3$

$$3\sigma_{np} = 3\sqrt{n\bar{p}(1-\bar{p})} - 3\sqrt{3(1-0.015)} = 5.157$$

$$UCL_{np} = n\bar{p} + 3\sigma_{np} = 3 + 5.157 = 8.157$$

$$UCL_{np} = n\bar{p} - 3\sigma_{np} = 3 - 5.157 = -2.157 = 0 \text{ (negative value will be replaced by zero)}$$

(b) From the Poisson distribution table (1000 × probability) shown below, the probability value of the twenty-first sample for which $np = 3$ and containing up to 7 defective components is taken as 0.988, and the probability of containing up to 6 defective components is 0.966.

r / np	0	1	....	....	6	7	....	10
0.02	980	1000	....	....	....	....	....	....
.								
.								
.								
3	050	199	.....	....	966	988	....	1000

Hence, the probability of the sample containing exactly 7 defective components is given as $0.988 - 0.966 = 0.022$.

(c) Probability of the sample containing 7 or more defective components is given as

$= 1 -$ Probability of sample containing up to 6 defective items

$= 1 - 0.966 = 0.034$

Example Problem 9: From a check of the mechanical defects observed on lots of 10 gear drives drawn from a production line lasting for 20 days, it is found that there were 50 defects.

(a) Find the control and warning limits of the c chart.
(b) What is the probability that a defect is found above the upper control limit, if the process average shifts to 5?
(c) What is the probability that a defect is found within the upper control limit and upper warning limit?

Solution: $$\bar{c} = \frac{\text{Total number of defects}}{\text{Number of samples}} = \frac{50}{20} = 2.5$$

(a) *Control limits:*

$$UCL_c = \bar{c} + 3\sqrt{\bar{c}} = 2.5 + 3\sqrt{2.5} = 7.24$$

$$LCL_c = \bar{c} - 3\sqrt{\bar{c}} = 2.5 - 3\sqrt{2.5} = -2.24 = 0 \text{ (due to negative value)}$$

Warning limits:

$$UWL_c = \bar{c} + 2\sqrt{\bar{c}} = 2.5 + 2\sqrt{2.5} = 5.66$$

$$LWL_c = \bar{c} - 2\sqrt{\bar{c}} = 2.5 - 2\sqrt{2.5} = -0.66 = 0 \text{ (due to negative value)}$$

(b) When $c = 5$, the probability of a defect falling within the control limit of 7.24 is calculated using the Poisson table as follows:

r / $c = np$	0	1	....	....	7	8	....	14
5			.....	....	867	932		

With $c = 5$, probability up to 7 defects = 0.867

With $c = 5$, probability up to 8 defects = 0.932

By the interpolation method,

With $c = 5$, probability for defects up to 7.24 is calculated as follows:

$$= 8.867 + (0.932 - 0.867)\ 0.24 = 0.8826 \approx 0.883$$

Probability of a defect sample falling above the UCL is $1 - 0.883 = 0.117$

(c) When c = 5, the probability of a defect falling within the UCL and UWL is calculated using the Poisson table as follows:

r / $c = np$	0	1	....	....	5	6	....	14
5			.....	....	616	762		

With $c = 5$, probability up to 5 = 0.616

With $c = 5$, probability up to 6 = 0.762

By the interpolation method,

With $c = 5$, probability up to 5.66 is calculated as follows:

$$= 0.616 + (0.762 - 0.616)\ 0.66 = 0.712$$

Probability of a defect sample falling within UCL and UWL is given as

$- 0.883 \quad 0.712 - 0.171$

Example Problem 10: A new iron rod manufacturing unit wants to prepare control parameters for its new production unit. The data collected from the initial inspection is as follows:

Inspection lot no.	Length of rods inspected in metres	No. of defects found
1	120	5
2	100	4
3	200	6
4	250	7
5	150	6
Total	820	28

Draw a suitable control chart and recommend a future value for the control parameter you are using.

Solution: As the sample size varies with the lot, it is appropriate to draw a u chart.

First, a standard lot size for the problem is to be fixed. As the quantity 150 is the value closest to all the given lot sizes, the basic lot size (k) is chosen as 150 in this case.

$$\bar{u} = \frac{28}{820} = 0.034; \; k\bar{u} = 150 \times 0.034 = 5.1$$

Also, the $k\bar{u}$ values are computed for all the given lots and tabulated as follows.

For example, for lot number 1: $k\bar{u} = 150 \times \frac{5}{120} = 6.25$

Lot no.	1	2	3	4	5
$k\bar{u}$	6.25	6	4.5	4.2	6

Control limits for each sample are calculated and tabulated as follows:

Inspection lot no.	Quantity, n	$3\sqrt{\frac{k\bar{u}}{n/k}}$	$UCL = k\bar{u} + 3\sqrt{\frac{k\bar{u}}{n/k}}$	$LCL = k\bar{u} - 3\sqrt{\frac{k\bar{u}}{n/k}}$
1	120	8.38	14.63	0
2	100	9	13.50	0
3	200	5.51	10.01	0
4	250	4.76	8.96	0
5	150	7.35	13.35	0

The control chart of the given lots is plotted as shown in Fig. 11.28.

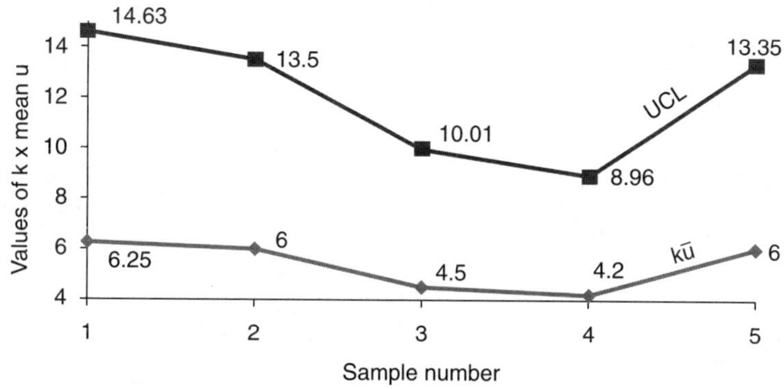

Figure 11.28 u chart for variable sample size

None of the points in the control chart is out of control, and hence, the kū value of 5.1 is assigned to k $\bar{u}_0$ and the existing UCL and LCL values are considered for future reference as well.

Example Problem 11: Determine the UCL and LCL for a u chart for the probability limits of 0.975 and 0.025, when n = 5 and $m_u = 2$.

Solution: To use the Poisson table values, $n \cdot \mu_u$ value is considered equivalent to μ_c (or) np.

$$\mu_c = n \,.\, \mu_u = 5 \times 2 = 10$$

For $\mu_c = 10$, the following values are obtained from the Poisson table:

r \ $\mu_c = np$	0	1	2	3	4	....	16	17
10			.....	010	029		973	986

By interpolation, the c value for the lower probability limit of 0.025 is given by

$$X = \frac{0.025 - 0.010}{0.029 - 0.010} = 0.789$$

$$c \text{ value} = 3 + 0.789 = 3.789$$

Similarly, by interpolation, the c value for the upper probability limit of 0.975 is computed as follows:

$$X = \frac{0.975 - 0.973}{0.986 - 0.973} = 0.154$$

c value = 16 + 0.154 = 16.154

$$\text{Lower control limit for the u chart} = \frac{c\ \text{value corresponding to lower limit}}{n}$$

$$= \frac{3.789}{5} = 0.758$$

$$\text{Upper control limit for the u chart} = \frac{c\ \text{value corresponding to upper limit}}{n}$$

$$= \frac{16.154}{5} = 3.231$$

Mean value for the u chart = $\mu_u = 2$

Example Problem 12: Draw a u chart for the following data. The sample size is 10.

Sample no.	1	2	3	4	5
No. of defects	10	12	9	8	10

Solution:

Sample no.	**Sample size, *n***	**No. of defects, *c***	**Average no. of defects per unit, $u = c/n$**
1	10	10	1.0
2	10	12	1.2
3	10	9	0.9
4	10	8	0.8
5	10	10	1.0
Total		49	4.9

$$\text{Central line, } \bar{u} = \frac{\Sigma c}{N} = \frac{49}{50} = 0.98$$

$$UCL_u = \bar{u} + 3\sqrt{\frac{\bar{u}}{n}} = 0.98 + 3\sqrt{\frac{0.98}{10}} = 192$$

$$LCL_u = \bar{u} - 3\sqrt{\frac{\bar{u}}{n}} = 0.98 - 3\sqrt{\frac{0.98}{10}} = 0.041$$

The control chart is shown in Fig. 11.29.

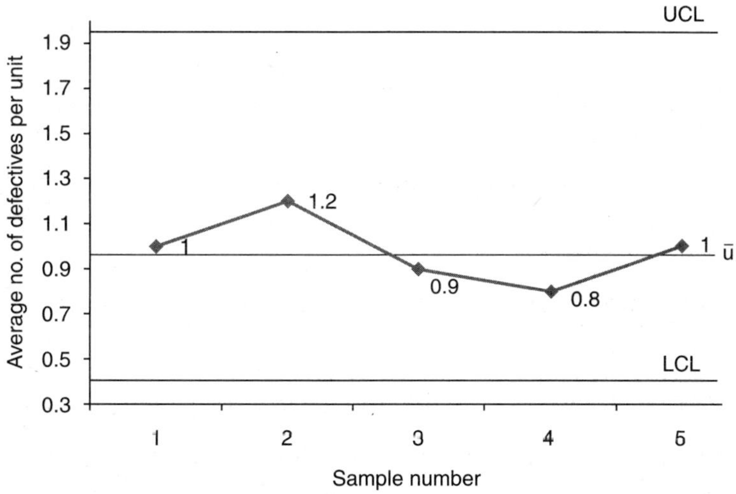

Figure 11.29 The *u* chart

All the points fall within the control limits, and hence, the defects are under control.

Example Problem 13: Determine the normal, tightened and reduced single inspection plans for a lot size of 5000 with AQL of 0.65% at a general inspection level of III.

Solution: From the following table giving the details of the sample size code letters, for the inspection level of III, the sample size code letter is found to be M for a lot size of 5,000.

Lot or batch size	Special inspection levels				General inspection levels		
	S-1	S-2	S-3	S-4	I	II	III
2 - 8	A	A	A	A	A	A	B
...							
3,200 ~ 10,000	C	D	F	G	J	L	M

For a normal inspection plan: From the following MIL STD 105 master table (single sampling plan for normal inspection), for sample size code letter M, the following details are obtained.

Sample size, $n = 315$

Sample size code letter	Sample size	Acceptable quality level (normal inspection)			
		0.010	....	0.65	......
		A_c R_e		A_c R_e	
M	315	.....		5 6	

For AQL at 0.65%, Acceptance number (A_c) = 5

Rejection number (R_e) = 6

Result: A random sample of 315 items is inspected and the number of defective items (d) is determined.

If $d \leq 5$, the lot is accepted; if $d \geq 6$, the lot is rejected.

For a tightened inspection plan: From the following MIL-STD-105 master table (single sampling plan for tightened inspection), for sample size code letter M, the following details are obtained.

Sample size, $n = 315$

Sample size code letter	Sample size	Acceptable quality level (tightened inspection)			
		0.010	....	0.65	......
		A_c R_e		A_c R_e	
M	315	.....		3 4	

For AQL at 0.65%, Acceptance number (A_c) = 3

Rejection number (R_e) = 4

Result: A random sample of 315 items is inspected and the number of defective items (d) is determined.

If $d \leq 3$, the lot is accepted; if $d \geq 4$, the lot is rejected.

For a reduced inspection plan: From the following MIL-STD-105 master table (single sampling plan for reduced inspection), for sample size code letter M, the following details are obtained.

Sample size, $n = 315$

Sample size code letter	Sample size	Acceptable quality level (reduced inspection)			
		0.010	....	0.65	......
		$A_c\ R_e$		$A_c\ R_e$	
M	125	.....		2 5	

For AQL at 0.65%, Acceptance number (A_c) = 2

Rejection number (R_e) = 5

Result: A random sample of 125 items is inspected and the number of defective items (d) is determined.

If $d \leq 2$, the lot is accepted; if $d \geq 5$, the lot is rejected.

If d is found to be between 2 and 5, the lot is accepted but the normal inspection plan is reinstated to take the final decision.

Example Problem 14: Determine the normal and reduced inspection plans for a double sampling plan to inspect a lot size of 3000 soft drink bottles for defects such as manufacturing defects or incorrect quantity of liquid, broken bottles, dust inside the bottle, improper capping, improper sealing, improper stickering, and so on. Any bottle containing one or more defects is found to be defective. An AQL of 10% and general inspection level of III is followed.

Solution: From the following table giving the details of sample size code letters, for the inspection level of III, the sample size code letter is found to be L for a lot size of 3000.

Lot or batch size	Special inspection levels				General inspection levels		
	S-1	S-2	S-3	S-4	I	II	III
2 ~ 8	A	A	A	A	A	A	B
...							
1,201 - 3,200	C	D	E	G	H	K	M

For a normal inspection plan: For sample size code letter L and AQL at 10%, the following data is obtained from the following MIL-STD-105 master table (double sampling plan for normal inspection).

Sample size code letter	Sample	Sample size	Cumulative sample size	Acceptable quality level (normal inspection)			
				0.010	....	10.0	......
				$A_c\,R_e$		$A_c\,R_e$	
K	First	80	80			11 16	
	Second	80	160			26 27	
L	First	125	125	.....		↑	
	Second	125	250				

The letter L has no A_c and R_e values, though sample sizes are available. However, the arrow pointing up indicates the applicable A_c and R_e values at the tip of the arrow.

$A_c = 11$, $R_e = 16$ for $n_1 = 80$ and $A_c = 26$, $R_e = 27$ for $n_2 = 80$ and the corresponding sample size code letter is K.

Hence, for AQL at 10%, the following decision (plan) is taken. The first sample of 80 units is drawn at random and inspected for the number of defectives (d_1). If $d_1 \le 11$, the lot is accepted. If $d_1 \ge 16$, the lot is rejected.

If d_1 is found to lie between 11 and 16, the second sample of 80 units is drawn at random and inspected for the number of defectives (d_2). If $(d_1 + d_2) \le 26$, the lot is accepted. If $(d_1 + d_2) \ge 27$, the lot is rejected.

For a reduced inspection plan: For sample size code letter L and AQL at 10%, the following data is obtained from the following MIL-STD-105 master table (double sampling plan for reduced inspection).

Sample size code letter	Sample	Sample size	Cumulative sample size	Acceptable quality level (reduced inspection)			
				0.010	....	10.0	......
				$A_c\,R_e$		$A_c\,R_e$	
K	First	32	32			5 10	
	Second	32	64			12 16	
L	First	50	50	.....		↑	
	Second	50	100				

The letter L has no A_c and R_e values, though sample sizes are available. However, the arrow pointing up indicates the applicable A_c and R_e values at the tip of the arrow.

$A_c = 5$, $R_e = 10$ for $n_1 = 32$ and $A_c = 12$, $R_e = 16$ for $n_2 = 64$ and the corresponding sample size code letter is K.

Hence, for AQL at 10%, the following decision (plan) is taken. The first sample of 32 units is drawn at random and inspected for the number of defectives (d_1). If $d_1 \leq 5$, the lot is accepted. If $d_1 \geq 10$, the lot is rejected.

If d_1 is found to lie between 5 and 10, the second sample of 32 units is drawn at random and inspected for the number of defectives (d_2). If $(d_1 + d_2) \leq 12$, the lot is accepted. If $(d_1 + d_2) \geq 16$, the lot is rejected.

Example Problem 15: Determine the normal inspection plan for a multiple sampling of a lot of 30 units of tubelights with an inspection level of II and AQL of 6.5%.

Solution: From the following table giving the details of sample size code letters, for the inspection level of III, the sample size code letter is found to be D for a lot size of 30.

Lot or batch size	Special inspection levels				General inspection levels		
	S-1	S-2	S-3	S-4	I	II	III
2 ~ 8	A	A	A	A	A	A	B
...							
26 ~ 50	A	A	B	C	C	D	E

For a normal inspection plan: For sample size code letter D and AQL at 6.5%, the following data is obtained from the MIL-STD-105 master table (multiple sampling plan for normal inspection).

Sample size code letter	Sample	Sample size	Cumulative sample size	Acceptable quality level (reduced inspection)			
				0.010 A_c R_e	....	6.5 A_c R_e	......
D	First	2	2	.....		– 2	
	Second	2	4			– 2	
	Third	2	6			0 2	
	Fourth	2	8			0 2	
	Fifth	2	10			1 3	
	Sixth	2	12			1 3	
	Seventh	2	14			2 3	

Hence, for AQL at 6.5%, the following decision (plan) is taken. The first sample and second sample of 2 units are drawn at random and inspected for the number of defectives, d_1 and d_2, separately. Since the A_c value is not given for the first and second samples, the lot is not accepted for the first two samples. The third sample of size 2 is drawn at random and inspected for the number of defectives, d_3. If $(d_1 + d_2 + d_3) \leq 0$, the lot is accepted. If $(d_1 + d_2 + d_3) \geq 2$, the lot is rejected.

If $(d_1 + d_2 + d_3)$ is found to be 1, a fourth sample of size 2 is drawn at random and inspected for the number of defectives, d_4. If $(d_1 + d_2 + d_3 + d_4) \leq 0$, the lot is accepted. If $(d_1 + d_2 + d_3 + d_4) \geq 2$, the lot is rejected.

If $(d_1 + d_2 + d_3 + d_4)$ is found to be 1, a fifth sample of size 2 is drawn at random and inspected for the number of defectives, d_5. If $(d_1 + d_2 + d_3 + d_4 + d_5) \leq 1$, the lot is accepted. If $(d_1 + d_2 + d_3 + d_4 + d_5) \geq 3$, the lot is rejected.

If $(d_1 + d_2 + d_3 + d_4 + d_5)$ is found to be 2, a sixth sample of size 2 is drawn at random and inspected for number of defectives, d_6. If $(d_1 + d_2 + d_3 + d_4 + d_5 + d_6) \leq 1$, the lot is accepted. If $(d_1 + d_2 + d_3 + d_4 + d_5 + d_6) \geq 3$, the lot is rejected.

If $(d_1 + d_2 + d_3 + d_4 + d_5 + d_6)$ is found to be 2, a seventh sample of size 2 is drawn at random and inspected for the number of defectives, d_7. If $(d_1 + d_2 + d_3 + d_4 + d_5 + d_6 + d_7) \leq 2$, the lot is accepted. If $(d_1 + d_2 + d_3 + d_4 + d_5 + d_6 + d_7) \geq 3$, the lot is rejected.

Note: In this problem, acceptance of the lot does not happen at first, second, fourth and sixth sample levels, even if $d_i = 0$, but rejection of the lot may happen or the next sample level of inspection may continue.

Example Problem 16: Which single sampling plan is to be used for a lot size of 6000, an LTPD of 5% with a consumer's risk of 0.10 and process average as 1.5% defective?

Solution: From the following table of the Dodge–Romig single sample lot tolerance, for a lot tolerance percent defective of 5% and consumer's risk of 0.10, for a lot size of 6000, the following data is obtained.

Lot size	**Process average, %**			
	0 – 0.04	......	1.01 – 1.50	......
	n c AOQL		*n c* AOQL	
5001 – 7000	.....		260 8 1.9	

Result: A sample of size 260 is to be drawn at random from a lot size of 6000 for inspection of the number of defectives, d_1. If $d_1 \leq 8$, the lot is accepted. If $d_1 > 8$, the lot is rejected.

Example Problem 17: For Example Problem 16, if the sampling plan is of double sampling, what should be the inspection plan?

Solution: From the following table of the Dodge–Romig double sample lot tolerance, for a lot tolerance percent defective of 5% and consumer's risk of 0.10, for a lot size of 6000, the following data is obtained.

Lot size	Process average, %								
	0 – 0.04	......	1.21 – 1.50						......
	....	......	Trial 1		Trial 2			AOQL, %	
5001 – 7000	.....	.....	n_1	c_1	n_2	$n_1 + n_2$	c_2		
	....	.....	120	2	260	380	12	2.1	

Result: The first sample size of 120 is drawn at random to check the number of defectives, d_1. If $d_1 \leq 2$, the lot is accepted. If $d_1 > 2$, draw a second sample of size 260 to check the number of defectives, d_2. If $(d_1 + d_2) \leq 12$, the lot is accepted. If $(d_1 + d_2) > 12$, the lot is rejected.

Example Problem 18: For Example Problem 16, the value of AOQL of 2% is given instead of LTPD of 5%. Find the single sampling plan. What is the LTPD value?

Solution: From the following table of the Dodge–Romig single sample AOQL, for AOQL of 2%, for a lot size of 6000, the following data is obtained.

Lot size	Process average, %					
	0 – 0.04	......	1.21 – 1.60			......
	n c		n	c	$100p_{0.10}$	
5001 – 7000	.....		185	6	5.6	

Result: A sample of size 185 is to be drawn at random from a lot size of 6000 for inspection of the number of defectives, d_1. If $d_1 \leq 6$, the lot is accepted. If $d_1 > 6$, the lot is rejected.

$$\text{LTPD} = \text{Value of } 100p_{0.10} = 5.6\%$$

Example Problem 19: Draw an OC for an inspection in which a sample of 10 items is drawn from a lot of 3000 items and the lot is accepted if the number of defectives found ≤ 1 and rejected if the number of defectives found is ≥ 2 in the sample. Find the probable number of defective items for the probability of accepting the lot at 0.6 from the OC for $c = 1$.

Solution: As the sample size is very small compared to the lot size, i.e., the sample size is less than 10%, it is reasonable to use the binomial distribution table to obtain the probabilities that a lot will be accepted for various lot qualities.

The following binomial table gives the required data for a sample size of 10 for acceptance numbers 1, 2, 3.

		Proportion defective, p											
n	x	.05	.10	.15	.20	.25	.30	.35	.40	.45	.50	.55	.60
10	0	.5987	.3487	.1969	.1074	.0563	.0282	.0135	.0060	.0025	.0010	.0003	.0001
$c = 1$	1	.9139	.7361	.5443	.3758	.2440	.1493	.0860	.0464	.0233	.0107	.0045	.0017
→	2	.9885	.9298	.8202	.6778	.5256	.3828	.2616	.1673	.0996	.0547	.0274	.0123
	3	.9990	.9872	.9500	.8791	.7759	.6496	.5138	.3823	.2660	.1719	.1020	.0548

From the table, it is clear that for number of defectives ≤ 1 (i.e., $c = 1$), the probability of accepting the lot (P_a) with 5% defectives ($p = 0.05$) is 0.9139. For a lot with 10% defectives (i.e., $p = 0.1$) for $c = 1$, the probability of accepting the lot (P_a) is 0.7361 and so on.

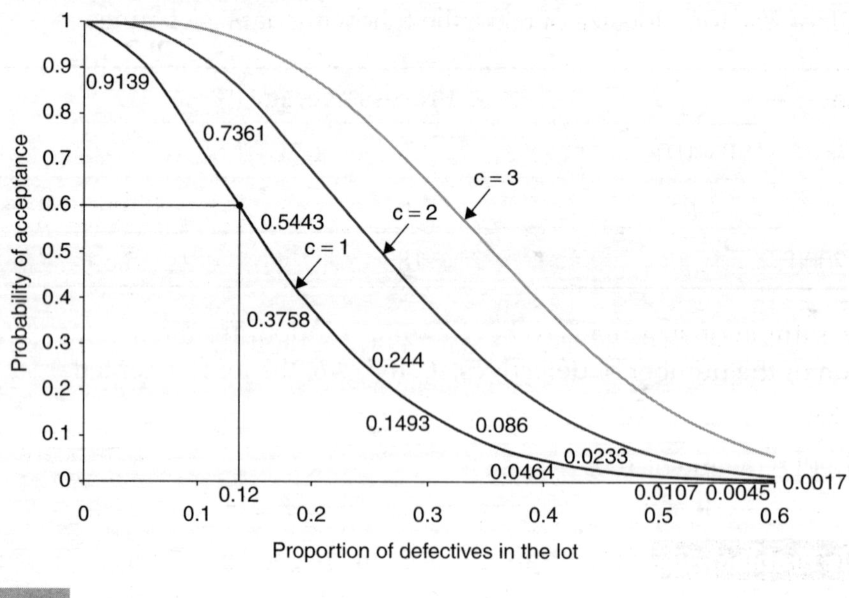

Figure 11.30 OC curve

By marking the points on a graph, with p values on the x axis and P_a values on the y axis, we obtain smooth curves for various c values. The graph (Fig. 11.30) shows the OC curves for $c = 1, 2$ and 3.

From the above OC curve, for $c = 1$, $P_a = 0.60$, the proportion of defectives is found to be 0.12. Hence, the number of defective items will be equal to $0.12 \times 3000 = 360$.

Example Problem 20: A single sampling inspection plan is planned for a large lot of mass-manufactured items. From each lot, draw and inspect a random sample of 50. If there are three or more defective items, inspect the whole lot and remove all the defective items. If less than three defective items are found, then accept the lot without further inspection. Evaluate P_a for $p = 0.01, 0.02, 0.03, 0.05, 0.07, 0.10, 0.15, 0.20, 0.30$.

Plot the OC and AOQ curves. Estimate: a) the producer's risk at p of 4% and b) the consumer's risk for LTPD of 8%.

Solution: Let P_a be the probability of accepting a lot containing a fraction 'p' of defective items.

P_0 is the probability of drawing a lot containing exactly 1 defective item.

P_1 is the probability of drawing a lot containing exactly 2 defective items.

P_2 is the probability of drawing a lot containing exactly 3 defective items.

Then, the probability of accepting a lot containing three or fewer defectives, $P_a = P_0 + P_1$

For a sample of 50 items, $P_0 = q^{50}$ where $q = 1 - p$

$$P_1 = C_1^{50}\, pq^{49} = 50\, pq^{49}$$

$$P_2 = C_2^{50}\, p^2\, q^{48} = 50 \times 49\, p^2q^{48} = 1225\, p^2q^{48}$$

$$P_a = q^{50} + 50\, pq^{49} + 1225\, p^2q^{48} = q^{48}\,(q^2 + 50\, pq + 1225\, p^2)$$

$$= (1-p)^{48}\,[(1-p)^2 + 50p(1-p) + 1225p^2] - (1-p)^{48}\,(1176p^2 + 48p + 1)$$

The P_a values for various fractions of defectives are calculated using the above formula for the inspection of 50 items and tabulated as follows:

p	0.01	0.02	0.03	0.05	0.07	0.10	0.15	0.20	0.25
P_a	0.986	0.922	0.811	0.541	0.311	0.112	0.014	0.001	0.000

The OC curve is drawn for the above p and P_a values, as shown in Fig. 11.31.

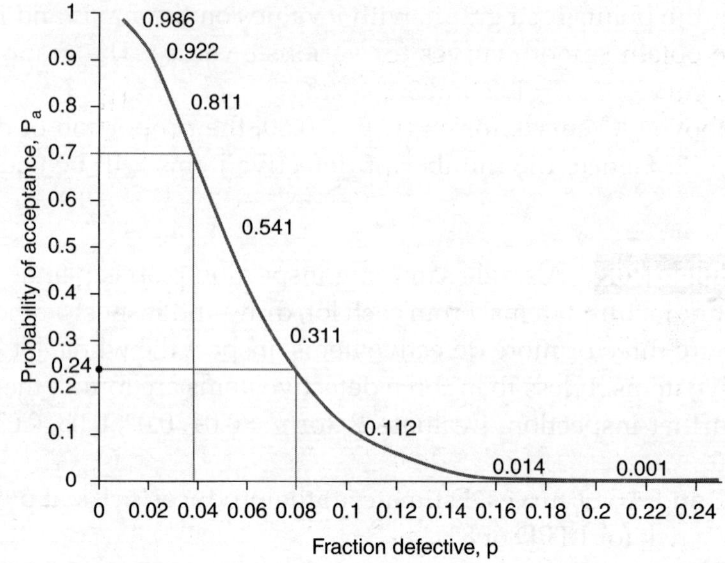

Figure 11.31 OC curve

From the above OC curve, it is found that $P_a = 0.7$, at $p = 4$ % (i.e., 0.04).

$$\text{Producer's risk} = 1 - P_a = 1 - 0.7 = 0.3 = 30\%$$

Similarly, $P_a \approx 0.24$ for $p = 8\%$ (i.e., 0.08)

$$\text{Consumer's risk} = P_a = 0.24 = 24\%$$

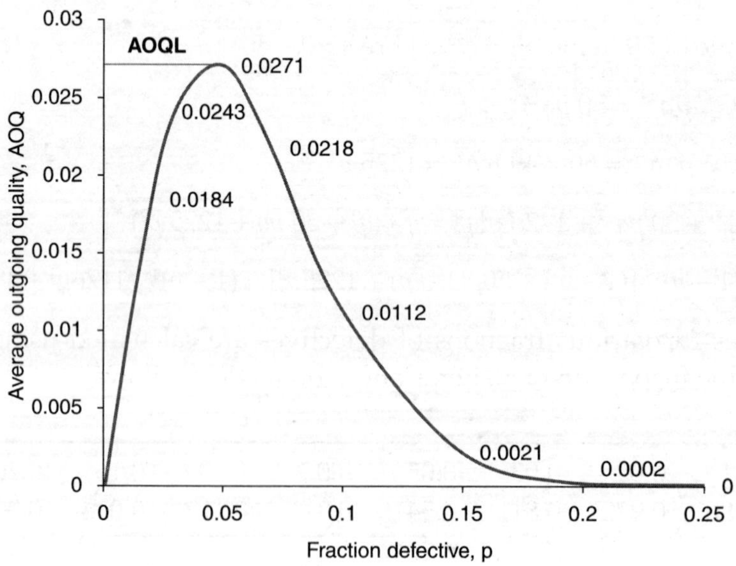

Figure 11.32 AOQ curve

The formula of AOQ = $P_a\,p$ is used to compute AOQ values for various 'p' values.

p	0.01	0.02	0.03	0.05	0.07	0.10	0.15	0.20	0.25
P_a	0.986	0.922	0.811	0.541	0.311	0.112	0.014	0.001	0.000
AOQ	0.0099	0.0184	0.0243	0.0271	0.0218	0.0112	0.0021	0.0002	0.000

Figure 11.32 shows the AOQ curve for these values.

The average outgoing quality limit (AOQL) is found to be 0.0271 (i.e., 2.71%) from the above AOQ curve.

Example Problem 21: Construct an AOQ curve for $N = 600$, $n = 10$, $c = 1$ using the formula AOQ = $P_a \cdot p\,(N\text{-}n)/N$. Find the AOQL from the curve.

Solution: Using the binomial table as the sample size is small (less than 10% compared with the lot size), the corresponding P_a values are taken for the increment value of 0.05 of p. As the P_a value is very close to 0 for the p values from 0.40, the tabulation is carried out up to a p value of 0.40.

p	0.05	0.10	0.15	0.20	0.25	0.30	0.35	0.40
P_a	0.9139	0.7361	0.5443	0.3758	0.2440	0.1493	0.0860	0.0464
AOQ	0.045	0.073	0.081	0.074	0.060	0.044	0.029	0.019

The following graph (Fig. 11.33) shows the AOQ curve for the given problem.

From the curve, the AOQL is found to be 0.082.

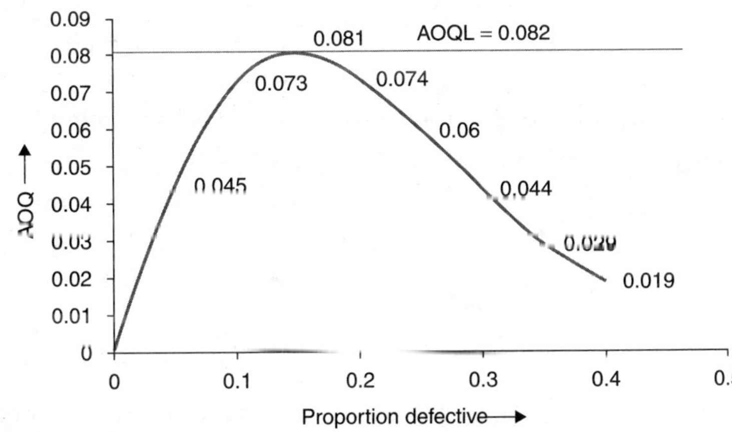

Figure 11.33 AOQ curve

SUMMARY

- Quality is defined as, "a degree to which a set of inherent characteristics fulfils the requirement".
- Quality control is the process of achieving quality in a product or service by implementation of standards and procedures.
- Quality assurance is considered a part of the quality management system focused on providing confidence that quality requirements will be fulfilled.
- Chance causes of variation occur due to poor design, poor maintenance of machines, poor working conditions, etc.
- Assignable causes of variation occur due to poor adjustment of equipment and machines, faulty controllers, malfunctioning of machines, operators' carelessness, poor batch of raw material, etc.
- A control chart is an online process-monitoring technique that utilises statistical calculations and graphical representations to understand the ongoing process in real time and to identify the process capability – whether the process is stable and predictable.
- Process capability gives the relationship between six-sigma limits (control limits) and the product specification limits.
- In acceptance sampling, the entire lot is divided into a number of smaller subgroups and a few samples of accepted numbers are selected at random for inspection. If the number of defectives is within the preset limit of some acceptance value, then the entire lot of products will be accepted as good, though it is an established fact that there would be a few defectives in the lot.
- Some benefits of sampling plan are reduced inspection cost, modern and scientific method of control, reduced wastage, quality assurance to all stakeholders, uniformity in production quality and better customer satisfaction.
- The OC curve is a graph indicating the probability that a particular lot with a certain percent of non-conforming items will be accepted or pass the inspection.
- Producer's risk (α) represents the probability of rejection of a conforming (good) lot.
- Consumer's risk (β) represents the probability of acceptance of a non-conforming (bad) lot.
- Acceptable quality level (AQL) represents the maximum percent of defectives that will make a lot easily acceptable.
- Rejectable quality level (RQL) represents the quality level unacceptable to the customer.

- Lot tolerance percent defective (LTPD) represents the upper limit of the percentage of defective items that a consumer is willing to tolerate in an accepted lot.

EXERCISES

Multiple Choice Questions

1. A control chart is a(an)
   a) Offline monitoring technique
   b) Online monitoring technique
   c) Pre-production monitoring technique
   d) All of the above

2. The data from a process that is a measured quantity in units is called
   a) Attribute
   b) Variable
   c) Constant
   d) Real

3. The data from a process that is a measured quantity in terms of defectives and defects is called
   a) Attribute
   b) Variable
   c) Constant
   d) Real

4. The distribution of measured data can be studied through
   a) $\overline{X}$ chart
   b) R chart
   c) C chart
   d) Both a and b

5. If the subgroup size of the variable data is between 2 and 10, which of the following control charts should be selected?
   a) $\overline{X}$ and mR charts
   b) $\overline{X}$ and R charts
   c) P and nP chart
   d) Both b and c

6. Which among the following is a kind of variable chart?
   a) U chart
   b) P chart
   c) C chart
   d) R chart

7. The moving average chart is selected if the subgroup size of the
   a) Attribute data is equal to 1
   b) Variable data is equal to 1
   c) Attribute data is more than 1
   d) Variable data is more than 1

8. A sample size of 2 m^2 is observed in which type of chart?
   a) P chart
   b) R chart
   c) $\overline{X}$ chart
   d) C chart

9. The reasons for chance causes
   a) Can be eliminated economically
   b) Cannot be eliminated economically
   c) Can be eliminated in less time
   d) None of the above

10. If the control chart is drawn for all the individual values of the subgroup, it is called
   a) $\overline{X}$ chart
   b) Long run chart
   c) Short run chart
   d) Tolerance chart

11. The data obtained from the inspection of items using a GO and NO GO gauge is called
   a) Attribute
   b) Variable
   c) Constant
   d) Real

12. The data obtained from the inspection of items using a micrometer is called
   a) Attribute
   b) Variable
   c) Constant
   d) Real

13. A defective item will
   a) Not have any defect
   b) Have only one defect
   c) Have many defects
   d) None of the above

14. The capability ratio is
   a) The ratio of process capability and sample size
   b) The ratio of specification range and process capability
   c) The ratio of the number of defects and the number of units inspected
   d) The ratio of the number of defectives and the number of units inspected

15. The most desirable condition for a process to be in control is
   a) $6\sigma < USL - LSL$
   b) $6\sigma = USL - LSL$
   c) $6\sigma > USL - LSL$
   d) $6\sigma < UCL - LCL$

16. When a process distribution is centred, then
   a) C_p will be less than C_{pk}
   b) C_p will be equal to 1
   c) C_p will be greater than C_{pk}
   d) C_p will be equal to C_{pk}

17. If C_p is less than 1, the process is said to be
   a) Capable
   b) Desirable condition
   c) Incapable
   d) None of the above

18. If the value C_{pk} is equal to 1, it means that the process is producing
   a) Smaller quantity of products
   b) Quality products
   c) Conforming products
   d) Non-conforming products

19. Process control is carried out

a) During production b) After production
c) Before production d) All of the above

20. Low cost, higher volume items require

a) No inspection b) Little inspection
c) Intensive inspection d) Complete inspection

21. High cost, low volume items require

a) No inspection b) Little inspection
c) Intensive inspection d) Complete inspection

22. The dividing lines between random and non-random variations from the mean of the distribution are called

a) Control limits b) Warning limits
c) Two-sigma limits d) Specification limits

23. The chart used to monitor variance is the

a) P chart b) C chart
c) R chart d) All of the above

24. The chart used to monitor the number of defects per unit is the

a) P chart b) C chart
c) R chart d) $\overline{X}$ chart

25. Process capability is

a) $(USL - LSL)/3\sigma$ b) $(USL - LSL)/6\sigma$
c) $(USL + LSL)/3\sigma$ d) $(USL + LSL)/6\sigma$

26. A 6σ process has a defect level below ____number of defects per million chances.

a) 4.3 b) 3.4
c) 5.6 d) 3.0

27. The moving average chart is a type of

a) $\overline{X}$ chart b) P chart
c) U chart d) S chart

28. Which of the following are required to construct the OC curve?

a) AOQL and LTPD b) α and β
c) Acceptance number and sample size d) AQL and LTPD

29. The value of AOQL should be

a) Less than the original proportion of non-conforming items
b) More than the original proportion of non-conforming items
c) Equal to the original proportion of non-conforming items
d) None of the above

30. The maximum number of allowable non-conforming items in the samples of a lot is the
   a) Rejection number
   b) Acceptance number
   c) AOQL
   d) LTPD

31. The number of samples to be inspected is not predetermined in
   a) Single sampling
   b) Double sampling
   c) Multiple sampling
   d) Item-by-item sampling

32. If the quality of the inspected lot shows a significant reduction, __________ inspection is considered.
   a) Normal
   b) Reduced
   c) Tightened
   d) None of the above

33. The Dodge–Romig sampling method is applicable for
   a) Single sampling plans
   b) Double sampling plans
   c) Multiple sampling plans
   d) Both single and double sampling plans

34. The Dodge–Romig sampling plan is designed for
   a) Minimum total inspection
   b) Minimum LTPD
   c) Minimum AOQL
   d) All the above

Answers

1. b	2. b	3. a	4. d	5. b
6. d	7. b	8. d	9. b	10. d
11. a	12. b	13. c	14. b	15. a
16. d	17. c	18. d	19. a	20. b
21. c	22. a	23. c	24. b	25. b
26. b	27. a	28. c	29. a	30. b
31. d	32. c	33. a	34. a	

Review Questions

1. Write about the importance of the statistical quality control process.
2. Write about the differences between quality assurance and quality control.
3. What is meant by control chart? Explain how a process is improved using a control chart.
4. Differentiate between variable data and attribute data.
5. Differentiate chance causes and assignable causes.
6. Explain the types of causes of variation with examples.

7. Differentiate chance causes from assignable causes of variation.
8. What do we mean by 'controllable errors'?
9. List the types of control charts and explain the selection method of control charts.
10. Write about the applications of various control charts.
11. What could be inferred from an $\overline{X}$ chart and an R chart?
12. Define 'run up' and 'run down' of a control chart.
13. State the advantages of control charts.
14. Why do we prefer sampling by attributes to sampling by variables?
15. Define 'non-conformity' and 'non-conforming'.
16. What is meant by process capability?
17. "The design tolerances specified for the product must be compatible with the process capability" – Comment on this statement.
18. What is the difference between the process capability index 'C_p' and 'C_{pk}'?
19. Explain in detail how the process capability index helps to predetermine the nature of the process.
20. What is acceptance sampling? What are its advantages?
21. "Manufacturer and Consumer strike a balance and come to an agreement with respect to the sampling inspection" – Explain this statement. What are the various types of sampling? Explain in brief.
22. Distinguish between 'sampling by variables' and 'sampling by attributes'.
23. What are the disadvantages of 100% sampling?
24. Define normal, tightened and reduced inspection.
25. Distinguish between the single and double sampling plans with the help of a diagram.
26. Explain the concept and characteristics of the double sampling plan.
27. Distinguish between the double and multiple sampling plans.
28. Explain in detail the Dodge–Romig lot-by-lot acceptance sampling method.
29. Explain the salient features of an ideal OC curve.
30. Discuss producer's risk and consumer's risk.
31. What is AQL?
32. What is meant by AOQL?
33. What is LTPD?
34. Six inspection samples of five amplifiers each are drawn from production. The table shows the power output obtained from each amplifier.

Lot no.	Amplifier				
	1	2	3	4	5
1	12	8	11	9	10
2	9	10	12	11	13

Lot no.	Amplifier				
	1	2	3	4	5
3	15	12	9	10	13
4	11	13	14	12	15
5	10	12	16	13	12
6	14	12	13	11	9

Draw the $\overline{X}$ and R charts. Check if the process meets the specification of 11.7 ± 2.5.

35. A subgroup of five items each is taken from a manufacturing process at regular intervals. A certain quality characteristic is measured and the $\overline{X}$ and R values are computed. After 25 subgroups, it is found that $\Sigma\overline{X} = 357.50$ and $\Sigma R = 8.80$. If the specification limits are 14.40 ± 0.40 and if the process is in statistical control, what conclusions can you draw about the ability of the process to produce items within specification?

36. The following table gives the daily inspection reports of 100 tubelights.

Sample number	Number of defectives	Sample number	Number of defectives
1	7	6	10
2	12	7	2
3	8	8	16
4	18	9	14
5	5	10	6

a) Determine the three-sigma control limits.
b) Does the chart prove that the process is in control?

37. If the specification limits are 2.230 to 2.550, $\overline{X} = 2.238$ and $s = 0.0065$, what percentage of measurements would be out of specification? If s is decreased to 0.0025, what percentage of the measurements would be out of specification?

38. Two samples of cloth (200 m in length) were inspected and minor defects were found, as listed below:

Sample no.	1	2	3	4	5	6	7	8	9	10
Defects	2	3	0	1	4	4	2	4	1	2

Prepare a control chart showing the control limits and warning limits. Does the chart need lower warning and control limits? If yes, why?

39. In a manufacturing process, 10 castings were inspected. Each casting was found to contain a certain number of defects, as given below:

Casting	1	2	3	4	5	6	7	8	9	10
No. of defects	2	4	1	5	5	6	3	4	0	7

Draw the appropriate control chart and check if the process is in control or not.

40. Construct an OC curve for a single sampling plan for $n = 100$ and $c = 2$. Also mark the important points. Use the Poisson distribution table for values of probability of acceptance at various 'p' values.

41. Draw a single sampling plan that satisfies a producer's risk of 5% for lots that are 1.8% non-conforming and a consumer's risk of 10% for lots that are non-conforming. Use the OC curve to determine a suitable sample size and acceptance number.

42. Draw an OC curve for $n = 250$ and $c = 2$. Use the following table values to construct the curve.

P_a	0.99	0.95	0.9	0.8	0.6	0.3	0.1
p_n	4.2	5.5	6.3	7.5	9.1	12	15

Appendix 1

Table of Constants for Control Charts for Variables

Sample size	For $\overline{X}$ chart		For *R* chart			For *S* chart		
	A_2	A_3	d_2	D_3	D_4	c_4	B_3	B_4
2	1.880	2.659	1.128	0	3.267	0.7979	0	3.267
3	1.023	1.954	1.693	0	2.574	0.8862	0	2.568
4	0.729	1.628	2.059	0	2.282	0.9213	0	2.266
5	0.577	1.427	2.326	0	2.114	0.9400	0	2.089
6	0.483	1.287	2.534	0	2.004	0.9515	0.030	1.970
7	0.419	1.182	2.704	0.076	1.924	0.9594	0.118	1.882
8	0.373	1.099	2.847	0.136	1.864	0.9650	0.185	1.815
9	0.337	1.032	2.970	0.184	1.816	0.9693	0.239	1.761
10	0.308	0.975	3.078	0.223	1.777	0.9727	0.284	1.716
11	0.285	0.927	3.173	0.256	1.744	0.9754	0.321	1.679
12	0.266	0.886	3.258	0.283	1.717	0.9776	0.354	1.646
13	0.249	0.850	3.336	0.307	1.693	0.9794	0.382	1.618
14	0.235	0.817	3.407	0.328	1.672	0.9810	0.406	1.594
15	0.223	0.789	3.472	0.347	1.653	0.9823	0.428	1.572
16	0.212	0.763	3.532	0.363	1.637	0.9835	0.448	1.552
17	0.203	0.739	3.588	0.378	1.622	0.9845	0.466	1.534
18	0.194	0.718	3.640	0.391	1.608	0.9854	0.482	1.518
19	0.187	0.698	3.689	0.403	1.597	0.9862	0.497	1.503
20	0.180	0.680	3.735	0.415	1.585	0.9869	0.510	1.490

Appendix 2

Table of Standard Normal Probabilities

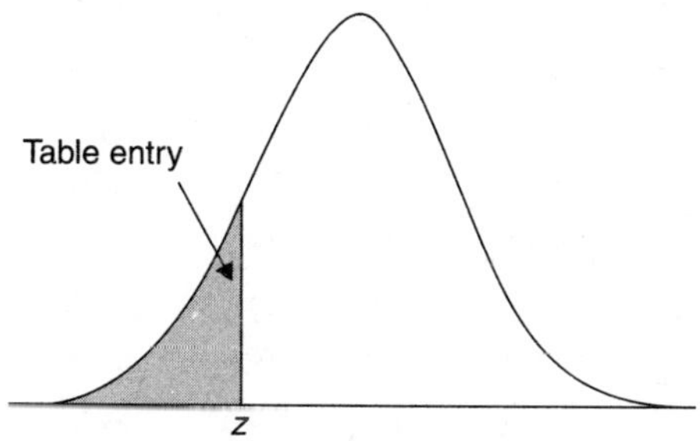

Table entry for *z* is the area under the standard normal curve to the left of *z*.

z	.00	.01	.02	.03	.04	.05	.06	.07	.08	.09
−3.4	.0003	.0003	.0003	.0003	.0003	.0003	.0003	.0003	.0003	.0002
−3.3	.0005	.0005	.0005	.0004	.0004	.0004	.0004	.0004	.0004	.0003
−3.2	.0007	.0007	.0006	.0006	.0006	.0006	.0006	.0005	.0005	.0005
−3.1	.0010	.0009	.0009	.0009	.0008	.0008	.0008	.0008	.0007	.0007
−3.0	.0013	.0013	.0013	.0012	.0012	.0011	.0011	.0011	.0010	.0010
−2.9	.0019	.0018	.0018	.0017	.0016	.0016	.0015	.0015	.0014	.0014
−2.8	.0026	.0025	.0024	.0023	.0023	.0022	.0021	.0021	.0020	.0019
−2.7	.0035	.0034	.0033	.0032	.0031	.0030	.0029	.0028	.0027	.0026
−2.6	.0047	.0045	.0044	.0043	.0041	.0040	.0039	.0038	.0037	.0036
−2.5	.0062	.0060	.0059	.0057	.0055	.0054	.0052	.0051	.0049	.0048
−2.4	.0082	.0080	.0078	.0075	.0073	.0071	.0069	.0068	.0066	.0064
−2.3	.0107	.0104	.0102	.0099	.0096	.0094	.0091	.0089	.0087	.0084
−2.2	.0139	.0136	.0132	.0129	.0125	.0122	.0119	.0116	.0113	.0110
−2.1	.0179	.0174	.0170	.0166	.0162	.0158	.0154	.0150	.0146	.0143
−2.0	.0228	.0222	.0217	.0212	.0207	.0202	.0197	.0192	.0188	.0183
−1.9	.0287	.0281	.0274	.0268	.0262	.0256	.0250	.0244	.0239	.0233
−1.8	.0359	.0351	.0344	.0336	.0329	.0322	.0314	.0307	.0301	.0294

(Continued)

Continued

z	.00	.01	.02	.03	.04	.05	.06	.07	.08	.09
–1.7	.0446	.0436	.0427	.0418	.0409	.0401	.0392	.0384	.0375	.0367
–1.6	.0548	.0537	.0526	.0516	.0505	.0495	.0485	.0475	.0465	.0455
–1.5	.0668	.0655	.0643	.0630	.0618	.0606	.0594	.0582	.0571	.0559
–1.4	.0808	.0793	.0778	.0764	.0749	.0735	.0721	.0708	.0694	.0681
–1.3	.0968	.0951	.0934	.0918	.0901	.0885	.0869	.0853	.0838	.0823
–1.2	.1151	.1131	.1112	.1093	.1075	.1056	.1038	.1020	.1003	.0985
–1.1	.1357	.1335	.1314	.1292	.1271	.1251	.1230	.1210	.1190	.1170
–1.0	.1587	.1562	.1539	.1515	.1492	.1469	.1446	.1423	.1401	.1379
–0.9	.1841	.1814	.1788	.1762	.1736	.1711	.1685	.1660	.1635	.1611
–0.8	.2119	.2090	.2061	.2033	.2005	.1977	.1949	.1922	.1894	.1867
–0.7	.2420	.2389	.2358	.2327	.2296	.2266	.2236	.2206	.2177	.2148
–0.6	.2743	.2709	.2676	.2643	.2611	.2578	.2546	.2514	.2483	.2451
–0.5	.3085	.3050	.3015	.2981	.2946	.2912	.2877	.2843	.2810	.2776
–0.4	.3446	.3409	.3372	.3336	.3300	.3264	.3228	.3192	.3156	.3121
–0.3	.3821	.3783	.3745	.3707	.3669	.3632	.3594	.3557	.3520	.3483
–0.2	.4207	.4168	.4129	.4090	.4052	.4013	.3974	.3936	.3897	.3859
–0.1	.4602	.4562	.4522	.4483	.4443	.4404	.4364	.4325	.4286	.4247
–0.0	.5000	.4960	.4920	.4880	.4840	.4801	.4761	.4721	.4681	.4641

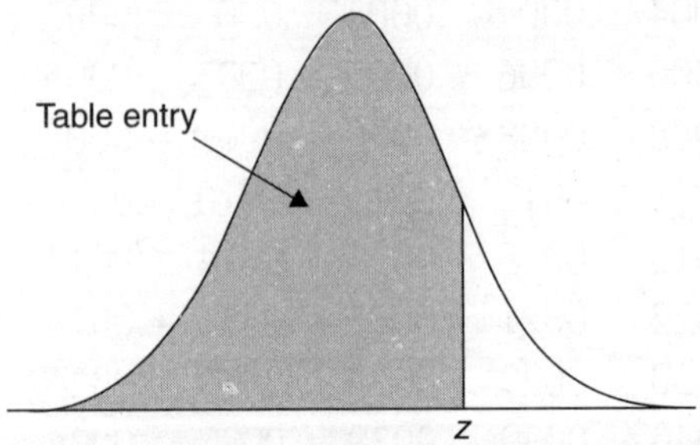

Table entry for *z* is the area under the standard normal curve to the left of *z*.

z	.00	.01	.02	.03	.04	.05	.06	.07	.08	.09
0.0	.5000	.5040	.5080	.5120	.5160	.5199	.5239	.5279	.5319	.5359
0.1	.5398	.5438	.5478	.5517	.5557	.5596	.5636	.5675	.5714	.5753
0.2	.5793	.5832	.5871	.5910	.5948	.5987	.6026	.6064	.6103	.6141
0.3	.6179	.6217	.6255	.6293	.6331	.6368	.6406	.6443	.6480	.6517
0.4	.6554	.6591	.6628	.6664	.6700	.6736	.6772	.6808	.6844	.6879
0.5	.6915	.6950	.6985	.7019	.7054	.7088	.7123	.7157	.7190	.7224
0.6	.7257	.7291	.7324	.7357	.7389	.7422	.7454	.7486	.7517	.7549

(Continued)

Continued

z	.00	.01	.02	.03	.04	.05	.06	.07	.08	.09
0.7	.7580	.7611	.7642	.7673	.7704	.7734	.7764	.7794	.7823	.7852
0.8	.7881	.7910	.7939	.7967	.7995	.8023	.8051	.8078	.8106	.8133
0.9	.8159	.8186	.8212	.8238	.8264	.8289	.8315	.8340	.8365	.8389
1.0	.8413	.8438	.8461	.8485	.8508	.8531	.8554	.8577	.8599	.8621
1.1	.8643	.8665	.8686	.8708	.8729	.8749	.8770	.8790	.8810	.8830
1.2	.8849	.8869	.8888	.8907	.8925	.8944	.8962	.8980	.8997	.9015
1.3	.9032	.9049	.9066	.9082	.9099	.9115	.9131	.9147	.9162	.9177
1.4	.9192	.9207	.9222	.9236	.9251	.9265	.9279	.9292	.9306	.9319
1.5	.9332	.9345	.9357	.9370	.9382	.9394	.9406	.9418	.9429	.9441
1.6	.9452	.9463	.9474	.9484	.9495	.9505	.9515	.9525	.9535	.9545
1.7	.9554	.9564	.9573	.9582	.9591	.9599	.9608	.9616	.9625	.9633
1.8	.9641	.9649	.9656	.9664	.9671	.9678	.9686	.9693	.9699	.9706
1.9	.9713	.9719	.9726	.9732	.9738	.9744	.9750	.9756	.9761	.9767
2.0	.9772	.9778	.9783	.9788	.9793	.9798	.9803	.9808	.9812	.9817
2.1	.9821	.9826	.9830	.9834	.9838	.9842	.9846	.9850	.9854	.9857
2.2	.9861	.9864	.9868	.9871	.9875	.9878	.9881	.9884	.9887	.9890
2.3	.9893	.9896	.9898	.9901	.9904	.9906	.9909	.9911	.9913	.9916
2.4	.9918	.9920	.9922	.9925	.9927	.9929	.9931	.9932	.9934	.9936
2.5	.9938	.9940	.9941	.9943	.9945	.9946	.9948	.9949	.9951	.9952
2.6	.9953	.9955	.9956	.9957	.9959	.9960	.9961	.9962	.9963	.9964
2.7	.9965	.9966	.9967	.9968	.9969	.9970	.9971	.9972	.9973	.9974
2.8	.9974	.9975	.9976	.9977	.9977	.9978	.9979	.9979	.9980	.9981
2.9	.9981	.9982	.9982	.9983	.9984	.9984	.9985	.9985	.9986	.9986
3.0	.9987	.9987	.9987	.9988	.9988	.9989	.9989	.9989	.9990	.9990
3.1	.9990	.9991	.9991	.9991	.9992	.9992	.9992	.9992	.9993	.9993
3.2	.9993	.9993	.9994	.9994	.9994	.9994	.9994	.9995	.9995	.9995
3.3	.9995	.9995	.9995	.9996	.9996	.9996	.9996	.9996	.9996	.9997
3.4	.9997	.9997	.9997	.9997	.9997	.9997	.9997	.9997	.9997	.9998

Index

U

V